Business Statistics for Competitive Advantage with Excel 2010

Cynthia Fraser

Business Statistics for Competitive Advantage with Excel 2010

Basics, Model Building, and Cases

Second Edition

 Springer

Cynthia Fraser
University of Virginia
Charlottesville, VA, USA
cfg8q@comm.virginia.edu

ISBN 978-1-4419-9856-9   e-ISBN 978-1-4419-9857-6
DOI 10.1007/978-1-4419-9857-6
Springer New York Dordrecht Heidelberg London

Library of Congress Control Number: 2011938464

Printed on acid-free paper

Springer is part of Springer Science+Business Media (www.springer.com)

To Len Lodish who taught me to include decision makers in the model building process

Contents

**Chapter 4   Quantifying the Influence of Performance Drivers
and Forecasting: Regression 109**

Preface

Exceptional managers know that they can create competitive advantages by basing decisions on performance response under alternative scenarios. To create these advantages, managers need to understand how to use statistics to provide this information. Statistics are created to make better decisions. Statistics are essential and relevant, and they must be easily and quickly produced using widely available software such as Excel. Then results must be translated into general business language and illustrated with compelling graphics to make them understandable and usable by decision makers. This book helps students master this process of using statistics to create competitive advantages as decision makers.

Statistics are essential, relevant, easy to produce, easy to understand, valuable, and fun when used to create competitive advantage.

The Examples, Assignments, and Cases Used to Illustrate Statistics for Decision Making Come from Business Problems

McIntire corporate sponsors and partners, such as Rolls-Royce, Procter & Gamble, Dell, and the industries that they do business in, provide many realistic examples. The book also features a number of examples of global business problems, including those from important emerging markets in China and India. Students are excited when statistics are used to study real and important business problems. This makes it easy to see how they will use statistics to create competitive advantages in their internships and careers.

Learning Is Hands On with Excel and Shortcuts

Each type of analysis is introduced with one or more examples. Following is an example of how to create the statistics in Excel and what the numbers mean in English.

Included in Excel sections are screenshots that allow students to easily master Excel. Featured are a number of popular Excel shortcuts, which are, themselves, a competitive advantage.

Powerful PivotTables and PivotCharts are introduced early and used throughout the book. Results are illustrated with graphics from Excel.

In each chapter, assignments or cases are included to allow students to practice using statistics for decision making and competitive advantage. Beginning in Chap. 9, Harvard Business School cases are suggested, which provide additional opportunities to use statistics to advantage.

Focus Is on What Statistics Mean to Decision Makers and How to Communicate Results

From the beginning, results are translated into English. In Chap. 5, results are condensed and summarized in PowerPoints and memos, the standards of communication in businesses. Later chapters include example memos for students to use as templates, making communication of statistics for decision making an easy skill to master.

Instructors, give your students the powerful skills that they will use to create competitive advantages as decision makers. Students, be prepared to discover that statistics are a powerful competitive advantage. Your mastery of the essential skills of creating and communicating statistics for improved decision making will enhance your career and make numbers fun.

New in the Second Edition

After reading The First Edition, students asked for more theory. The second edition includes more explanation of hypothesis tests and confidence intervals, and how t, F, and chi square distributions behave.

Excel pages feature Excel 2010 and include more shortcuts and fewer cell references to help form linkages between Excel menus more easily.

In Chap. 5, the introduction of effective PowerPoints borrows heavily from Cliff Atkinson's effective *Beyond Bullet Points* (Microsoft Press, 2007). Presenting statistical results in PowerPoints requires specific skill. Chapter 5 offers explanations and guidelines for effective PowerPoint file creation, design, and organization.

An introduction to analysis of variance (ANOVA) has been added to Chap. 10. Because regression with indicator variables is closely related to ANOVA, they are compared in Chap. 10.

The financial and economic events of 2008–2010 changed business dramatically. Examples have been updated to illustrate how the impacts of recent changes can be acknowledged to build powerful valid models.

The Data Files, Solution Files, and Chapter PowerPoints

The data files for text examples, cases, lab problems and assignments are stored on Blackboard and may be accessed using this link:

https://blackboard.comm.virginia.edu/webapps/portal/frameset.jsp

Instructors can gain access to the files, as well as solution files and chapter PowerPoints by registering on the Springer site:

http://www.springer.com/statistics/business%2C+economics+%26+finance/book/978-1-4419-9856-9?changeHeader

Business people can gain access to the files by emailing the author cfg8q@virginia.edu.

Acknowledgments

Preliminary and first editions of *Statistics for Decision Making & Competitive Advantage* were used at The McIntire School, University of Virginia, and I thank the many bright, motivated, and enthusiastic students who provided comments and suggestions.

Cynthia Fraser
Charlottesville, VA

Chapter 1
Statistics for Decision Making and Competitive Advantage

In the increasingly competitive global arena of business in the twenty-first century, a select few business graduates distinguish themselves by enhanced decision making backed by statistics. No longer is the production of statistics confined to quantitative analysis and market research divisions in firms. Statistics are useful when they are applied to improve decision making and so are used daily by managers in each of the functional areas of business. Excel and other statistical software live in our laptops, providing immediate access to statistical tools that can be used to improve decision making.

1.1 Statistical Competences Translate into Competitive Advantages

The majority of business graduates can create descriptive statistics and use Excel. Fewer have mastered the ability to frame a decision problem so that information needs can be identified and satisfied with statistical analysis. Fewer can build powerful and valid models to identify performance drivers, compare decision alternative scenarios, and forecast future performance. Fewer can translate statistical results into general business English that is easily understood by everyone in a decision making team. And fewer still have the ability to illustrate memos with compelling and informative graphics. Each of these competences provides a competitive advantage to those few who have mastery. This book will help you to attain these competences and the competitive advantages they promise.

Most examples in this book are taken from real businesses and concern real decision problems. A number of examples focus on decision making in global markets. By reading about how executives and managers successfully use statistics to increase information and improve decision making in a variety of case applications, you will be able to frame a variety of decision problems in your firm, whether small or multinational. The end of chapter assignments will give you practice framing diverse problems, practicing statistical analyses, and translating results into easily understood reports or presentations.

Many examples in this book feature bottom line conclusions. From the statistical results, you read what managers would conclude with those results. These conclusions and implications are written in general business English, rather than statistical jargon, so that anyone in a decision team will understand. Assignments ask you to feature bottom line conclusions and general business English.

Translation of statistical results into general business English is necessary to ensure their effective use. If decision makers, our audience for statistical results, don't understand the conclusions and implications from statistical analysis, the information created by the analysis will not be used. An appendix is devoted to writing memos that your audience will read and understand and to effective PowerPoint slide designs for effective presentation of results. Memos and PowerPoints are the predominant forms of communication in businesses. Decision making is compressed and information must be distilled, well written, and illustrated. Decision makers read memos. Use memos to make the most of your analyses, conclusions, and recommendations.

C. Fraser, *Business Statistics for Competitive Advantage with Excel 2010: Basics, Model Building, and Cases*,
DOI 10.1007/978-1-4419-9857-6_1, © Springer Science+Business Media, LLC 2012

In the majority of examples, analysis includes graphics. Seeing data provides an information dimension beyond numbers in tables. You need to see the data and their shape and dispersion in order to understand a market or population well. To become a master modeler, you need to be able to see how change in one variable is driving a change in another. Graphics are essential to solid model building and analysis. Graphics are also essential to effective translation of results. Effective memos and PowerPoint slides feature key graphics that help your audience digest and remember results. We feature PivotTables and PivotCharts in Chap. 7. These are routinely used in business to efficiently organize and display data. When you are at home in the language of PivotTables and PivotCharts, you will have a competitive advantage; practice using them to organize financial analyses and market data. Form the habit of looking at data and results whenever you are considering decision alternatives.

1.2 The Path toward Statistical Competence and Competitive Advantage

This book assumes basic statistical knowledge and reviews basics quickly. Basics form the foundation for essential model building. Chapters 2 and 3 present a concentrated introduction to data and their descriptive statistics, samples, and inference. Learn how to efficiently describe data and how to infer population characteristics from samples.

Model building with simple regression begins in Chap. 4 and occupies the focus of the remaining chapters. To be competitive, business graduates must have competence in model building and forecasting. A model building mentality focused on performance drivers and their synergies is a competitive advantage. Practice thinking of decision variables as drivers of performance and also that performance is driven by decision variables. Performance will improve if this linkage becomes second nature.

The approach to model building is steeped in logic and begins with logic and experience. Models must make sense in order to be useful. When you understand how decision variables drive performance under alternate scenarios, you can make better decisions, enhancing performance. Model building is an art that begins with logic.

Model building chapters 11, 12, and 13 include nonlinear regression and logit regression. Nearly all aspects of business performance behave in nonlinear ways. We see diminishing or increasing changes in performance in response to changes in drivers. It is useful to begin model building with the simplifying assumption of constant response, but it is essential to be able to grow beyond simple linear models to realistic models that reflect nonconstant response. Logit regression, appropriate for the analysis of bounded performance measures such as market share and probability of trial, has many useful applications in business and is an essential tool for managers. Resources and markets are limited, and responses to decision variables are also necessarily limited, as a consequence. Visualize the changing pattern of response when you consider decision alternatives and the ways they drive performance.

1.3 Use Excel for Competitive Advantage

This book features widely available Excel software, including many commonly used shortcuts. Excel is powerful, comprehensive, and user friendly. Appendices with screenshots follow each chapter to make software interactions simple. Re-create the chapter examples by following the steps in the Excel sections. This will give you confidence using the software. Then forge ahead and generalize your analyses by working through end of chapter assignments. The more often you use the statistical tools and software, the easier the analysis becomes.

1.4 Statistical Competence Is Powerful and Yours

Statistics and their potential to alter decisions and improve performance are important to you. With more and better information from statistical analysis, you will be equipped to make superior decisions and outperform the competition. You will find that the competitive advantages from statistical competence are powerful and yours.

Chapter 2
Describing Your Data

This chapter introduces *descriptive* statistics, center, spread, and distribution shape, which are almost always included with any statistical analysis to characterize a dataset. The particular descriptive statistics used depend on the *scale* that has been used to assign numbers to represent the characteristics of entities being studied. When the distribution of continuous data is bell shaped, we have convenient properties that make description easier. This chapter looks at dataset types and their description.

2.1 Describe Data with Summary Statistics and Histograms

We use numbers to measure aspects of businesses, customers, and competitors. These measured aspects are *data*. Data become meaningful when we use statistics to describe patterns within particular *samples* or collections of businesses, customers, competitors, or other entities.

Example 2.1 Yankees' Salaries: Is It a Winning Offer?

Suppose that the Yankees want to sign a promising rookie. They expect to offer $1M, and they want to be sure they are neither paying too much nor too little. What would the General Manager need to know to decide whether this is the right offer?

He might first look at how much the other Yankees earn. Their 2005 salaries are listed in Table 2.1.

Table 2.1 Yankees' salaries (in $M) in alphabetical order

Crosby	$.3	Johnson	$16.0	Posada	$11.0	Sierra	$1.5
Flaherty	.8	Martinez	2.8	Rivera	10.5	Sturtze	.9
Giambi	1.3	Matsui	8.0	Rodriguez	21.7	Williams	12.4
Gordon	3.8	Mussina	19.0	Rodriguez F	3.2	Womack	2.0
Jeter	19.6	Phillips	.3	Sheffield	13.0		

What should he do with these data?

Data are more useful if they are ordered by the aspect of interest. In his case, the General Manager would re-sort the data by salary (Table 2.2).

C. Fraser, *Business Statistics for Competitive Advantage with Excel 2010: Basics, Model Building, and Cases*, DOI 10.1007/978-1-4419-9857-6_2, © Springer Science+Business Media, LLC 2012

Table 2.2 Yankees sorted by salary (in $M)

Rodriguez	$21.7	Williams	$12.4	Rodriguez F	$3.2	Sturtze	$.9
Jeter	19.6	Posada	11.0	Martinez	2.8	Flaherty	.8
Mussina	19.0	Rivera	10.5	Womack	2.0	Crosby	.3
Johnson	16.0	Matsui	8.0	Sierra	1.5	Phillips	.3
Sheffield	13.0	Gordon	3.8	Giambi	1.3		

Now he can see that the lowest Yankee salary, the *minimum*, is $300,000 and the highest salary, the *maximum*, is $21.7M. The difference between the maximum and the minimum is the *range* in salaries, which is $21.4M, in this example. From these statistics, we know that the salary offer of $1M falls in the lower portion of this range. In addition, however, he needs to know just how unusual the extreme salaries are to better assess the offer.

He'd like to know whether the rookie would be in the better paid half of the team. This could affect morale of other players with lower salaries. The *median,* or middle, salary is $3.8M. The lower paid half of the team earns between $300,000 and $3.8M, and the higher paid half of the team earns between $3.8 and $21.7M. Thus, the rookie would be in the bottom half. The General Manager needs to know more to fully assess the offer.

Often, a *histogram* and a *cumulative distribution plot* are used to visually assess data, as shown in Figs. 2.1 and 2.2. A histogram illustrates central tendency, dispersion, and symmetry. The histogram of team salaries shows us that a large proportion, more than 40%, earn more than $400,000, but less than the average, or *mean*, salary of $8M.

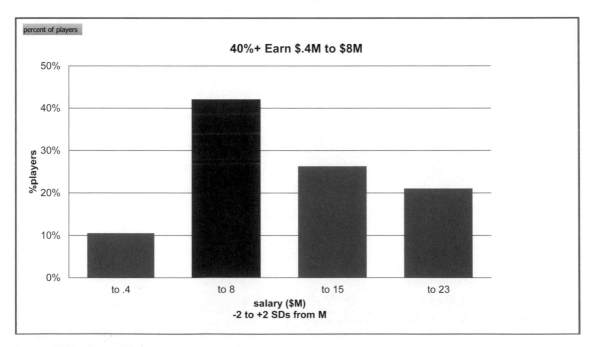

Fig. 2.1 Histogram of Yankee salaries

The cumulative distribution makes it easy to see the *median*, or 50th percentile, which is one measure of central tendency. It is also easy to find the *interquartile range*, the range of values that the middle 50% of the datapoints occupy, providing a measure of the data dispersion.

The cumulative distribution reveals that the *interquartile range*, between the 25th and 75th percentiles, is more than $10M. A quarter earn less than $1.4M, the 25th percentile, about half earn between $1.5 and $13M, and a quarter earn more than $13M, the 75th percentile. Half of the players have salaries below the median of $4M and half have salaries above $4M.

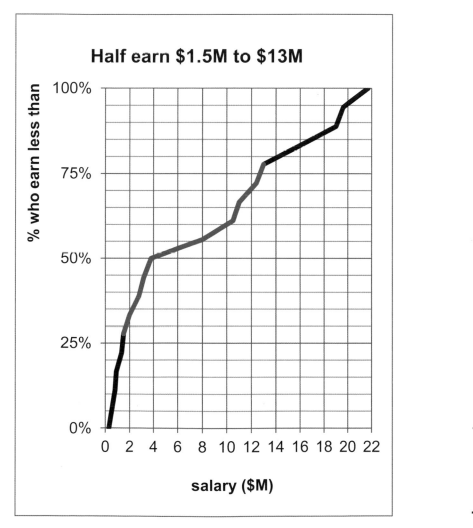

Salary ($M)	
25%	1.4
Median	4
75%	13

Fig. 2.2 Cumulative distribution of salaries

2.2 Outliers Can Distort the Picture

Outliers are extreme elements and considered unusual when compared with other sample elements. Because they are extraordinary, they can distort descriptive statistics.

Example 2.2 Executive Compensation: Is the Board's Offer on Target?
The board of a large corporation is pondering the total compensation package of the CEO, which includes salary, stock ownership, and fringe benefits. Last year, the CEO earned $2,000,000. For comparison, the Board consulted *Forbes*' summary of the total compensation of the 500 largest corporations. The histogram, cumulative frequency distribution, and descriptive statistics are shown in Figs. 2.3 and 2.4.

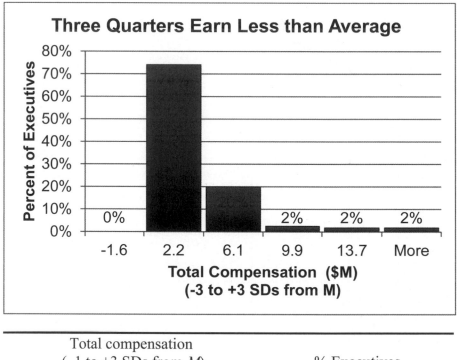

Total compensation (-1 to $+3$ SDs from M)	% Executives
2	74
6	20
10	2
14	2
More	2

Fig. 2.3 Histogram of executive compensation

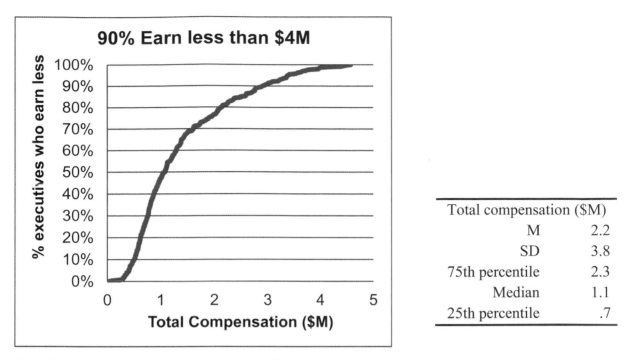

Fig. 2.4 Cumulative distribution of total compensation

Total compensation ($M)	
M	2.2
SD	3.8
75th percentile	2.3
Median	1.1
25th percentile	.7

The average executive compensation in this sample of large corporations is $2.2M. Half of the sample of 447 executives earn $1.1M (the median) or less. One quarter earn less than $.7M, the middle half, or *interquartile range,* earn between $.7 and $2.3M, and one quarter earn more than $2.3M.

Why is the *mean*, $2.2M, so much larger than the median, $1.1M? There is a group of eight *outliers*, shown as more than three standard deviations above the mean in Fig. 2.4, who are compensated extraordinarily well with each collecting a compensation package of more than $14M.

When we exclude these 8 outliers, 11 additional outliers emerge. This cycle repeats, because the distribution is highly skewed. When we remove outliers, the new mean is adjusted, making other executives appear to be more extreme. In this case, removing this 10%, or the 44 best compensated executives, gives us a better picture of what "typical" compensation is, which is shown in Fig. 2.5.

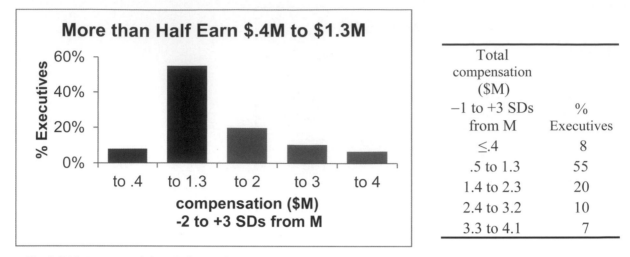

Total compensation ($M)	
−1 to +3 SDs from M	% Executives
≤.4	8
.5 to 1.3	55
1.4 to 2.3	20
2.4 to 3.2	10
3.3 to 4.1	7

Fig. 2.5 Histogram and descriptive statistics with 44 outliers excluded

Ignoring the 44 outliers, the average compensation is about $1.3M and the median compensation is about $1M as shown in Fig. 2.6. The mean and median are closer. With this more representative description of executive compensation in large corporations, the board has an indication that the $2M package is well above average. More than three quarters of executives earn less. Because extraordinary executives exist, the original distribution of compensation is skewed, with relatively few exceptional executives being exceptionally well compensated.

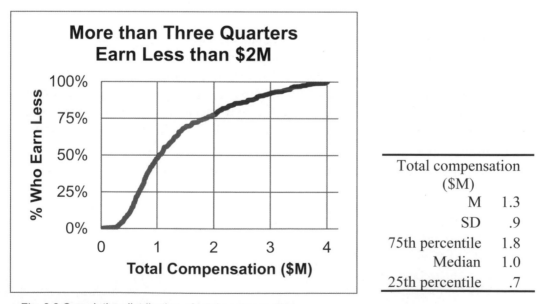

Total compensation ($M)	
M	1.3
SD	.9
75th percentile	1.8
Median	1.0
25th percentile	.7

Fig. 2.6 Cumulative distribution of total compensation

2.3 Round Descriptive Statistics

In the examples above, output statistics from statistical packages are presented with many decimal points of accuracy. The Yankee General Manager in Example 2.1 and the board considering executive compensation in Example 2.2 will most likely be negotiating in hundred thousands. It would be distracting and unnecessary to report descriptive statistics with more than two or three significant digits. In the *Yankees* example, the average salary is $8,000,000 (not $7,797,000). In the *executive compensation* example, the average total compensation is $1,300,000 (not $1,349,970). It is deceptive to present results with many significant digits, creating an illusion of precision. In addition to being honest, statistics in two or three significant digits are much easier for decision makers to process and remember.

2.4 Share the Story that Your Graphics Illustrate

Use your graphics to support the conclusion you have reached from your analysis. Choose a "bottom line" title that shares with your audience what it is that they should be able to see. Often this title should relate specifically to your reasons for analyzing data. In the executive compensation example, the board is considering a $2M offer. The chart titles capture board interest by highlighting this critical value. The bottom line, that a $2M offer is relatively high when compared with similar firms, makes the illustrations relevant.

Many have the unfortunate and unimaginative habit of choosing chart titles that name the type of chart. "Histogram of Executive Salaries," tells the audience little, beyond the obvious realization that they must form their own independent conclusions from the analysis. Choose a bottom line title so that decision makers can take away your conclusion from the analysis. Develop the good habit of titling your graphics to enhance their relevance and interest.

2.5 Central Tendency, Dispersion, and Skewness Describe Data

The baseball salaries and executive compensation examples focused on two measures of *central tendency*: the *mean*, or average, and the *median*, or middle. Both examples also refer to a measure of *dispersion* or variability: the *range* separating the minimum and maximum. *Skewness* reflects distribution symmetry. To describe data, we need statistics to assess central tendency, dispersion, and skewness. The statistics we choose depends on the scale that has been used to code the data we are analyzing.

2.6 Data Are Measured with Quantitative or Categorical Scales

If the numbers in a dataset represent amount, or magnitude of an aspect, and if differences between adjacent numbers are equivalent, the data are *quantitative* or *continuous*. Data measured in dollars (i.e., revenues, costs, prices, and profits) or percentages (i.e., market share, rate of return, and exam scores) are continuous. Quantitative numbers can be added, subtracted, divided, or multiplied to produce meaningful results.

With quantitative data, report central tendency with the mean, *M:*

$$\mu = \frac{\sum x_i}{N} \quad \text{for describing a population and}$$

$$\overline{X} = \frac{\sum x_i}{N} \quad \text{for describing a sample from a population,}$$

where x_i are datapoint values and N is the number of datapoints that we are describing.

The median can also be used to assess central tendency, and the range, variance, and standard deviation can be used to assess dispersion. The *variance* is the average squared difference between each of the datapoints and the mean:

$$\sigma^2 = \frac{\sum(x_i - \mu)^2}{N} \quad \text{for a population and}$$

$$s^2 = \frac{\sum(x_i - \overline{X})^2}{(N-1)} \quad \text{for a sample from a population.}$$

The *standard deviation SD*, σ for a population and s for a sample, is the square root of the variance, which gives us a measure of dispersion in the more easily interpreted original units, rather than squared units.

To assess distribution symmetry, assess its skewness:

$$\text{Skewness} = \frac{n}{(n-1)(n-2)} \sum \left(\frac{x_i - \overline{x}}{s} \right)^3.$$

Skewness of zero indicates a symmetric distribution, and skewness between −1 and +1 is evidence of an approximately symmetric distribution.

If numbers in a dataset are arbitrary and used to distinguish categories, the data are *nominal* or *categorical*. Football jersey numbers and your student ID are nominal. A larger number doesn't mean that a player is better or a student is older or smarter. Categorical numbers can be tabulated to identify the most popular number, occurring most frequently, the *mode*, to report central tendency. Categorical numbers cannot be added, subtracted, divided, or multiplied.

Quantitative measures convey more information, including direction and magnitude, whereas categorical measures convey less, sometimes direction and sometimes merely category membership. One more informative type of categorical data is *ordinal* scales used to rank order data, or to convey direction, but not magnitude. With ordinal data, an element (which could be a business, a person, or a country) with the most or best is coded as "1," second place as "2," and so on. With ordinal numbers, or rankings, data can be sorted, but not added, subtracted, divided, or multiplied. As with other categorical data, the mode represents the central tendency of ordinal data.

When focus is on membership in a particular category, the *proportion* of sample elements in the category is a continuous measure of central tendency. Proportions are quantitative and can be added, subtracted, divided, or multiplied, although they are bounded by zero, below, and by one, above.

2.7 Continuous Data Are Sometimes Normal

Continuous variables are often *normally distributed*, and their histograms resemble symmetric, bell-shaped curves, with the majority of datapoints clustered around the mean. Most elements are "average" with values near the mean; fewer elements are unusual and far from the mean. If continuous data are normally distributed, we need only the mean and standard deviation to describe these data and description is simplified.

Example 2.3 Normal SAT Scores
Standardized tests, such as the SAT, capitalize on normality. Math and verbal SATs are both specifically constructed to produce normally distributed scores with mean $M = 500$ and standard deviation SD − 100 over the population of students (Fig. 2.7).

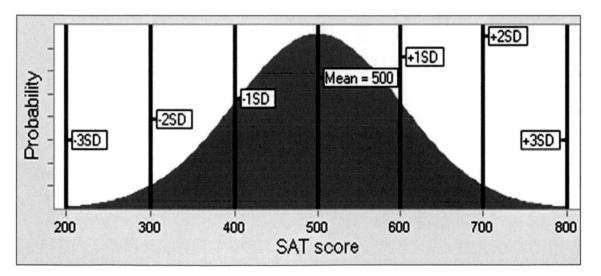

Fig. 2.7 Normally distributed SAT scores

2.8 The Empirical Rule Simplifies Description

Normally distributed data have a very useful property described by the Empirical Rule:

2/3 of the data lie within one standard deviation of the mean.
95% of the data lie within two standard deviations of the mean.

This is a powerful rule! If data are normally distributed, data can be described with just two statistics: the mean and the standard deviation.

Returning to SAT scores, if we know that the average score is 500 and the standard deviation is 100, we also know that

> 2/3 of SAT scores will fall within 100 points of the mean of 500, or between 400 and 600.
> 95% of SAT scores will fall within 200 points of the mean of 500, or between 300 and 700.

Example 2.4 Class of '10 SATs: This Class Is Normal and Exceptional

Descriptive statistics and a histogram of math SATs of a third year class of business students reveal an interquartile range from 640 to 730, with mean of 690 and standard deviation of 70, as shown in Fig. 2.8. Skewness is −.5, indicating approximate symmetry.

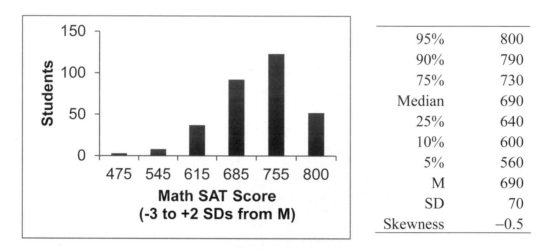

95%	800
90%	790
75%	730
Median	690
25%	640
10%	600
5%	560
M	690
SD	70
Skewness	−0.5

Fig. 2.8 Histograms and descriptive statistics of class '10 math SATs

Are Class '10 math SATs Normally distributed? Approximately. Class '10 scores are bell shaped. There are "too many" perfect scores of 800.

The Empirical Rule would predict that 2/3 of the class would have scores within 1 standard deviation of 70 points of the mean of 690, or within the interval 620–760. There are actually 67% (=37% + 30%).

The Empirical Rule would also predict that only 2.5% of the class would have scores more than two standard deviations below or above the mean of 690: scores below 550 and above 830. We find that 4% actually do have scores below 530, although none score above 830 (inasmuch as perfect SAT score is 800). This class of business students has math SATs that are nearly normal, but not exactly normal.

To summarize Class '10 students' SAT scores, report the following.

> Class '10 students' math SAT scores are approximately normally distributed with mean of 690 and standard deviation of 70.

Relative to the larger population of all SAT takers, the smaller standard deviation in Class '10 students' math SAT scores, 70 versus 100, indicates that Class '06 students are a more homogeneous group than the more varied population.

2.9 Describe Categorical Variables Graphically: Column and PivotCharts

Numbers representing category membership in nominal, or categorical, data are described by tabulating their frequencies. The most popular category is the *mode*. Visually, we show our tabulations with a *Pareto* chart, which orders categories by their popularity.

Example 2.5 Who Is Honest and Ethical?
Figure 2.9 shows a Column chart of results of a survey of 1,014 adults by Gallup.

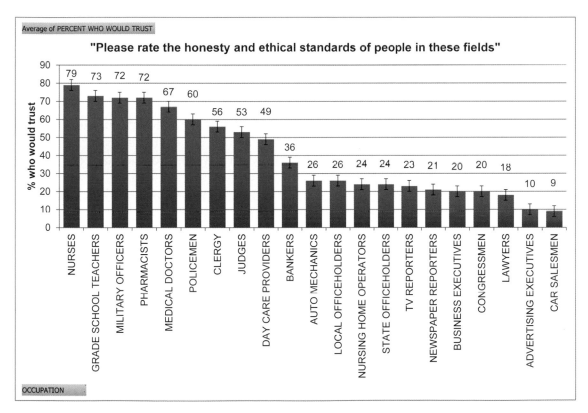

Fig. 2.9 Pareto charts of the percentages who judge professions honest

More Americans trust and respect nurses (79%, the *modal* response) than people in other professions, including doctors, clergy, and teachers. Although a small minority judge business executives (20%) and advertising professionals (10%) as honest and ethical, most do not judge people in those fields to be honest (which highlights the importance of ethical business behavior in the future).

2.10 Descriptive Statistics Depend on the Data and Your Packaging

Descriptive statistics, graphics, central tendency, and dispersion, depend upon the type of scale used to measure data characteristics (i.e., quantitative or categorical). Table 2.3 summarizes the descriptive statistics (graph, central tendency, dispersion, shape) used for both types of data.

Table 2.3 Descriptive statistics (central tendency, disperson, graphics) for two types of data

	Quantitative	*Categorical*
Central tendency	Mean	Mode
	Median	Proportion
Dispersion	Range	
	Standard deviation	
Symmetry	Skewness	
Graphics	Histogram	Pareto chart
	Cumulative distribution	Pie chart
		Column chart

If continuous data are normally distributed, a dataset can be completely described with just the mean and standard deviation. We know from the Empirical Rule that 2/3 of the data will lie within one standard deviation of the mean and that 95% of the data will lie within two standard deviations of the mean.

Effective results are those that are remembered and used to improve decision making. Your presentation of results will influence whether decision makers remember and use your results. Round statistics to two or three significant digits to make them honest, digestible, and memorable. Title your graphics with the bottom line to guide and facilitate decision makers' conclusions.

Excel 2.1 Produce Descriptive Statistics and Histograms

Executive Compensation
We describe executive compensation packages by producing descriptive statistics, a histogram, and cumulative distribution.

First, freeze the top row of **Excel 2.1 Executive Compensation.xls** so that column labels are visible when you are at the bottom of the dataset.

Select the first cell, **A1**, and then use Excel shortcuts **Alt WFR.** (The shortcuts, activated with **Alt**, select the vie**W** menu , the **F**reeze panes menu, and then freeze **R**ows.)

Select **B1**, then use shortcuts to move to the end of the file where we add descriptive statistics. **Cntl+down arrow** scrolls through all the cells containing data in the same column and stops at the last filled cell.

Descriptive statistics: In the first empty cell in the column, below the data, use shortcuts to find the sample mean: Alt **MUA**.

Use the following:

STDEV(*array***)** function to find the standard deviation
PERCENTILE(*array***, .75)** and **PERCENTILE(***array***, .25)** functions to find the 75th and 25th percentile values
MEDIAN(*array***)** function to find the median
SKEW(*array***)** function to find skewness

B454		f_x	=SKEW(B2:B448)

	A	B	C
1		*Total Compensation (MM$)*	
446		20.658	
447		32.582	
448		53.111	
449	M	2.22	
450	SD	3.84	
451	75%	2.26	
452	median	1.13	
453	25%	0.72	
454	skew	7.44	

Histograms: To make a histogram of salaries, Excel needs to know what ranges of values to combine. Set these *bins*, or categories, to differences from the sample mean that are in widths of standard deviations.

Histogram bins.xls uses formulas to find cutoff values for histogram bins of three standard deviations below the mean to three standard deviations above the mean using a default mean of 0 and standard deviation of 1. Change these to the sample mean and standard deviation.

Open **histogram bins.xls**, select **A1:E9**, then copy **Cntl+C**.

	A	B	C	D	E
1	M	SD	*SDs from M*	*distribution % if Normal*	*SDs from M*
2	0	1	-3	0.1%	3 ≤ sds below mean (outliers)
3			-2	2.1%	2 ≤ sds < 3 below mean
4			-1	13.6%	1 ≤ sds < 2 below mean
5			0	34.1%	0 ≤ sds < 1 below mean
6			1	34.1%	0 < sds ≤ 1 above mean
7			2	13.6%	1 < sds ≤ 2 above mean
8			3	2.1%	2 < sds ≤ 3 above mean
9				0.1%	3 < sds above mean (outliers)

In the executive compensation file, select **C448**, **[Enter]**, to paste the histogram bins formulas.

	A	B	C	D	E	F	G	H	I
1		Total Compensation (MM$)							
446		20.658							
447		32.582							
448		53.111	M	SD	SDs from M	distribu tion % if Normal	SDs from M		
449	M	2.22	0	1	-3	0.1%	3 ≤ sds below mean (outliers)		
450	SD	3.84			-2	2.1%	2 ≤ sds < 3 below mean		
451	75%	2.26			-1	13.6%	1 ≤ sds < 2 below mean		
452	median	1.13			0	34.1%	0 ≤ sds < 1 below mean		
453	25%	0.72			1	34.1%	0 < sds ≤ 1 above mean		
454	skew	7.44			2	13.6%	1 < sds ≤ 2 above mean		
455					3	2.1%	2 < sds ≤ 3 above mean		
456						0.1%	3 < sds above mean (outliers)		

In **C449**, replace the mean of 0 with the sample mean by entering **=B449**.

In **D449**, replace the standard deviation of 1 with the sample standard deviation by entering **=B450**.

D449	▼	*fx*	=B450

◢	A	B	C	D
1		*Total Compensation (MM$)*		
446		20.658		
447		32.582		
448		53.111 *M*		*SD*
449	M	2.22	2.22	3.84
450	SD	3.84		

To see the distribution of *total compensation*, use shortcuts **Alt AY11 H** to request a histogram. (**Alt AY11** selects the dat**A** menu and the data Anal**Y**sis menu.)

◢	A	B	C	D	E	F	G
1		*Total Compensation (MM$)*					
441		14.652					
442		14.925					
443		15.706					
444		15.915					
445		16.172					
446		20.658					
447		32.582					
						distribution %	
					SDs	*if*	*SDs*
448		53.111 *M*		*SD*	*from M*	*Normal*	*from M*
449	M	2.22	2.22	3.84	-9.299	0.1%	3 ≤ sds below mean (outliers)
450	SD	3.84			-5.461	2.1%	2 ≤ sds < 3 below mean
451		75%	2.26		-1.623	13.6%	1 ≤ sds < 2 below mean
452	median	1.13			2.2153	34.1%	0 ≤ sds < 1 below mean
453		25%	0.72		6.0533	34.1%	0 < sds ≤ 1 above mean
454	skew	7.44			9.8914	13.6%	1 < sds ≤ 2 above mean
455					13.729	2.1%	2 < sds ≤ 3 above mean

Histogram

Input
Input Range: $B$1:$B$448
Bin Range: $E$448:$E$455
☑ Labels

Output options
○ Output Range:
⦿ New Worksheet Ply:
○ New Workbook

☐ Pareto (sorted histogram)
☐ Cumulative Percentage
☑ Chart Output

OK
Cancel
Help

For **Input Range**, enter the *total compensation* cells, **B1:B448**; for **Bin Range,** enter the *histogram bin* cells, **E448:E455**, with **L**abels**,** and **C**hart Output.

Select **A2:A8** and use shortcuts **Alt H9** to reduce the unnecessary decimals. (**H** selects the **H**ome menu and **9** selects the reduce decimals function of the Number menu.)

	A	B	C	D	E	F	G	H	I
1	SDs from M	Frequency							
2	-9.3	0							
3	-5.5	0							
4	-1.6	0							
5	2.2	331							
6	6.1	89							
7	9.9	11							
8	13.7	8							
9	More	8							
10									
11									

Excel 2.2 Sort to Produce Descriptives Without Outliers

Outliers are executives whose total compensation is more than three standard deviations greater than the mean. There are eight such executives in this sample, tabulated in the **More** histogram bin, and each earns more than $14M.

To easily remove outliers, sort the rows from lowest to highest *total compensation* ($M), select *total compensation* data in column **B** (but not statistics below the data), then use shortcuts to sort: **Alt ASA,** **C**ontinue with the current selection, **S**ort. (**A** selects the dat**A** menu, **S** selects the Sort menu, and **A** specifies **A**scending.)

B	C	D
Total Compensation (MM$)		
0.029		
0.052		
0.102		
0.222		

Sort Warning

Microsoft Excel found data next to your selection. Since you have not selected this data, it will not be sorted.

What do you want to do?

○ Expand the selection
◉ Continue with the current selection

[Sort] [Cancel]

◢	A	B
1		Total Compensation (MM$)
439		12.838
440		13.071
441		**14.652**
442		14.925

Scroll up from the end of **B** to identify the eight rows that are less than 14.

Recalculate the mean, standard deviation, 25th percentile, median, 75th percentile, and skewness, including only rows with *total compensation* less than $14M, by changing the end of the array in each Excel function. (The histogram bins formulas will automatically update bin cutoffs with your new mean and standard deviation.)

B454	▾	f_x	=SKEW(B2:B440)		
	A	B		C	D
		Total Compensation (MM$)			
		20.658			
		32.582			
		53.111	*M*		*SD*
M		1.84		1.84	2.01
SD		2.01			
	75%	2.15			
median		1.12			
	25%	0.71			
skew	◈	**3.07**			

Rerun the histogram tabulation, excluding the eight outliers by changing the array end in **Input Data**.

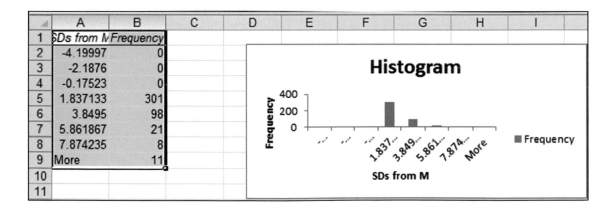

	A	B	C	D	E	F	G	H	I
1	SDs from M	Frequency							
2	-1.4	0							
3	-0.5	0							
4	0.4	32							
5	1.3	222							
6	2.3	80							
7	3.2	42							
8	4.1	27							
9	More	0							
10									
11									

Repeat this process to continue excluding outliers until there are no outliers. Because the distribution of total compensation is highly skewed, *outliers* will continue to appear.

Including only executives whose total compensation is less than $4.1M, the descriptive statistics become more representative.

	A	B	C	D	E	F	G	H	I
1	SDs from M	Frequency							
2	-4.19997	0							
3	-2.1876	0							
4	-0.17523	0							
5	1.837133	301							
6	3.8495	98							
7	5.861867	21							
8	7.874235	8							
9	More	11							
10									
11									

Excel 2.3 Plot a Cumulative Distribution

To see the cumulative distribution of total compensation, choose **Rank and Percentile, Alt AY11, Rank and Percentile.**

For convenience, select and delete column **C, Alt HDC**. (**H** selects the **H**ome menu, **D** selects the **D**elete menu, and **C** deletes the **C**olumn.)

Plot *Total Compensation* in **B** by *Percent* in **C** to see the cumulative distribution plot, **Alt ND**.

Excel 2.4 Find and View Distribution Percentages with a PivotChart

Class of '10 Math SATs

	A	B
1		*MathSAT*
313		490
314		470
315		450
316		430
317	*mean*	685
318	*sd*	70
319	*skewness*	-0.5

Descriptive statistics: At the end of the dataset, add the *mean,* the *standard deviation*, and *skewness.*

To further assess normality, we want to see the sample percentages that are –3 to +3 standard deviations from the sample mean. Make the descriptive statistics and histogram tabulation.

Histogram tabulation: Copy and paste the histogram bins.xls formulas into the Excel 2.4 **SATs.xls** file in columns **E**, **F**, and **G**. Then, change the mean and standard deviation to those from the sample.

	F317		f_x	=B318		
	A	B	C	D	E	F	G
1		**MathSAT**	**VerbalSAT**	**TotalSAT**			
311		510	550	1060			
312		500	520	1020			
313		490	590	1080			
314		470	590	1060			
315		450	570	1020			
316		430	510	940 *M*		*SD*	*SDs from M*
317	M	685			685	70	475
318	SD	70					545
319	skewness	-0.5					615
320							685
321							755
322							825
323							895

Order the histogram tabulation.

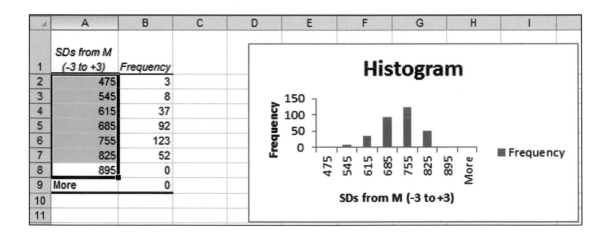

	A	B
1	SDs from M (-3 to +3)	Frequency
2	475	3
3	545	8
4	615	37
5	685	92
6	755	123
7	825	52
8	895	0
9	More	0
10		
11		

	A	B	C
1	SDs from M (-3 to +3)	Frequency	distribution % if Normal
2	475	3	0.1%
3	545	8	2.1%
4	615	37	13.6%
5	685	92	34.1%
6	755	123	34.1%
7	825	52	13.6%
8	895	0	2.1%
9	More	0	0.1%

PivotTable and PivotChart of a distribution in percents: Reduce decimals in A2:A7, copy *distribution % if Normal* from the **SATs** sheet and paste into the histogram sheet.

Select **A1:C8** and make a PivotTable, **Alt NVT**. (**N** selects the i**N**sert menu, **V** selects the Pi**V**ot menu, and **T** inserts a Pivot**T**able.)

To set up your PivotTable, drag *SDs from M* to **ROW** and *Frequency* to Σ **values**.

Change the table to percents using shortcuts: **Alt JTAG**. (**JT** accesses the Options menu, **A** accesses Show Values As, and **G** corresponds to *% **G**rand total*.)

To compare the sample distribution with Normal, drag the *distribution % if Normal* to the Σ **values** box and then change to percents and reduce decimals.

Make the **PivotChart** by clicking inside the table, and then **Alt JTC**. (**C** requests insertion of a **C**hart.)

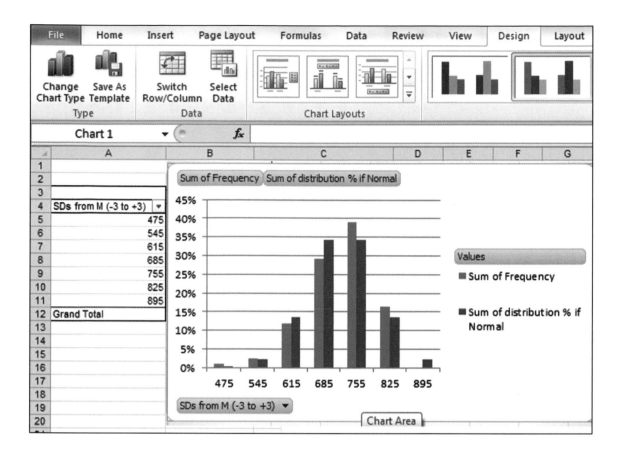

Excel 2.5 Produce a Column Chart of a Nominal Variable

A firm is targeting customers who consult a news source daily. Management wants to compare the popularity of news sources. To facilitate comparisons, we make a PivotChart from a Gallup Poll of 992 Americans. Data are in **Excel 2.5 News Sources.xls**.

Open **Excel 2.3 News Sources.xls**, select **A1:B11**, and insert a Column chart **Alt NC. (N** activates I**N**sert, and **C** inserts a **C**olumn chart.)

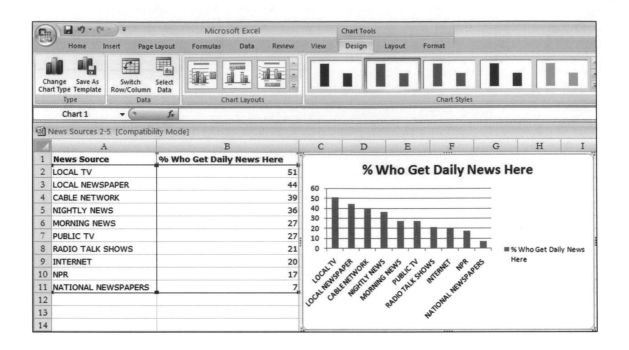

To add the vertical margin of error bars, using a click inside a column, **Alt JARM, Tab, F**(ixed value), **Tab,** 3 (the approximate margin of error). **(JA** selects the Layout menu, **R** selects the Erro**R** Bar menu, and **M** selects the custo**M** Error Bar menu.)

Choose **Design Chart Layout 6, Alt JCL, Tab** to layout 6.

Type in a bottom line title and a vertical axis title and add data labels.

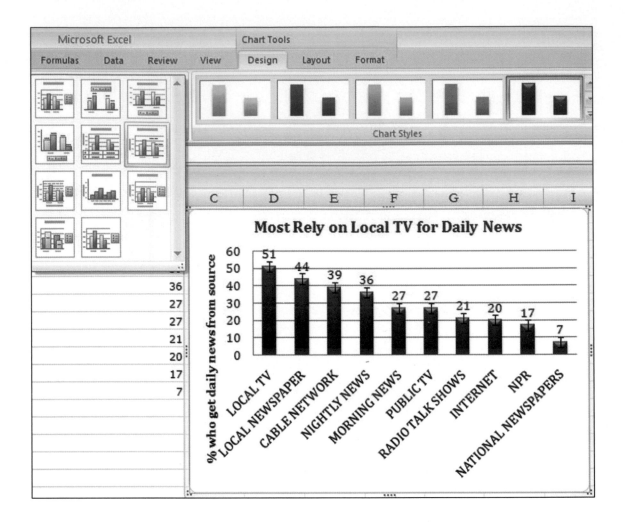

Excel Shortcuts at Your Fingertips by Shortcut Key

Alt activates the shortcuts menus, linking keyboard letters to Excel menus. Press **Alt**, then release and press letters linked to the menus you want.

Press **Alt H**ome, then

9 to select the reduce decimals function

IC to select the **I**nsert function and to insert a **C**olumn to the left

Press **Alt H**, then

> **DC** to select the **D**elete function to delete **C**olumn(s)

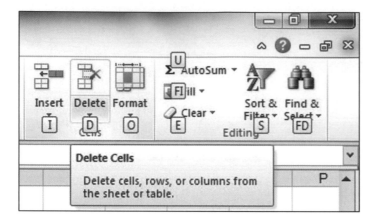

Press **Alt** Dat**A**, then

> **SS** to select the **S**ort menus

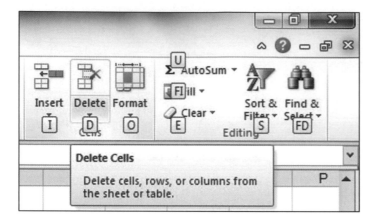

> **Y11** to select the Data Anal**Y**sis menus

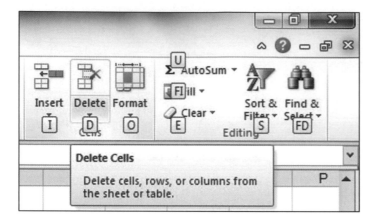

Press **Alt I**N**sert**, then

   VT to insert a Pivot**T**able

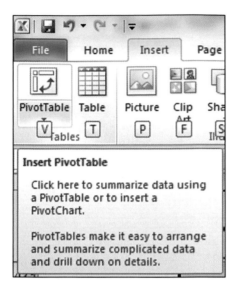

   C to insert a **C**olumn chart

Press **Alt N**, then

Q to insert a Pie chart

D to insert a scatterplot

Press **Alt N**, then

   X to insert a te**X**t box

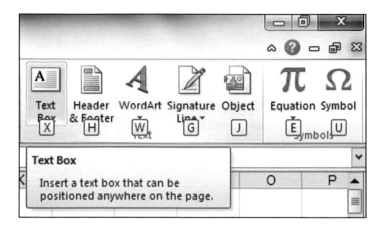

Press **Alt** Vie**W**, then **FR** to select the **F**reeze panes menus and to freeze **R**ows

Press **Alt JA**, then

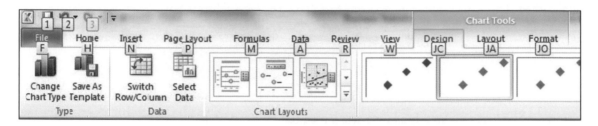

B to select the Dat**A** La**B**els menus.

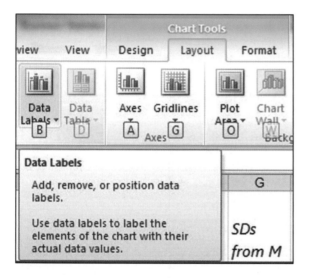

Press **Alt JA**, then

RM to select the Erro**R** Bar and the custo**M** error bar menus.

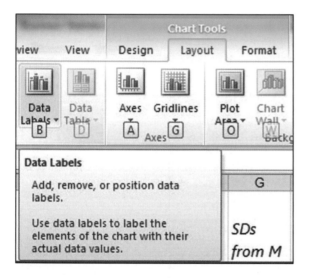

Shift+arrow selects cells scrolled over.

Cntl+

<u>C</u>opies

down arrow scrolls through all cells in the same column that contain data

R fills in values of empty cells in a <u>R</u>ow using a formula from the first cell in a selected array.

Shift+down arrow selects all filled cells in the column.

Significant Digits Guidelines

The number of significant digits in a number conveys information. Significant digits include the following:

1. All nonzero numbers
2. Zeros between nonzero numbers
3. Trailing zeros

Zeros acting as placeholders aren't counted.

The number 2,061 has four significant digits, whereas the number 2,610 has three, because the zero is merely a placeholder. The number 0.0920 has three significant digits, "9," "2," and the final, trailing "0." The first two zeros are placeholders that aren't counted.

In rare cases, it is not clear whether zero is a placeholder or a significant digit. The number 40,000 could represent the range 39,500–40,499. In that case, the number of significant digits is 1, and the zeros are placeholders. Alternatively, 40,000 could represent the range 39,995–40,004. In this latter case, the number of significant digits is 4, because the zeros convey meaning. When in doubt, a number could be written in scientific notation, which is unambiguous. For one significant digit, 40,000 becomes $4 \times E^4$. For four significant digits, 40,000 becomes $4.000 \times E^4$.

Lab 2 Descriptive Statistics

A Typical Executive's Compensation
Help the Board of a firm in the financial industry evaluate the $2MM compensation package that they expect to offer the CEO. Summarize the Forbes' data on executive compensation in **Lab 2 Executive Compensation.xls**.

1. Find the sample mean, standard deviation, and skewness of *compensation*, and then make a histogram of *compensation* in financial firms.

   Average compensation: _____ Standard deviation: _____ Skewness: _____
   How many executives earn an unusually high or low package (more than 3 SDs above or below the average)? _____

2. Remove outliers and rerun, repeating this process until outliers no longer appear.

   Average compensation: _____ Standard deviation: _____ Skewness: _____

3. Compare the distribution of compensation in the financial sector with a Normal distribution with the same mean and standard deviation.

   How does the actual distribution differ from Normal? _____

   _____

4. Find the 75%, median, and 25% compensation values, excluding outliers, and then plot the cumulative distribution of executive compensation.

   In large financial firms,
   25% of executives earn more than _____

   Half of executives earn less than _____

   25% of executives earn less than _____

   Half of executives earn between _____ and _____

5. What can the Board say to the CEO to describe the $2MM package proposal?

   _____

6. One Board member has heard rumors that American Express, a competitor, may try to hire the CEO. Will the $2MM package be competitive?   Y or N

Hollywood Politics

Managers of a political campaign are considering launch of an effort to attract Hollywood celebrity endorsements.

Summarize public opinion of celebrity endorsements reported in a CBS News/New York Times poll. Data are in **Lab 2 Hollywood Politics.xls**. Be sure to round your answers to two or three significant digits.

1. What percent of Republicans prefer celebrities to stay out of politics? ___ to ____ %.

2. What percent of Democrats prefer celebrities to stay out of politics? ____ to ____%.

3. Make a PivotTable and PivotChart (Column chart) comparing the percentages of Republicans, Democrats, and Independents who prefer celebrities to stay out of politics. Add a bottom line title that summarizes poll results.

Assignment 2-1 Procter & Gamble's Global Advertising

Procter & Gamble spent $5,960,000 on advertising in 51 global markets. These data from Advertising Age, Global Marketing, are in **Assignment 2-1 P&G Global Advertising.xls.** P&G Corporate is reviewing the firm's global advertising strategy, which is the result of decisions made by many brand management teams. Corporate wants to be sure that these many brand level decisions produce an effective allocation when viewed together.

Describe Procter & Gamble's advertising spending across the 51 countries that make up the global markets.

1. Identify *countries* that are **outliers.**

   - Find the sample mean and standard deviation, then use these to make a histogram.
   - Sort the *countries* by *advertising*, then recalculate the sample mean and standard deviation and make a second histogram, *excluding outliers*.
   - Repeat the process of removing outliers and updating the sample mean, standard deviation, and histogram until there are *no more outliers*.

2. Illustrate advertising levels in countries that are not outliers. Add a "bottom line" chart title.

3. Summarize your analysis by describing *P&G's advertising* in *countries* around the world, excluding outliers.

   Include
   - One or more measures of central tendency, such as the mean and median
   - One or more measures of dispersion, such as the standard deviation and range
   - The similarity of the distribution to a Normal distribution
     Be sure to round your answers to two or three significant digits.

4. Considering the entire sample, which advertising strategy describes the P&G strategy better: (1) advertise at a moderate level in many global markets or (2) advertise heavily to a small number of key markets and spend a little in many other markets?

Assignment 2-2 Best Practices Survey

Firm managers use statistics to advantage. Sometimes when results are lackluster, more significant digits are used, because readers will spend less time digesting results and results with more significant digits are less likely to be remembered. Sometimes when results are impressive, fewer significant digits are used to motivate readers to digest and remember.

Choose an Annual Report and cite the firm and the year.

1. In the body of the report, what range of significant digits are used to report numerical results? Cite two examples, one with the smallest number of significant digits and one with the largest number of significant digits.

2. In the Financial Exhibits at the end, what range of significant digits are used? Cite two examples, one with the smallest number of significant digits and one with the largest number of significant digits.

3. Survey the graphics. Cite an example where the "bottom line" is used to help readers interpret. Cite an example where the title could be more effective, and provide a suggestion for a better title.

Assignment 2-3 Shortcut Challenge

Complete the steps in the first Excel page of Lab 2 (find descriptive statistics, make a histogram, sort to identify and remove outliers, then compare with Normal in a PivotChart) and record your time. Extra points will be added for a time less than 5 min.

CASE 2-1 VW Backgrounds

Volkswagen management commissioned background music for New Beetle commercials. The advertising message is that the New Beetle is unique . . . "round in a world of squares." To be effective, the background music must support this message.

Thirty customers were asked to write down the first word that came to mind when they listened to the music. The clip is in **Case 2-1 VW background.MP3** and words evoked are contained in **Case 2-1 VW background.xls.** Listen to the clip and then describe the market response.

Create a PivotTable of the percentage who associated each image with the music and sort rows so that the modal image is first.

Create a PivotChart to illustrate the images associated with the background music.
(Add a "bottom line" title and round percentages to two significant digits.)

What is the modal image created by the VW commercial's background music?

Is this music a good choice for the VW commercial? Explain.

Chapter 3
Hypothesis Tests, Confidence Intervals, and Simulation to Infer Population Characteristics and Differences

Samples are collected and analyzed to estimate population characteristics. Chapter 3 explores the practice of *inference*: how *hypotheses* about what may be true in the population are tested and how population parameters are estimated with *confidence intervals*. Included in this chapter are tests of hypotheses and confidence intervals for

1. A population mean from a single sample
2. The difference between means of two populations, or segments from two independent samples
3. The mean difference within one population between two time periods or two scenarios from two matched or paired samples

In some cases, it is useful to simulate random samples using decision makers' assumptions about a population to estimate demand and its sensitivity to those assumptions. Monte Carlo simulation is introduced in this chapter.

3.1 Sample Means Are Random Variables

The descriptive statistics from each sample of a population are unique. In the example that follows, each team in a New Product Development class collected a sample from a population to estimate population demand for their concept. Each of the team's statistics is unique, but predictable, inasmuch as the sample statistics are random variables with a predictable sampling distribution. If many random samples of a given size are drawn from a population, the means from those samples will be similar and their distribution will be Normal and centered at the population mean.

Example 3.1 Thirsty on Campus: Is There Sufficient Demand?
An enterprising New Product Development class has an idea to sell on campus custom-flavored, enriched bottles of water from dispensers that would add customers' desired vitamins and natural flavors to each bottle. To assess profit potential, they need an estimate of demand for bottled water on campus. If demand exceeds the break even level of seven bottles per week per customer, the business would generate profit.

The class translated break even demand into hypotheses that could be tested using a sample of potential customers. The entrepreneurial class needs to know whether demand exceeds seven bottles per consumer per week, because below this level of demand, revenues wouldn't

C. Fraser, *Business Statistics for Competitive Advantage with Excel 2010: Basics, Model Building, and Cases*, DOI 10.1007/978-1-4419-9857-6_3, © Springer Science+Business Media, LLC 2012

cover expenses. Hypotheses are formulated as *null* and *alternative*. In this case, the null hypothesis states a limiting conclusion about the population mean. This default conclusion cannot be rejected unless the data indicate that it is highly unlikely. The null hypothesis is that of insufficient demand, which would lead the class to stop development:

H_0: Campus consumers drink no more than seven bottles of water per week on average:

$$\mu \leq 7.$$

Unless sample data indicate sufficient demand, the class will stop development.

In this case, the alternative hypothesis states a conclusion that the population mean exceeds the qualifying condition. The null hypothesis is rejected only with sufficient evidence from a sample that it is unlikely to be true.

In **Thirsty**, the alternate hypothesis supports a conclusion that population demand is sufficient and would lead to a decision to proceed with the new product's development:

H_1: Campus consumers drink more than seven bottles of water per week on average:

$$\mu > 7.$$

Given sufficient demand in a sample, the class would reject the null hypothesis and proceed with the project.

Sample statistics are used to determine whether the population mean is likely to be less than 7, using the sample mean as the estimate. To test the hypotheses regarding mean demand in the population of customers on campus, each of the 15 student teams in the class independently surveyed a random sample of 30 consumers from the campus. The distribution of means of many "large" ($N \geq 30$) random samples is Normal and centered on the unknown population mean. If the null hypothesis is true, and mean demand less than or equal to seven bottles per customer per week, each team's sample mean should be close to 7.

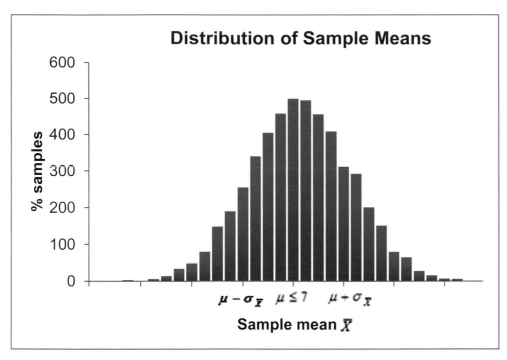

Fig. 3.1 Distribution of sample means under the null hypothesis

On average, across all random samples of the same size N, the average difference between sample means and the population mean is the standard error of sample means:

$$\sigma_{\bar{X}} = \sigma / \sqrt{N},$$

where σ is the standard deviation in the population and N is the sample size. The standard error is larger when there is more variation in the population and when the sample size is smaller.

With random samples of 30, population mean $\mu = 10.2$, and standard deviation $\sigma = 4.0$, the sampling standard error would be

$$s_{\bar{X}} = \sigma / \sqrt{30} = 4 / 5.5 = .7.$$

From the empirical rule introduced in Chap. 2, we would expect 2/3 of the teams' sample means to fall within one standard error of the population mean:

$$\mu - s_{\bar{X}} \leq \bar{X} \leq \mu + s_{\bar{X}}$$

$$10.2 - .7 \leq \bar{X} \leq 10.2 + .7$$

$$9.5 \leq \bar{X} \leq 10.9,$$

and we expect 95% of the teams' *sample means* to fall within two standard errors of the population mean:

$$\mu - 2s_{\bar{X}} \leq \bar{X} \leq \mu + 2s_{\bar{X}}$$

$$10.2 - 2(.7) \leq \bar{X} \leq 10.2 + 2(.7)$$

$$8.8 \leq \bar{X} \leq 11.6.$$

Nearly all of sample means can be expected to fall within three standard errors of the mean, 8.1–12.3.

Each team calculated the sample mean and standard deviation from their sample. Team 1, for example, found that average demand in their sample was 11.2 bottles per week, with a standard deviation of 4.5 bottles. Each team's descriptive statistics from the 15 samples are shown in Fig. 3.2.

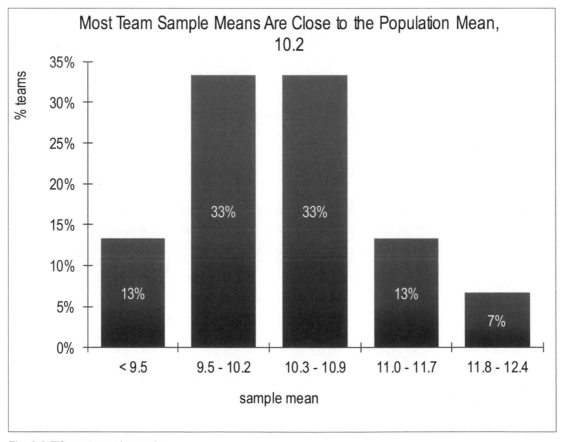

Fig. 3.2 Fifteen teams' samples

Team	Sample statistics Average demand per consumer per week, $\overline{X}$	Standard deviation, s_i
1	11.2	4.5
2	10.9	4.0
3	10.6	4.3
4	9.5	3.4
5	9.0	3.9
6	10.8	4.6
7	9.6	3.8
8	9.9	4.1
9	9.7	3.7
10	10.7	4.2
11	9.0	3.8
12	9.8	3.6
13	10.5	3.1
14	12.2	4.9
15	11.6	4.2

Sample means across the 15 teams ranged from 9.0 to 12.2 bottles per week per consumer. Each team's sample mean, $\overline{X}$, is close to the true unknown population mean, $\mu = 10.2$, and not as close to the hypothetical population mean of 7. Each of the sample standard deviations is close to the true, unknown population standard deviation $\sigma = 4$. In addition, each team's sample statistics are unique.

Because the population standard deviation is almost never known, but estimated from a sample, the standard error is also estimated from a sample using the estimate of the population standard deviation s:

$$s_{\overline{X}} = s/\sqrt{N}$$

When the standard deviation is estimated from a sample (which is nearly always), the distribution of standardized sample means $\overline{X}/s_{\overline{X}}$ is distributed as *Student t*, which is approximately Normal.

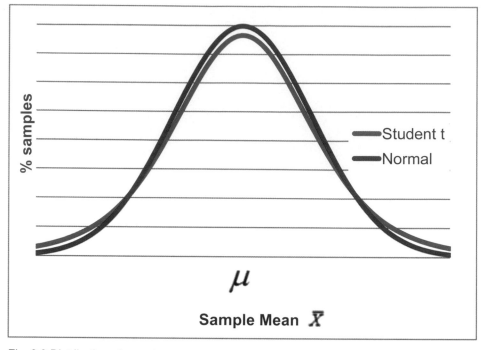

Fig. 3.3 Distribution of sample means

Student t has slightly fatter tails than Normal, because we are estimating the standard deviation. How much fatter the tails are depends on the sample size. *Student t* is a family of distributions indexed by sample size. The difference from Normal is more if a sample size is small. For sample sizes of about 30 or more, there is little difference between *Student t* and Normal. An estimate of the standard deviation from the sample is close to the true population value if the sample size meets or exceeds 30.

3.2 Infer Whether a Population Mean Exceeds a Target

Each team asks, "How likely is it that we would observe this sample mean, were the population mean seven or less?" From the Empirical Rule, sample means are expected to fall within approximately two standard errors of the population mean 95% of the time.

Rearranging the empirical rule formula, we see that *Student t* counts the standard errors between a sample mean and the population mean:

$$\left| \overline{X} - \mu \right| / s_{\overline{X}} = t_{N-1}.$$

A difference between a sample mean and the break even level of 7 that is more than approximately two standard errors ($t > 2$) is a signal that population demand is unlikely to be 7 or less. In this case, the sample mean would lie to the extreme right in the hypothetical distribution of sample means with the center at the hypothetical population mean of 7, where fewer than 5% of sample means are expected.

In the **Thirsty** example, each team calculated the number of standard errors by which their sample mean exceeded 7. Next, each referred to a table of *Student t* values or used statistical software to find the area under the right distribution tail, called the p Value. Were true demand less than 7, it would be unusual to observe a sample mean more than $t_{2\alpha=.1;\ 29}=1.7$ standard errors greater than 7. The larger a t value, the smaller the corresponding p Value will be, and the less likely the sample statistics would be observed were the null hypothesis true:

p Value > .05: If the null hypothesis were true, it would not be unusual to observe the data.
      The conclusion of insufficient demand H_0 cannot be rejected.
      The team recommends halting development.

p Value ≤ .05: If the null hypothesis were true, it would be unusual to observe the data.
      Reject the null hypothesis.
      The team recommends proceeding with development.

Each team used software to test the hypothesis that demand exceeds 7. Analyses of Team 8 are illustrated in Fig. 3.4, as an example.

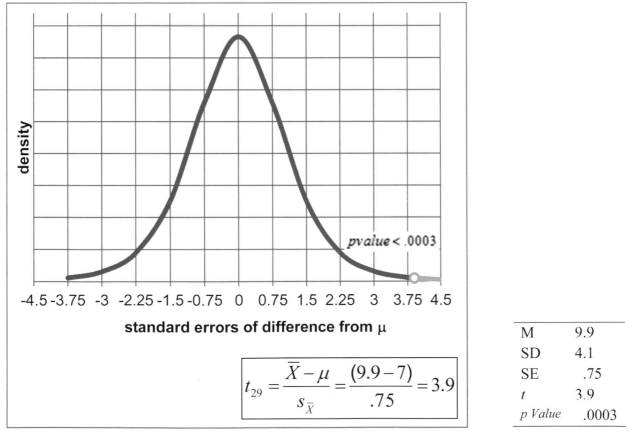

M	9.9
SD	4.1
SE	.75
t	3.9
p Value	.0003

$$t_{29}=\frac{\overline{X}-\mu}{s_{\overline{X}}}=\frac{(9.9-7)}{.75}=3.9$$

Fig. 3.4 t Test of the hypothesis that population demand is 7 or less

Reviewing these results, Team 8 would conclude the following:

> Demand in our sample of 30 ranged from 0 to 19 bottles per person per week, averaging 9.9 bottles per person per week. With this sample of 30, the standard error is .75 bottles per week. Our sample mean is 3.9 standard errors greater than breakeven of 7. (The *t* statistic is 3.9.) Were population demand 7 or less, it would be unusual to observe demand of 9.9 in a sample of 30. The *p Value* is .0003. We conclude that demand is not 7 or less.

In a test of the level of demand for bottles of water, each team used a "one tail" test. Regardless of how much demand exceeded seven bottles per consumer per week, a team would vote to proceed with development as long as they could be reasonably sure demand would exceed breakeven. They require only that the chance of observing the data be less than 5%, the *critical p Value*, were true demand less than 7. Thus, it is only the area under the right tail that concerns them.

3.3 Confidence Intervals Estimate the Population Mean

The class of entrepreneurs in the **Thirsty** example doesn't know that the population mean is 10.2 bottles per customer per week; therefore, each team will estimate this mean using their sample data. Rearranging the formula for a *t test*, we see that each team can use its sample standard error, the *Student t* value for their sample size, and the desired level of confidence to estimate the range that is likely to contain the true population mean:

$$\overline{X} - t_{\alpha, N-1} s_{\overline{X}} < \mu < \overline{X} + t_{\alpha, N-1} s_{\overline{X}},$$

where α is the chance that a sample is drawn from one of the sample distribution tails, and $t_{\alpha, (N-1)}$ is the *critical Student t* value for a chosen level of certainty $(1 - \alpha)$ and sample size N.

The *confidence level* $(1 - \alpha)$ allows us to specify the level of certainty that an interval will contain the population mean. Generally, decision makers desire a 95% level of confidence $(\alpha = 0.05)$, ensuring that in 95 out of 100 samples, the interval would contain the population mean. The *critical Student t* value for 95% confidence with a sample of 30 $(N = 30)$ is $t_{\alpha/2, (N-1)=29} = 2.05$. In 95% of random samples of the 30 drawn, we expect the sample means to be no further than 2.05 standard errors from the population mean:

$$\overline{X} - 2.05 s_{\overline{X}} \leq \mu \leq \overline{X} + 2.05 s_{\overline{X}}.$$

Each team's sample standard error, margin of error, and 95% confidence interval from the **Thirsty** example are shown in Table 3.1.

Table 3.1 Confidence intervals from each team's sample

$Team_i$	Average/demand/ consumer/week $\overline{X}_i$	Standard deviation, s_i	Standard error $s_{\overline{X}}$	Margin of error $2.05s_{\overline{X}}$	95% Confidence interval $\overline{X} \pm 2.05s_{\overline{X}}$	
1	11.2	4.5	.84	1.7	9.5	12.9
2	10.9	4.0	.74	1.5	9.4	12.4
3	10.6	4.3	.80	1.6	9.0	12.2
4	9.5	3.4	.63	1.3	8.2	10.8
5	9.0	3.9	.72	1.5	7.5	10.5
6	10.8	4.6	.85	1.7	9.1	12.5
7	9.6	3.8	.71	1.5	8.1	11.1
8	9.9	4.1	.75	1.5	8.4	11.4
9	9.7	3.7	.69	1.4	8.3	11.1
10	10.7	4.2	.78	1.6	9.1	12.3
11	9.0	3.8	.71	1.5	7.5	10.5
12	9.8	3.6	.67	1.4	8.4	11.2
13	10.5	3.1	.58	1.2	9.3	11.7
14	12.2	4.9	.91	1.9	10.3	14.1
15	11.6	4.2	.78	1.6	10.0	13.2

In practice, 15 samples would not be collected. A single sample would be selected, just as each individual team did in their market research. Team 8's analysis is shown in Fig. 3.5 as an example.

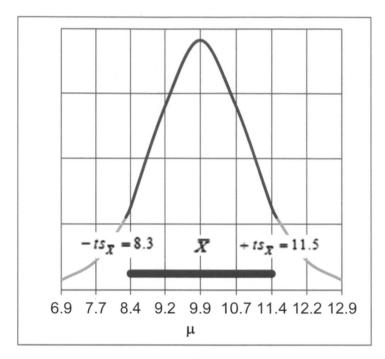

M	9.9
SD	.75
Critical t	2.1
Margin of error	1.5
95% lower	8.3
95% upper	11.5

Fig. 3.5 Confidence interval for bottled water demand μ

Team 8 would conclude the following:

Average demand in our sample of 30 is 9.9 bottles per person per week, with a margin of error of 1.5 bottles. It is likely that average campus demand is between 8.3 and 11.5 bottles per person per week.

3.4 Calculate Approximate Confidence Intervals with Mental Math

When the sample size is "large," $N \geq 30$, we can use an approximate $t \cong 2.0$ to produce approximate confidence intervals with mental math. Using $t \cong 2$ for an approximate 95% level of confidence, the 15 student teams each calculated the likely ranges for bottled water demand in the population, as shown in Table 3.2.

Table 3.2 Each team's approximate confidence interval

Team$_i$	Average customer demand/week $\overline{X}_i$	Standard error $s_{\overline{X}}$	Margin of error $2.05s_{\overline{X}}$	95% Confidence interval $\overline{X} \pm 2.05s_{\overline{X}}$		Approximate margin of error $2s_{\overline{X}}$	Approximate 95% confidence interval $\overline{X} \pm 2s_{\overline{X}}$	
1	11.2	.84	1.7	9.5	12.9	1.7	9.5	12.9
2	10.9	.74	1.5	9.4	12.4	1.5	9.4	12.4
3	10.6	.80	1.6	9.0	12.2	1.6	9.0	12.2
4	9.5	.63	1.3	8.2	10.8	1.3	8.2	10.8
5	9.0	.72	1.5	7.5	10.5	1.4	7.6	10.4
6	10.8	.85	1.7	9.1	12.5	1.7	9.1	12.5
7	9.6	.71	1.5	8.1	11.1	1.4	8.2	11.0
8	9.9	.75	1.5	8.4	11.4	1.5	8.4	11.4
9	9.7	.69	1.4	8.3	11.1	1.4	8.3	11.1
10	10.7	.78	1.6	9.1	12.3	1.6	9.1	12.3
11	9.0	.71	1.5	7.5	10.5	1.4	7.6	10.4
12	9.8	.67	1.4	8.4	11.2	1.3	8.5	11.1
13	10.5	.58	1.2	9.3	11.7	1.2	9.3	11.7
14	12.2	.91	1.9	10.3	14.1	1.8	10.4	14.0
15	11.6	.78	1.6	10.0	13.2	1.6	10.0	13.2

With the approximation, Team 8's conclusion remains the following: expected demand will range from 8.4 to 11.4 bottles per week per customer.

3.5 Margin of Error Is Inversely Proportional To Sample Size

The larger a sample N is, the smaller the *95% confidence interval* is

$$\overline{X} - 2s_{\overline{X}} \leq \mu \leq \overline{X} + 2s_{\overline{X}},$$

because the standard error $s_{\overline{X}}$ and *margin of error*, roughly $2s_{\overline{X}}$, are inversely proportional to the square root of our size N, as shown in Fig. 3.6.

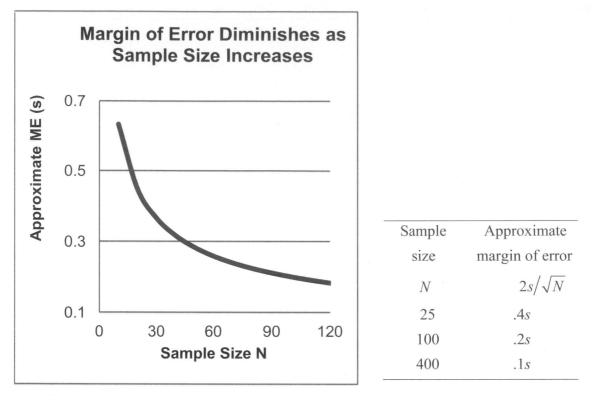

Sample size	Approximate margin of error
N	$2s/\sqrt{N}$
25	$.4s$
100	$.2s$
400	$.1s$

Fig. 3.6 Margin of error, given sample size

To double precision, the sample size must be quadrupled. Gains in precision become increasingly more expensive.

3.6 Samples Are Efficient

Sample statistics are used to estimate population statistics because it is often neither possible nor feasible to identify and measure the entire population. The time and expense involved in identifying and measuring all population elements are prohibitive. To survey the bottled water consumption of each faculty member, student, and staff member on campus would take many hours. An estimate of demand is inferred from a random, representative sample including these people. Although sample estimates will not be exactly the same as population statistics because of sampling error, samples are amazingly efficient if properly drawn and representative of the population.

3.7 Use Monte Carlo Simulation Samples to Incorporate Uncertainty and Quantify Implications of Assumptions

The Team 8 partners were concerned that they might either pass up a profitable opportunity or invest in an unprofitable business. Their estimate of average bottles of water demanded per customer per week seemed promising, although there was a fairly large difference between breakeven and the profit they felt necessary to warrant the investment.

Demand depended on bottles per customer as well as share of bottles sold on campus. They were unsure whether they would be successful in capturing 10% share of bottled water sold on campus, but this was the best estimate. The team constructed a demand worksheet, shown in Table 3.3, from the best guess of market share, as well as the key performance measure, estimated bottles sold, assuming 30K customers on campus, with sales 40 weeks in a year.

Table 3.3 Worksheet for bottled water demand

Bottles/customer/week	9.90
Market share	10%
Bottles sold (K)	1,188

Sales of approximately 1.2M bottles were attractive; however, they realized that this was simply the best guess and depended on uncertain demand as well as uncertain market share. Sales were likely to be either substantially higher or lower, which increased the risk of their investment. They really needed to know the chances that demand would exceed 800K bottles in the first year, because demand less than 800K would not justify their investment.

To quantify the risks, the team decided to use Monte Carlo simulation to incorporate both demand and share uncertainty and their assumptions into their forecast and decision. Results would show the outcomes and their likelihoods under the team's assumptions.

Demand assumptions: The team used their sample statistics to specify assumptions about demand.

- Demand for bottled water was Normally distributed, 99, and the sample mean was the most likely level.
- The standard error was .75.
- There was a 95% chance that demand would be greater than 8.3 and less than 11.5.

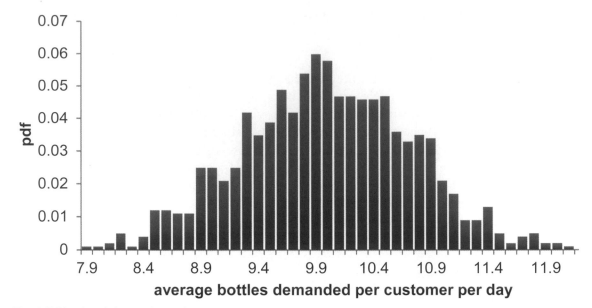

Fig. 3.7 Simulated demand sample

Share assumptions: The team thought that market share could be as low as 5% or as high as 15%, but they thought any share within this range was equally likely. Excel offers a choice of Normal or Uniform distribution to match assumptions. When sample data, judgment, or expert opinion is used to specify assumptions, a Normal distribution can often be assumed. With less information, a Uniform distribution can be used, specifying only the assumed *minimum* and *maximum*. Lacking sample data or experience, they assumed that potential shares were Uniform. Their assumptions produced the simulated sample of 1,000 shares as shown in Fig. 3.8.

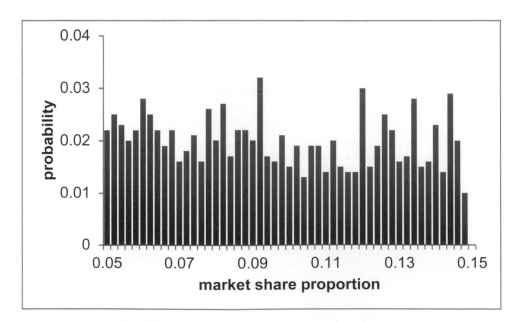

Fig. 3.8 Simulated sample of market shares

The simulated samples of bottles per customer per week, the market share, the number of customers, and the number of weeks were multiplied to find the simulated distribution of bottles sold per customer per year, as shown in Fig. 3.9.

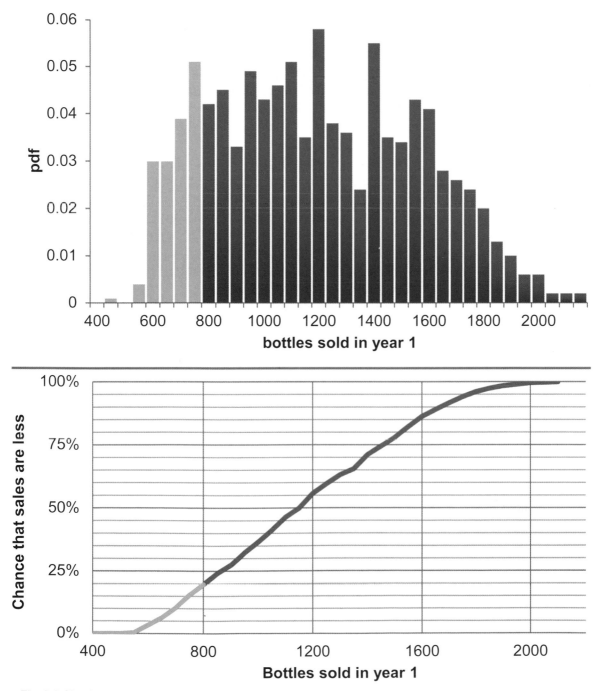

Fig. 3.9 Simulated distribution of demand

If the assumptions are valid, there is an 80% chance that the team will sell more than 800K bottles in the first year. Because sales are likely to exceed this minimally acceptable level of 800K bottles, given team assumptions, the team was more confident that the potential demand warranted their investment.

3.8 Determine Whether Two Segments Differ with *Student t*

Example 3.2 SmartScribe: Is Income a Useful Base for Segmentation?
SmartScribe, marketers of a brand of smart pens, would like to identify the demographic segment with the highest demand for its new concept. Smart pens record presentation notes onto a file that can be downloaded. The new pens were being sold at a relatively high price; thus, the adopters may have higher incomes. To test this hypothesis, customers at an office supply retail store were sorted into SmartScribe purchasers, which management refers to as the adopters, and nonadopter customers. Random samples from these two segments were drawn and offered a store coupon in exchange for completion of a short survey, which included a measure of annual household income. The survey was completed by 56 SmartScribe pen adopters and 41 nonadopters. SmartScribe needs to determine whether income is a useful demographic indicator of interest.

The null hypothesis states the conclusion that the average annual household income of adopters is not greater than that of nonadopters, as illustrated in Fig. 3.10:

H_0: Average annual household incomes of adopters is equal to or less than that of non-adopters of the new pen:

$$\mu_{Adopters} \leq \mu_{Nonadopters}$$

or

$$\mu_{Adopters} - \mu_{Nonadopters} \leq 0.$$

Alternatively,

H_1: Average annual household incomes of adopters exceed that of nonadopters of the new pen:

$$\mu_{Adopters} > \mu_{Nonadopters}$$

or

$$\mu_{Adopters} - \mu_{Nonadopters} \leq 0.$$

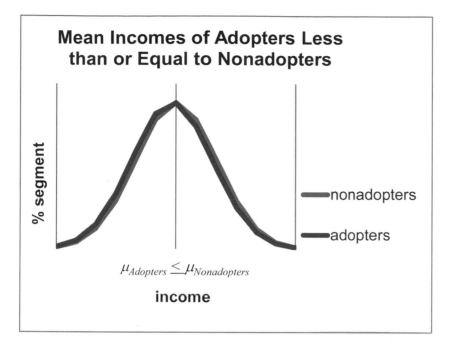

Fig. 3.10 The null hypothesis adopters earn less or equivalent incomes to nonadopters

If there is no difference in incomes between the two segment samples, or if adopters earn lower incomes, the null hypothesis cannot be rejected based on the sample evidence.

A test of the significance of the difference between the two segments' average annual household incomes is based on the difference between the two sample means

$$\overline{X}_{Adopters} - \overline{X}_{Nonadopters}$$

and the standard error of the difference

$$s_{\overline{X}_{Adopters} - \overline{X}_{Nonadopters}}.$$

The standard error of average difference in annual household income (in thousands) is

$$s_{\overline{X}_{Adopters} - \overline{X}_{Nonadopters}} = \sqrt{s^2_{\overline{X}_{Adopters}} \Big/ N_{Adopters} + s^2_{\overline{X}_{Nonadopters}} \Big/ N_{Nonadopters}}$$

$$= \sqrt{[2,300 / 41 + 2,670 / 56]} = \$10.2..$$

This estimate for the standard error of the difference between segment means assumes that the two segment standard deviations may differ. It is usually not known whether the segment standard deviations are equivalent; thus, this is a conservative assumption.

The number of standard errors of difference between sample means is measured with *Student t*:

$$t_{(N-1)} = (\overline{X}_{Adopters} - \overline{X}_{Nonadopters}) \Big/ s_{\overline{X}_{Adopters} - \overline{X}_{Nonadopters}}.$$

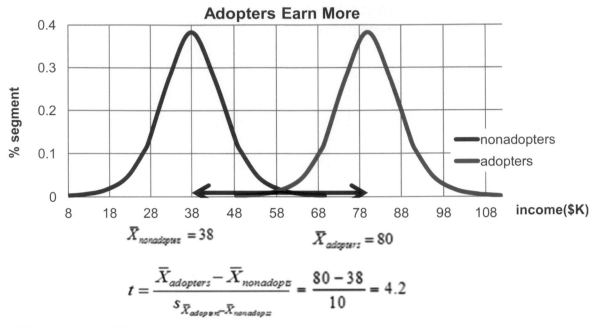

Fig. 3.11 *t test* of difference between segment means

From the *t test* of difference between segment incomes, shown in Fig. 3.11, SmartScribe management could conclude the following:

> In segment samples of 56 adopters and 41 nonadopters, the corresponding average segment sample incomes are $80K and $38K, a difference of $42K. Were there no difference in segment mean incomes in the population, it would be unusual to observe this difference in segment average incomes in the segment samples. Based on sample evidence, we conclude that average incomes of adopters cannot be less than or equal to the average incomes of nonadopters. Income is a useful basis for segmentation.

3.9 Estimate the Extent of Difference Between Two Segments

From the sample data, SmartScribe managers estimated the average annual household income difference (in thousands) between adopters and nonadopters:

$$\bar{X}_{Adopters} - \bar{X}_{Nonadopters} = \$80.1 - \$38.5 = \$41.6.$$

The approximate 95% confidence interval of the difference in annual household incomes between adopters and nonadopters is

$$\left(\overline{X}_{Adopters} - \overline{X}_{Nonadopters}\right) - 2s_{\overline{X}_{Adopters} - \overline{X}_{Nonadopters}} \leq \left(\mu_{Adopters} - \mu_{Nonadopters}\right)$$

$$\leq \left(\overline{X}_{Adopters} - \overline{X}_{Nonadopters}\right) + 2s_{\overline{X}_{Adopters} - \overline{X}_{Nonadopters}}$$

$$\$41.6 - 2(\$10.2) \leq (\mu_{Adopters} - \mu_{Nonadopters}) \leq \$41.6 + 2(\$10.2)$$

$$\$21.2 \leq (\mu_{Adopters} - \mu_{Nonadopters}) \leq \$62.0.$$

Thus, the firm estimates that the average difference in annual household income between adopters and nonadopters is $21,000–$62,000, as shown in Fig. 3.12. Management will conclude that annual household income can be used to differentiate the two market segments and that adopters are wealthier than nonadopters.

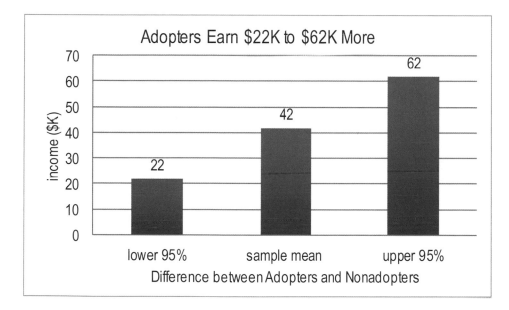

Fig. 3.12 95% Confidence interval of the difference between segments

In our samples of 56 adopters and 41 nonadopters, the corresponding average difference in income between segment samples is $42K, and the margin of error of the difference is $20K. Relative to nonadopters, we estimate that adopters earn $21K–$62K more on average each year. To construct confidence intervals for the difference in means of two samples, we assume that either (1) both segments' characteristics are bell shaped (distributed approximately Normal) and we've randomly sampled both segments or (2) "large" random samples from both segments have been collected.

The *approximate t* is used in the example, instead of the *t* that corresponds to a confidence interval for the difference between segments. The degrees of freedom calculation for the confidence interval for a difference between segments uses an approximation formula that Excel and other software provide. In practice, Excel would be used to find the confidence interval that corresponds to the two sample sizes.

3.10 Confidence Intervals Complement Hypothesis Tests

Confidence intervals and hypothesis tests are consistent and complementary, but are used to make different decisions. If a decision maker needs to make a qualitative Yes/No decision, a hypothesis test is used. If a decision maker instead requires a quantitative estimate, such as level of demand, confidence intervals are used. Hypothesis tests tell us whether demand exceeds a critical level or whether segments differ. Confidence intervals quantify demand or magnitude of differences between segments.

3.11 Estimate a Population Proportion from a Sample Proportion

Example 3.3 Guinea Pigs
A pharmaceutical company gauges reactions to their products by applying them to animals. An animal rights activist has threatened to start a campaign to boycott the company's products if the animal testing doesn't stop. Concerned managers have hired four public opinion polling organizations to learn whether medical testing on animals is accepted or not.

Four independent pollsters each surveyed 30 Americans and found the proportions shown in Table 3.4 agree that medical testing on animals is morally acceptable:

Table 3.4 Sample approval proportions by poll

Poll	Sample approval proportion
1	$P_1 = 16 / 30 = .53$
2	$P_2 = 19 / 30 = .63$
3	$P_3 = 17 / 30 = .57$
4	$P_4 = 21 / 30 = .70$

If numerous random samples are taken, sample proportions P will be approximately Normally distributed around the unknown population proportion $\pi = .6$, as long as this true proportion is not close to either 0 or 1.

The standard deviation of the sample proportions P, the *standard error of the sample proportion*, measures dispersion of samples of size N from the population proportion π:

$$\sigma_\pi = \sqrt{\pi(1-\pi)/N},$$

which is estimated with the sample proportion P:

$$s_P = \sqrt{P(1-P)/N}.$$

The four poll organizations each estimated the proportion of Americans who agree that medical testing on animals is morally acceptable, as shown in Table 3.5.

Table 3.5 Confidence interval of approval proportion by poll, $N = 30$

Poll i	Sample proportion, P_i	Standard error, s_{P_i} (N=30)	Margin of error for 95% confidence, $zs_{P_i} = 1.96s_{P_i}$	Interval containing the population proportion with 95% confidence, $P_i \pm zs_{P_i}$
1	.57	.090	.18	.39 to .75
2	.61	.089	.17	.44 to .78
3	.58	.090	.18	.40 to .76
4	.63	.088	.16	.47 to .79

With samples of just 30, margins of error are relatively large, and we are uncertain whether a minority or a sizeable majority approves. In practice, polling organizations use much larger samples, which shrink margins of error and corresponding confidence intervals. Had samples of 1,000 been collected instead, the poll results would be as shown in Table 3.6.

Table 3.6 Confidence interval of approval proportion by poll, $N = 1,000$

Poll i	Sample proportion, P_i	Standard error, s_{P_i} (N=1,000)	Margin of error for 95% confidence, $zs_{P_i} = 1.96s_{P_i}$	95% Confidence interval, $P_i \pm zs_{P_i}$
1	.57	.016	.031	.54 to .60
2	.61	.015	.029	.58 to .64
3	.58	.016	.031	.55 to .61
4	.63	.015	.029	.60 to .66

With much larger samples and correspondingly smaller margins of error, it becomes clear that the majority approves of medical testing on animals.

The second polling organization would report the following:

The majority of a random sample of 1,000 Americans approves of medical testing on animals. 61% believe medical testing on animals is morally acceptable, with a margin of error of 3%.

3.12 Conditions for Assuming Approximate Normality

It is appropriate to use the Normal distribution to approximate the distribution of possible sample proportions if sample size is "large" ($N \geq 30$) and both $N \times P \geq 5$ and $N \times (1 - P) \geq 5$. When the true population proportion is very close to either 0 or 1, we cannot reasonably assume that the distribution of sample proportions is Normal. A rule of thumb suggests that $P \times N$ and $(1 - P) \times N$ ought to be at least 5 in order to use Normal inferences about proportions. For a sample of 30, the sample proportion P would need to be between .17 and .83 to use Normal inferences. For a sample of 1,000, the sample proportion P would need to be between .01 and .99. Drawing larger samples allows more precise inference of population proportions from samples.

3.13 Conservative Confidence Intervals for a Proportion

Polling organizations report the sample proportion and margin of error rather than a confidence interval. For example, "61% approve of medical testing on animals. (The margin of error from this poll is 3% points.)" A 95% level of confidence is the industry standard. Because the true proportion and its standard deviation are unknown, and because pollsters stake their reputations on valid results, a *conservative* approach, which assumes a true proportion of .5, is used. This conservative approach

$$s_P = \sqrt{.5(1 - .5)/N}$$

yields the largest possible standard error for a given sample size and makes the margin of error ($z \times s_P$) a simple function of the square root of the sample size N.

   With this conservative approach and samples of $N = 1,000$, the pollsters' results are shown in Table 3.7.

Table 3.7 Conservative confidence intervals for approval proportions, $N = 1,000$

Poll_i	Sample proportion, P	Conservative margin of error for 95% confidence, $zs_P = 1.96s_P$	Conservative 95% confidence interval $P - zs_P \leq \pi \leq P + zs_P$	
1	.57	.031	.54	.60
2	.61	.031	.58	.64
3	.58	.031	.55	.61
4	.63	.031	.60	.66

   An effective display of proportions or shares is a *Pie chart*. The second poll organization used Excel to create this illustration of their survey results, as shown in Fig. 3.13.

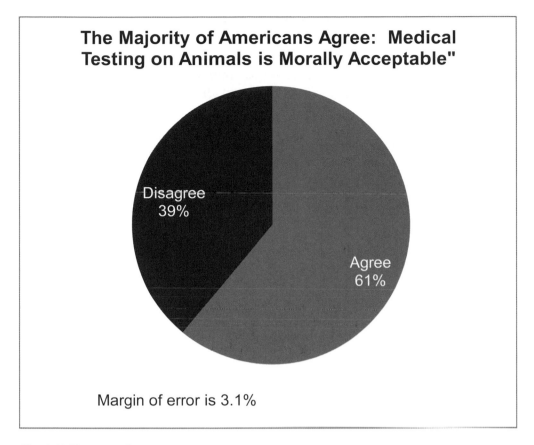

Fig. 3.13 Pie chart of approval percentage

The second polling organization would report the following:

Sixty-one percent of American adults agree that medical testing on animals is morally acceptable. Poll results have a margin of error of 3.1% points. The majority of Americans supports medical testing on animals.

Other appropriate applications for confidence intervals to estimate population proportions or shares include

- Proportion who prefer a new formulation to an old one in a taste test
- Share of retailers who offer a brand
- Market share of a product in a specified market
- Proportion of employees who call in sick when they're well
- Proportion of new hires who will perform exceptionally well on the job

3.14 Assess the Difference Between Alternate Scenarios or Pairs

Sometimes management is concerned with the comparison of means from a single sample taken under varying conditions – at different times or in different scenarios – or comparison of sample pairs, such as the difference between an employee's opinion and the opinion of the employee's supervisor:

- Financial management might be interested in comparing the reactions of a sample of investors to "socially desirable" stock portfolios, excluding stocks of firms that manufacture or market weapons, tobacco, or alcohol, versus alternate portfolios that promise similar returns at similar risk levels, but are not "socially desirable."
- Marketing management might be interested in comparing taste ratings of sodas that contain varying levels of red coloring: do redder sodas taste better to customers?
- Management might be interested in comparing satisfaction ratings following a change that allows employees to work at home.

These examples compare *repeated samples*, where participants have provided multiple responses that can be compared:

- Financial management might also be interested in comparing the risk preferences of husbands and wives.
- Marketing management might want to compare children and parents' preferences for red sodas.
- Management might also be interested in comparing the satisfaction ratings of those employees with their supervisors' satisfaction ratings.

In these examples, interest is in comparing means from *matched pairs*.

In either case of repeated or matched samples, a *t test* can be used to determine whether the difference is nonzero. Testing hypotheses that concern a difference between pairs is equivalent to a *one sample t test*. The difference is tested in the same way that a characteristic mean is tested, using a *one sample test*.

Example 3.4 Are "Socially Desirable" Portfolios Undesirable?

An investment consulting firm's management believes that they have difficulty selling "socially desirable" portfolios because potential investors assume those funds are inferior investments. Socially desirable funds exclude stocks of firms that manufacture or market weapons, tobacco, or alcohol. There may be a perceived sacrifice associated with socially desirable investments, which causes investors to avoid portfolios labeled with that name.

The null hypothesis is

H_0: Investors rate "socially desirable" portfolios at least as attractive as equally risky conventional portfolios promising equivalent returns:

$$\mu_{Socially\ Desirable} - \mu_{Conventional} \geq 0.$$

If investors do not penalize "socially desirable" funds, the null hypothesis cannot be rejected. The alternative hypothesis is

H_1: Investors rate "socially desirable" portfolios as less attractive than other equally risky portfolios promising equivalent returns:

$$\mu_{Socially\ Desirable} - \mu_{Conventional} < 0.$$

Thirty-three investors were asked to evaluate two stock portfolios on a scale of attractiveness (-3 = "Not At All Appealing" to 3 = "Very Appealing"). The two portfolios promised equivalent returns and were equally risky. One contained only socially desirable stocks, whereas the other included stocks from companies that sell tobacco, alcohol, and arms. These are shown in Table 3.8.

Table 3.8 Paired ratings of other and socially desirable portfolios

Appeal of conventional portfolio	Appeal of socially desirable portfolio	Difference	Appeal of conventional portfolio	Appeal of socially desirable portfolio	Difference
−3	1	−4	2	−1	3
−3	2	−5	2	−1	3
−3	3	−6	2	−2	4
−3	3	−6	2	2	0
0	−1	1	2	1	1
0	1	−1	2	2	0
1	−3	4	2	2	0
1	−3	4	2	3	−1
1	−1	2	3	−3	6
1	−1	2	3	−3	6
1	−1	2	3	−3	6
1	1	0	3	−1	4
1	1	0	3	−1	4
1	2	−1	3	−3	6
2	−3	5	3	3	0
2	−3	5	3	3	0
2	2 −	4			

From a random sample of 33 investors' ratings of conventional and socially desirable portfolios of equivalent risk and return, the average difference is 1.5 points on a 7 point scale of attractiveness:

$$\overline{X}_{dif} = \overline{X}_{SD} - \overline{X}_{C} = -.2 - 1.3 = -1.5.$$

With this sample of 33, the standard error of the difference is .6:

$$s_{\overline{X}_{dif}} = s_{dif} / \sqrt{N} = 3.4 / \sqrt{33} = .6.$$

The average difference in attractiveness between the conventional and the socially desirable portfolio is 2.5 standard errors:

$$t = \overline{X}_{dif} / s_{\overline{X}_{dif}} = -1.5 / .6 = -2.5.$$

The *p Value* for $t = -2.5$ for a sample size of 33 is .009. Were the socially desirable portfolio at least as attractive as the conventional portfolio with equivalent risk and return, it would be unusual to observe such a large sample mean difference in ratings. Based on sample evidence, shown in Fig. 3.14, we conclude that a "socially desirable" label reduces portfolio attractiveness.

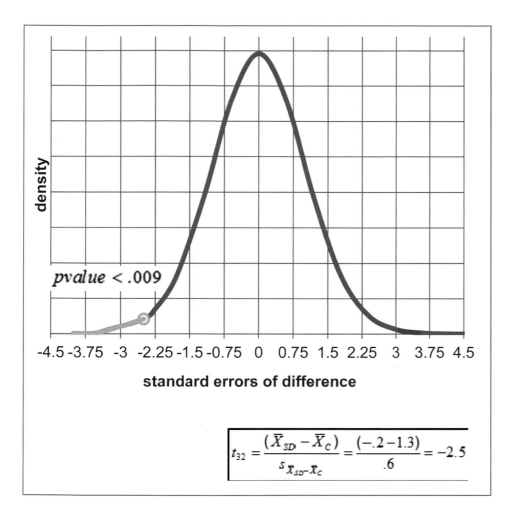

$$t_{32} = \frac{(\bar{X}_{SD} - \bar{X}_c)}{s_{\bar{X}_{SD} - \bar{X}_c}} = \frac{(-.2 - 1.3)}{.6} = -2.5$$

Fig. 3.14 *t test* of differences between paired ratings of socially desirable and conventional portfolios

The 95% confidence interval for the difference is

$$\bar{X}_{dif} \pm t_{\alpha/2, N-1} s_{\bar{X}_{dif}}$$

$$-1.5 \pm 2.04(.6)$$

or -2.7 to $-.3$ on the 7 point scale, shown in Fig. 3.15.

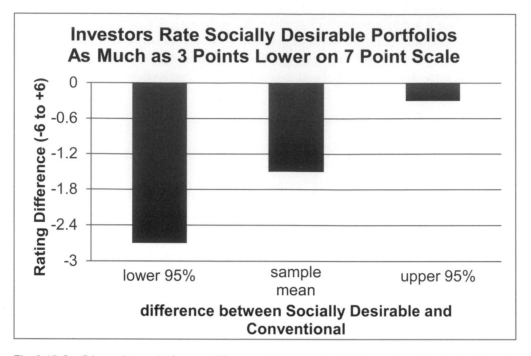

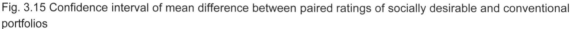

Fig. 3.15 Confidence interval of mean difference between paired ratings of socially desirable and conventional portfolios

The investment consultants would conclude the following:

A "socially desirable" label reduces investors' judged attractiveness ratings. Investors downgrade the attractiveness of "socially desirable" portfolios by about 1–3 points on a 7 point scale, relative to equivalent, but conventional, portfolios.

3.15 Inference from Sample to Population

Managers use sample statistics to infer population characteristics, knowing that inference from a sample is efficient and reliable. Because sample standard errors are approximately normally distributed, we can use the empirical rule to build confidence intervals to estimate population means and to test hypotheses about population means with *t tests*. We can determine whether a population mean is likely to equal, be less than, or exceed a target value, and we can estimate the range that is likely to include a population mean.

Our certainty that a population mean will fall within a sample based confidence interval depends on the amount of population variation and on the sample size. To double precision, sample size must be quadrupled, because the margin of error is inversely proportional to the square root of sample size.

Differences are important to managers, because differences drive decision making. If customers differ, segments are targeted in varying degrees. If employee satisfaction differs between alternate work environments, the workplace may be altered. Inference about differences between two populations is similar and relies on differences between two independent samples. A *t test* can be used to determine whether there is a likely difference between two population means, and with a confidence interval, we can estimate the likely size of difference.

Excel 3.1 Test the Level of a Population Mean with a *One Sample t Test*

Thirsty on Campus
Team 8 wants to know whether the demand for bottled water exceeds a break even level of seven bottles per day. To compare the level of demand to this critical level, we will use a *t test* of *Bottles* purchased per day.

Open **Excel 3.1 Bottled Water Demand.xls.**

Find the sample *mean* and *standard deviation.*

Find the standard error by dividing the sample standard deviation by the square root of sample size, **30.**

	B34	f_x	=B33/SQRT(30)
	A		B
1			**bottles**
29			15
30			16
31			19
32	M		9.9
33	SD		4.11
34	standard error		0.75

Find *t* by finding the difference between the sample mean and the critical value, **7,** and then divide that difference by the standard error.

Find the *p Value* for this *t* using the Excel function **TDIST.RT(*t,df*).** For degrees of freedom, *df,* enter the sample size minus 1, **29** (= 30–1).

	B36	f_x	=T.DIST.RT(B35,29)
	A		B
1			**bottles**
34	standard error		0.75
35	t		3.869
36	pvalue		0.0003

Excel 3.2 Make a Confidence Interval for a Population Mean

Determine the range likely to contain average demand in the population. Construct the 95% confidence interval for the population mean *Bottles* demanded.

B37		f_x	=T.INV.2T(0.05,29)
	A		B
1			**bottles**
31			19
32	M		9.9
33	SD		4.11
34	standard error		0.75
35	t		3.9
36	pvalue		0.0003
37	critical t		2.05

Use the Excel function **T.INV.2T**(*probability, df*) to find the critical *t* value for 95% confidence. For *probability*, enter **.05** for a 95% level of confidence. For *df*, enter the sample size minus one, **29**.

Find the margin of error by multiplying the *critical t* with the standard error.

B40		f_x	=B32+B38
	A		B
1			**bottles**
31			19
32	M		9.9
33	SD		4.11
34	standard error		0.75
35	t		3.9
36	pvalue		0.0003
37	critical t		2.05
38	margin of error		1.5
39	lower 95%		8.4
40	upper 95%		11.4

Add and subtract the *margin of error* from the sample *mean* to find the 95% *upper* and *lower* confidence interval limits.

Excel 3.3 Illustrate Confidence Intervals with Column Charts

T-Mobile's Service

T-Mobile managers have conducted a survey of customers in 32 major metropolitan areas to assess the quality of service along three key areas: coverage, absence of dropped calls, and static. Customers rated T-Mobile service along each of these three dimensions using a 5 point scale (1 = poor to 5 = excellent). Management's goal is to be able to offer service that is not perceived as inferior. This goal translates into mean ratings that exceed 3 on the 5 point scale in the national market across all three service dimensions. Make a 95% confidence interval to estimate the average perceived quality of service.

Open **Excel 3.3 t-mobile.xls.**

95% confidence intervals: Find the sample *mean*, *standard deviation*, *standard error*, *critical t margin of error*, and *lower and upper 95%* confidence interval bounds for coverage, dropped calls, and static ratings.

C34	f_x =AVERAGE(C2:C33)			
	A	B	C	D	E
			coverage rating (1=Poor to 5=Excellent)	dropped calls rating (1=Poor to 5=Excellent)	static rating (1=Poor to 5=Excellent)
1	city	service			
30	las vegas	tmobile	3	4	4
31	kansas city	tmobile	3	4	3
32	miami	tmobile	2	3	4
33	raleigh	tmobile	1	3	2
34		M	2.3	3.4	2.9
35		SD	0.98	0.61	0.56
36		standard error	0.17	0.11	0.10
37		critical t	2.0	2.0	2.0
38		margin of error	0.4	0.2	0.2
39		lower 95%	1.9	3.2	2.7
40		upper 95%	2.6	3.6	3.1

Column chart of confidence intervals: To see the confidence intervals for all three service dimension ratings, first select and copy row **1**, containing labels, **Cntl+C**, and then paste above the *lower* 95% confidence interval bounds.

Select row **39, Alt HIE**. (**Alt** activates shortcuts, **H** selects the **H**ome menu, **I** selects **I**nsert menu, and **E** inserts copied cells.)

	A	B	C coverage rating (1=Poor to 5=Excellent)	D dropped calls rating (1=Poor to 5=Excellent)	E static rating (1=Poor to 5=Excellent)
1	city	service			
30	las vegas	tmobile	3	4	4
31	kansas city	tmobile	3	4	3
32	miami	tmobile	2	3	4
33	raleigh	tmobile	1	3	2
34		M	2.3	3.4	2.9
35		SD	0.98	0.61	0.56
36		standard error	0.17	0.11	0.10
37		critical t	2.0	2.0	2.0
38		margin of error	0.4	0.2	0.2
39	city	service	coverage rating (1=Poor to 5=Excellent)	dropped calls rating (1=Poor to 5=Excellent)	static rating (1=Poor to 5=Excellent)
40		lower 95%	1.9	3.2	2.7
41		upper 95%	2.6	3.6	3.1

To make a Column chart, select the labels and 95% confidence interval bounds, and then use shortcuts

Alt NC. (N invokes the I**N**sert menu and **C** specifies a **C**olumn chart.)

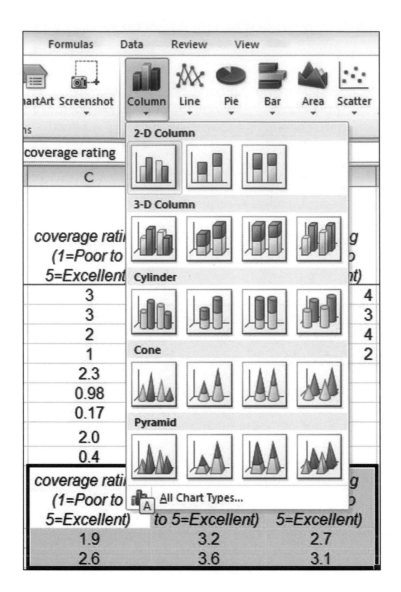

Choose **Design, Chart Layout 6** to add a vertical axis label.

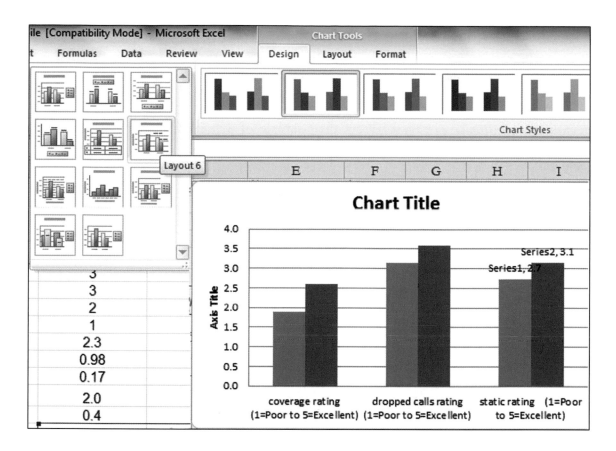

Adjust axes and type in axis label and chart title:

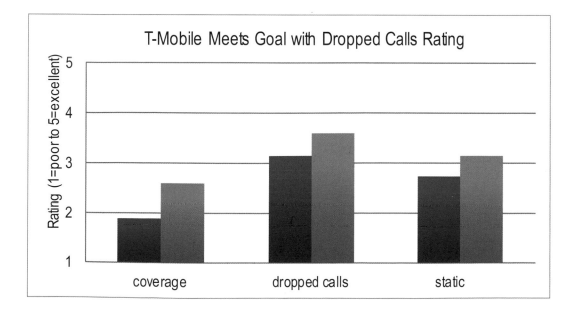

Excel 3.4 Conduct a Monte Carlo Simulation

Use Team 8's sample statistics and market share assumptions to assess market potential for custom-enhanced bottled water on campus.

Average demand per customer per week: From their sample of 30, Team 8 estimates that mean bottles demanded per customer per week will be 10.
The team also believes that

- There is only a 2.5% chance that average customer demand could be less than 8 bottles per customer per week.
- There is only a 2.5% chance that average customer demand could be more than 12 bottles per customer per week.

To see possible mean demand levels and their likelihoods, use Excel to simulate a distribution of average bottles demanded per customer per week.

Alt AY11 Random Number Generation

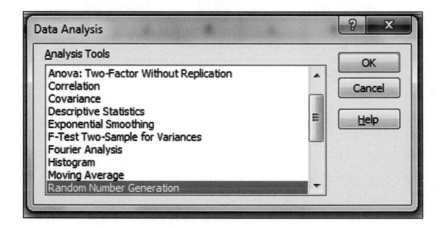

Request 1 sample of 1,000 observations, with Normal distribution, mean 10, standard deviation 1 (equal to the range that is expected to contain average demand, divided by 4).

Add a label and make a histogram to see the probability distribution.

To see the simulated distribution clearly, use about 30 bins. For the range from 7 to 13, use bin widths of .2: 7, 7.2, 7.4, …, 12.8, 13.0.

	A	B	C	D	E	F	G	H	I
1	*bins*	*Frequency*							
2	7	0							
3	7.2	1							
4	7.4	2							
5	7.6	2							
6	7.8	3							
7	8	12							
8	8.2	13							
9	8.4	10							
10	8.6	21							
11	8.8	40							

Market share: The team believes that they will achieve a market share of approximately 10%, and that

- There is just a 2.5% chance that share would be lower than 5%.
- There is only a 2.5% chance that share would be higher than 15%.

Use random number generation to create a simulated sample of 1,000 market shares. Specify one sample of 1,000 with Normal distribution with mean .1 and standard deviation .025.

Add a label and make a histogram to see the simulated distribution of market shares.

To see the distribution of simulated market shares clearly, create about 30 bins ranging from .025 to .175 of width .005: .025, .030, …, .170, .175.

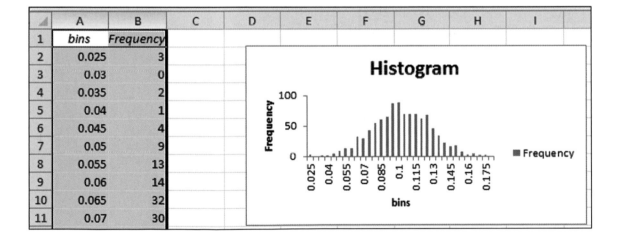

	A	B	C
1	demand	bins	share
2	11.295	7	0.1117
3	9.8558	7.2	0.1233
4	9.9246	7.4	0.0644
5	9.3373	7.6	0.1517
6	9.682	7.8	0.1088
7	11.188	8	0.1712
8	9.6768	8.2	0.0727
9	9.9441	8.4	0.1349

Copy and paste the simulated sample of *market share* into the sheet with simulated *mean demand* per customer per week.

Find the distribution of *bottles sold* by multiplying total customers on campus, 30K, weeks per year, 40, *mean demand* per customer per week, and *market share*.

D2				f_x	=30*40*A2*C2	
	A	B	C	D	E	
1	demand	bins	share	bottles sold (K)		
2	11.295	7	0.1117	1514		
3	9.8558	7.2	0.1233	1458		
4	9.9246	7.4	0.0644	767		
5	9.3373	7.6	0.1517	1699		
6	9.682	7.8	0.1088	1264		
7	11.188	8	0.1712	2299		
8	9.6768	8.2	0.0727	844		

Make a histogram of *bottles sold* to see possible outcomes and their likelihoods.

To easily identify the chance that sales will exceed 800K, the break even level, set the bin range at the minimum and maximum of simulated annual sales, which would be 100K and 2,300K in this case. Make bins of width 100(K), from the minimum to the maximum.

D1004				f_x	=MAX(D2:D1001)	
	A	B	C	D	E	
1	demand	bins	share	bottles sold (K)	bins	
1001	9.4685		0.0762	866		
1002						
1003	7.1427	7	0.0109	136	minimum	
1004	13.842	13	0.1712	2311	maximum	

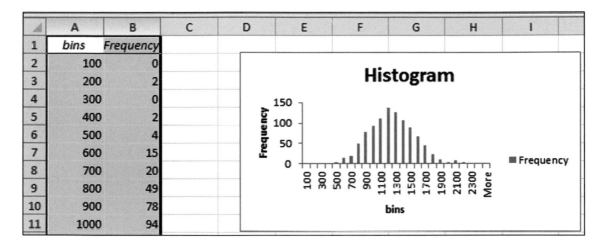

Use Rank and Percentile to find the chance that bottles sold will exceed 1000K.

	A	B	C	D
1	Point	bottles sold (K)	Rank	Percent
738	282	1000	737	26.30%
739	593	1000	738	26.20%
740	290	1000	739	26.10%
741	198	999	740	26.00%
742	18	998	741	25.90%

Excel 3.5 Test the Difference Between Two Segment Means with a *Two Sample t Test*

Pampers Preemies

Procter & Gamble management would like to know whether household income is a good base for segmentation in the market for their new preemie diaper. Test the hypothesis that average income is greater in the segment likely to try the new diapers than in the segment unlikely to try.

Open **Excel 3.5 Pampers Segment Income.xls.** The first column **A** contains *Likely Trier income* ($K) and the second column **B** contains *Unlikely Trier income* ($K).

Use the Excel function **T.TEST**(*array1,array2,tails,type*) to find the *p Value* from a *t test* of the difference between average incomes of the two segments.
For *array1*, enter the sample *Likely Trier income* values.
For *array2*, enter the sample *Unlikely Trier income* values.
For *tails*, enter **1** for a *one tail test*, and for *type*, enter **3** to signal a *two sample t test* that allows the standard deviations to differ between segments.

A59		f_x	=T.TEST(A2:A57,B2:B42,1,3)	
	A	B	C	D	E
1	Likely Triers Income	Unlikely Triers Income			
56	141				
57	156				
58					
59	0.00005	pvalue			

Excel 3.6 Construct a Confidence Interval for the Difference Between Two Segments

Estimate the difference in incomes between the Unlikely and Likely Trier segments.

	A	B	C
1		Likely Triers Income	Unlikely Triers Income
54		132	
55		139	
56		141	
57		156	
58			
59	p value	0.00005	
60	M	80.1	38.5
61	SD	51.7	48.0

At the end of the **Excel 3.6 Pampers Segment Income.xls** dataset, find the segment sample means and standard deviations.

Find the difference between segment means and the standard error of the difference from the square root of the sum of segment variances (equal to the squared standard deviations), each divided by the segment sample size.

G60		f_x =SQRT(B61^2/56+C61^2/41)					
	A	B	C	D	E	F	G
1		Likely Triers Income	Unlikely Triers Income				
54		132					
55		139					
56		141					
57		156					
58							
59	p value	0.00005					
60	M	80.1	38.5 difference		41.6 SE		10.2
61	SD	51.7	48.0				

Find the approximate margin of error, which will be twice the standard error.

Make the 95% confidence interval for the difference by adding and subtracting the margin of error from the mean difference.

E61		f_x	=E60+G61			
A	**B**	**C**	**D**	**E**	**F**	**G**
1	Likely Triers Income	Unlikely Triers Income				
54	132					
55	139					
56	141					
57	156					
58						
59 p value	0.00005		lower 95%	21.2		
60 M	80.1	38.5	difference	41.6	SE	10.2
61 SD	51.7	48.0	upper 95%	62.0	approximate margin of error	20.4

Excel 3.7 Illustrate the Difference Between Two Segment Means with a Column Chart

Illustrate the difference between average incomes of Likely and Unlikely Triers.

Select the segment sample means, the *lower* and *upper 95%* confidence interval bounds, and their labels, and then use shortcuts to insert a Column chart: **Alt NC.**

Choose a <u>S</u>tyle with **Alt JCS.**

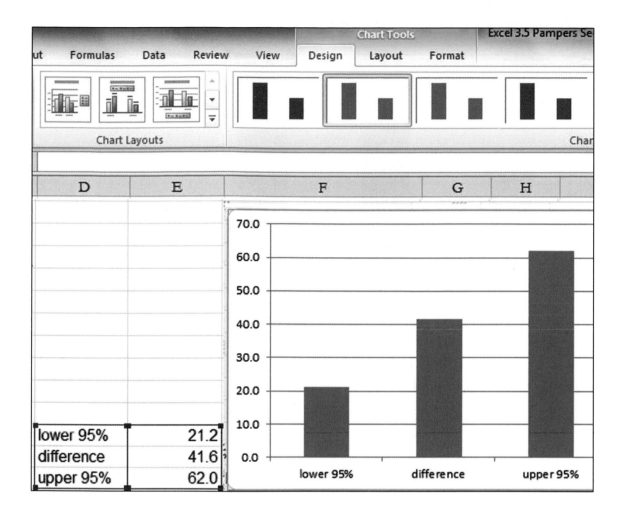

Choose **L**ayout 9, using **Alt JCL**. (**JC** selects the Design menu and **L** selects Chart **L**ayouts.)

Add data la**B**els: **Alt JAB**. (**JA** invokes the Layout menus, and **B** selects the data La**B**els menu.)

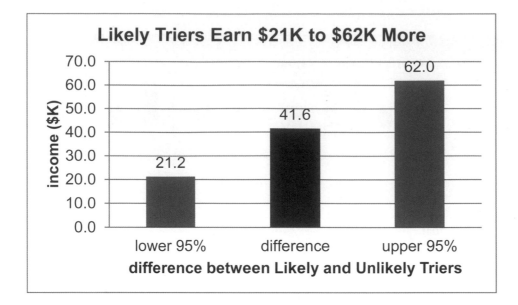

Excel 3.8 Construct a Pie Chart of Shares

Moral Acceptance of Medical Testing on Animals

Construct a Pie chart to illustrate how sample ratings of the acceptability of medical testing on animals are split.

Open a new workbook and type in two new columns, *segment* and *%surveyed.*

In the *segment* column, type in *acceptable* and *unacceptable.*

	A	B	C
1	Segment	% Surveyed	
2	Acceptable	61%	
3	Unacceptable	39%	

In the *%surveyed* column, type in the sample proportions that found medical testing on animals acceptable, 61%, and unacceptable, 39%.

B5		f_x	=0.5*SQRT(1/1000)		
	A	B	C	D	E
1	Segment	% Surveyed			
2	Acceptable	61%			
3	Unacceptable	39%			
4					
5	SE	0.016			

Find the conservative standard error of the proportion from $P = .5$ and sample size of 1,000.

B6		f_x	=1.96*B5

	A	B	C	D
1	Segment	% Surveyed		
2	Acceptable	61%		
3	Unacceptable	39%		
4				
5	SE	0.016		
6	Margin of error	0.031		

Find the margin of error from the *critical Z* for 95% confidence (1.96) and the *conservative standard error of the proportion.*

To make a Pie chart, select the six label and data cells, and then use shortcuts to insert a Pie chart: **Alt NQ**.

Choose a **Chart Style** from the **Design** menu.

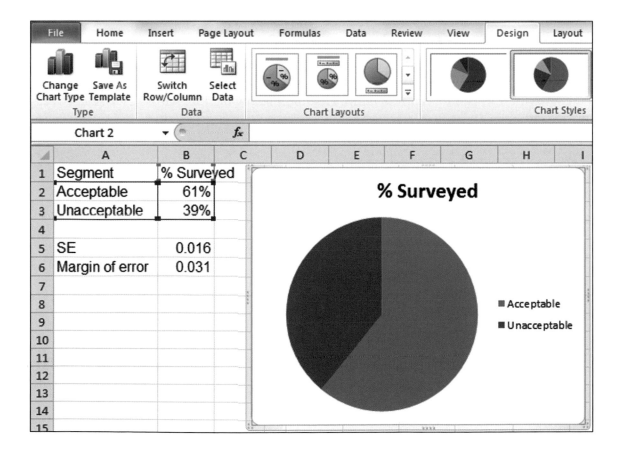

Choose **Design, Chart Layout 1**.

To add the margin of error, use shortcuts to insert a text box below the Pie: **Alt NX**. (**X** selects Te**X**t box from the I**N**sert menu.)

Type in *Margin of error: 3.1%*.

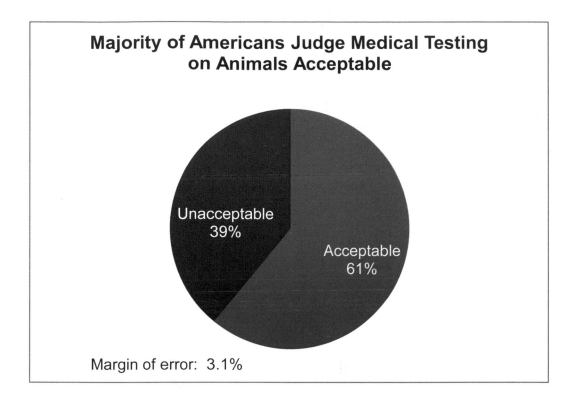

Excel 3.9 Test the Difference in Between Alternate Scenarios or Pairs with a *Paired t Test*

Difference Between Conventional and Socially Desirable Portfolio Ratings
Test the hypothesis that the average difference between ratings of a Conventional portfolio and ratings of a socially desirable portfolio is greater than 0.

Open **Excel 3.9 SD Portfolio.xls.**

Use the Excel function **T.TEST**(*array1,array2,tails,type*) to calculate a *paired t test*.

For *array1*, enter the *socially desirable portfolio ratings*.
For *array2*, enter the *conventional portfolio ratings*.
(It does not matter which array is first.)
For *tails*, enter **1** for a *one tail test*, and for *type*, enter **1** to specify a *paired t test*.

	C35		f_x	=T.TEST(A2:A34,B2:B34,1,1)	

	A	B	C	D
1	socially desirable portfolio	conventional portfolio rating	difference	
32	2	-3	5	
33	3	-3	6	
34	3	-3	6	
35		pvalue	0.0097	

Excel 3.10 Construct a Confidence Interval for the Difference Between Alternate Scenarios or Pairs

To estimate the population difference in investors' ratings of socially desirable and conventional portfolios from sample data, construct a confidence interval of the average rating difference.

Find the mean and standard deviation of the *difference.*

Use the standard deviation of the difference to find the margin of error of the difference with **CONFIDENCE.T**(*alpha, standard deviation, sample size*).
For a 95% confidence interval, enter **.05** for *alpha.*
Enter the *sample size*, **33**. (Do not subtract 1; Excel will do this.)

	C38		f_x	=CONFIDENCE.T(0.05,C37,33)	

	A	B	C	D
1	socially desirable portfolio	conventional portfolio rating	difference	
33	3	-3	6	
34	3	-3	6	
35		pvalue	0.0097	
36		M	-1.5	
37		SD	3.4	
38		margin of error	1.2	

	A	B	C
1	*socially desirable portfolio*	*conventional portfolio rating*	*difference*
33	3	-3	6
34	3	-3	6
35		pvalue	0.0097
36		M	-1.5
37		SD	3.4
38		margin of error	1.2
39		lower 95%	-2.7
40		upper 95%	-0.3

Subtract and add the margin of error from the mean difference to find the 95% confidence interval bounds for the difference.

Lab Practice 3 Inference

Cingular's Position in the Cell Phone Service Market

Cingular's managers have conducted a survey of customers in 21 major metropolitan areas to assess the quality of service along three key areas: *coverage, absence of dropped calls*, and *static*. Customers rated Cingular service along each of these three dimensions using a 5 point scale (1 = poor to 5 = excellent). Data are in **Lab Practice 3 cingular.xls**

Management's goal is to be able to offer service that is not perceived as inferior. This goal translates into mean ratings that are greater than 3 on the 5 point scale in the national market across all three service dimensions.

Based on this sample, average ratings in all major metropolitan areas are

_____ to _____ for *coverage*

_____ to _____ for absence of *dropped calls*

_____ to _____ for *static*, with 95% confidence

Management can conclude that they have achieved their goal along _____ *coverage*, _____*dropped calls*, _____ *static*.

Value of a Nationals Uniform

The Nationals General Manager is concerned that his club may not be paying competitive salaries. He has asked you to compare Nationals' salaries with salaries of players for the closest team in the National League East, the Phillies. He suspects that the Phillies may win more games because they are attracting better players with higher salary offers. Data are in **Lab Practice 3 Nationals.xls.**

This is a _____*tail t test.*

p Value from *one tail t test* of difference in team *salary* means _____

The General Manager can conclude that, relative to the Phillies, the Nationals are paid _____ _____less_____the same.

Extra Value of a Phillies Uniform

If you conclude that the Phillies do earn higher salaries, estimate the average difference at a 95% level of confidence.

On average, players for the Phillies earn _____ to _____ more than players for the Nationals.

The pooled standard error of the difference in mean salaries is _____

Illustrate the 95% confidence interval for the difference between the two teams' salaries with a Column chart.

Confidence in Chinese Imports

Following the recall of a number of products imported from China, the Associated Press–Ipsos Poll asked 1,005 randomly selected adults about the perceived safety of products imported from China. Poll results are shown below.

"When it comes to the products that you buy that are made in China, how confident are you that those products are safe . . . ?"

Confident	Not confident	Unsure
%	%	%
42	57	1

Use these data to construct a *conservative 95% confidence interval* for the proportion *Not Confident* that Chinese imports are safe.

_____ to _____ % are not confident that products made in China are safe.

Illustrate your result with a Pie chart which includes the margin of error in a text box. Add a bottom line title.

Lab 3 Inference: Dell PDA Plans

Managers at Dell are considering a joint venture with a Chinese firm to launch a new PDA equipped with Qwerty keyboard and loaded with Microsoft Office.

I. Estimate of percent of PDA owners who will replace their PDAs

In a concept test using a random sample of 1,000 PDA owners, 20% indicated that they would probably or definitely replace their PDA with the new product within the next quarter. Norms from past market research indicate that 80% who indicate intent to replace actually will.

1. *Expected Dell PDA share* = 80% × sample proportion = _____.

2. Use Excel to create a sample of 1,000 *Dell PDA share* proportions from a Normal distribution with mean equal to the sample proportion expected to purchase.

   To choose a standard deviation for your simulation

   (a)   Use subjective judgment to estimate the lowest possible *Dell PDA share* (which would be too high only 2.5% of the time): _____.

   (b)   Use subjective judgment to estimate the highest possible *Dell PDA share* (which would be too low only 2.5% of the time): _____.

   (c)   Estimate the standard deviation of *Dell PDA share* by dividing the likely range, a to b, by 4, following the empirical rule: _____.

   Create a histogram to illustrate the distribution of potential *Dell PDA shares.*

II. Simulate Dell Shipments to Assess Potential, Given Assumptions

Managers want to know the likelihood that *shipments of Dell PDA* (which is the product of *Dell PDA share* and *World shipments*) will exceed 80K in the third quarter of 2011.

Build a spreadsheet linking *Dell shipments* to *world shipments* and *Dell market share*:

$$\text{Dell shipments}_t = \text{Dell PDA proportion} \times \text{world shipments}_t.$$

The world PDA market declined in the first two quarters of 2011, down 40% from shipments in 2010.

Management assumes that potential *World shipments* in the third quarter of 2011 are normally distributed and most likely to be 600K.

Managers believe that there is only a 2.5% chance that *World shipments* would be less than 400K in 2011 and that there is only a 2.5% chance that *World shipments* would be greater than 800K in 2011.

Use Excel to create a sample of 1,000 *world shipments* to produce a distribution of potential *World shipments*.

Use your samples of potential *Dell PDA proportion* and *World PDA shipments* to find the distribution of possible *Dell shipments*.

Given managers' assumptions, what is the chance that *Dell shipments* will exceed **80,000 in the third quarter of 2011**? _____%

III. Distinguish Likely Dell PDA Adopters

Those PDA owners who indicated that they were likely to switch to the Dell PDA may be more price conscious than other PDA owners. In the concept test, 100 participants were asked to rate the importance of several PDA attributes, including price. These data are in **Lab 3 Inference Dell PDA.xls.**

1. Do Likely Adopters rate price higher in importance than Unlikely Adopters?

   p Value: _____ Conclusion: Y or N

2. What is the expected difference in *price importance* between Likely and Unlikely Adopters? _____

3. Approximate 95% confidence interval of the difference in *price importance* between Likely and Unlikely Adopters: _____ to _____

4. Make a Column chart to illustrate your results.

Assignment 3-1 The Marriott Difference

There are 51 branded hotels in Washington, DC, owned or managed by Marriott or competitors. The hotel industry in Washington, DC, is representative of the hotel industry in cities throughout the United States. Differences in quality and price distinguish the hotels. Marriott would like to claim that its hotels offer higher average quality lodging than competing hotels and that Marriott's average *starting room price* is equivalent to competitors' average *starting room price*. The dataset **Assignment 3-1 DC Hotels.xls** contains *Guest rating*, a measure of quality, and *starting room price* for Marriott hotels and for competitors' hotels.

1. Can Marriott claim that Marriott hotels are rated higher in quality than competitors' hotels? (Assume a 95% level of confidence.)
   - (a) State the null and alternative hypotheses.
   - (b) State your conclusion in <u>one sentence</u> with words that a technically savvy manager would understand.
   - (c) State your conclusion in <u>one sentence</u> with words that a manager, not necessarily statistically savvy, would understand.
2. Can Marriott claim that Marriott hotels are priced equivalently to competitors? (Assume a 95% level of confidence.)
   - (a) State the null and alternative hypotheses.
   - (b) State your conclusion in <u>one sentence</u> that a technically savvy manager would understand.
   - (c) State your conclusion in <u>one sentence</u> with words that a manager, not necessarily statistically savvy, would understand.

Assignment 3-2 Bottled Water Possibilities

The students in Team 8, Stephanie, Shawn, Erica, and Tyler, want to know how their assumptions regarding

Demand for bottled water
Market share
affect the chances that *bottles sold* will **exceed 800,000**.

Conduct a Monte Carlo simulation of bottles sold with the following assumptions:

- *Average demand for bottled water* will be 9.9 bottles per customer per week.
- There is a 2.5% chance that *average demand for bottled water* will be less than 8 bottles per customer per week.

- There is a 2.5% chance that *average demand for bottled water* will be greater than 12 bottles per customer per week.
- *Market share* that Team 8 could achieve with their custom bottled water dispensers could be **as low as 5%** and **as high as 15%**, and the *market share* possibilities within this range are equally likely or **uniformly distributed**.

What are the chances that Team 8 could sell **at least 800,000** bottles in the first year, given these assumptions?

Include the distribution of bottles sold to illustrate your answer.

Assignment 3-3 Immigration in the United States

The FOX News/Opinion Dynamics Poll, July 11–12, 2006, of (*N=*) 900 registered voters nationwide, reports public opinion concerning immigrants and proposed immigration legislation:

"In general, do you think immigrants who come to the United States today join society and give to the country or stay separate from society and take from the country?"

	Join Society/Give	Stay Separate/Take	Depends (vol.)	Unsure
	%	%	%	%
7/11–12/06	41	36	17	6

"Do you think the United States should increase or decrease the number of legal immigrants allowed to move to this country?"

	Increase	Decrease	No Change (vol.)	Unsure
	%	%	%	%
7/11–12/06	24	51	17	8

Use these data to construct **conservative** 95% *confidence intervals* for the *proportions* who (1) agree that immigrants joint society/give and (2) agree that the United States should increase the number of legal immigrants.

Briefly summarize the opinions of all registered voters using language that American adults would understand.

Illustrate your summary with Pie charts embedded in your report.
Be sure to include the margins of error in your Pie charts.

Assignment 3-4 McLattes

McDonalds recently sponsored a blind taste test of lattes from Starbucks and their own McCafes. A sample of 30 Starbucks customers tasted both lattes from unmarked cups and provided ratings on a –3 (= worst latte I've ever tasted) to +3 (= best latte I've ever tasted) scale. These data are in **Assignment 3-4 Latte.xls**.

Can McDonalds claim that their lattes taste every bit as good as Starbucks' lattes? (Please use 95% confidence.)

What evidence allows you to reach this conclusion?

Assignment 3-5 A Barbie Duff in Stuff

Mattel recently sponsored a test of their new Barbie designed by Hillary Duff. The Duff Barbie is dressed in Stuff, Hillary Duff clothing designs, and resembles Hillary Duff. Mattel wanted to know whether the Duff Barbie could compete with rival MGA Entertainment's Bratz dolls.
A sample of thirty 7 year old girls attended Barbie parties, played with both dolls, then rated both on a –3 (= Not At All Like Me) to +3 (= Just Like Me) scale. These data are in **Assignment 3-5 Barbie.xls**.

Do the 7 year olds identify more strongly with the Duff Barbie in Stuff than the Bratz? (Please use 95% confidence.)

What evidence allows you to reach this conclusion?

CASE 3-1 Yankees Versus Marlins: The Value of a Yankee Uniform[1]

The Marlins General Manager is disgruntled because two desirable rookies accepted offers from the Yankees instead of the Marlins. He believes that Yankee salaries must be noticeably higher; otherwise, the best players would join the Marlins organization. Is there a difference in salaries between the two teams? If the typical Yankee is better compensated, the general manager is planning to chat with the owners about sweetening the Marlins' offers. He suspects that the owners will argue that the typical Yankee is older and more experienced, justifying some difference in salaries.
Data are in **Case 3-1 Yankees v Marlins Salaries.xls**.

---

[1] This example is a hypothetical scenario using actual data.

Determine

- Whether Yankees earn more on average than Marlins
- Whether players for the Yankees are older on average than players for the Marlins

If you find a difference in either case, construct a *95% confidence interval* of the expected difference in any season.

Briefly summarize your results using language that the General Manager and owners would understand, and illustrate with a Column chart.

CASE 3-2 Gender Pay

The Human Resources Manager of Sam's Club was shocked by the revelations of gender discrimination by Wal-Mart ("How Corporate America is Betraying Women," *Fortune*, January 10, 2005) and wants to demonstrate that there is no gender difference in average salaries in his firm. He also wants to know whether levels of responsibility (measured with the Position variable) and experience differ between men and women, inasmuch as this could explain a difference in salaries.
Case 3-2 GenderPay.xls contains *salaries, positions*, and *experience* of men and women from a random sample of the company records.

Determine

- Whether the sample supports a conclusion that men and women are paid equally
- Whether average level of *responsibility* differs across genders
- Whether average *experience* differs across genders

If you find that the data lead to rejection of the null hypothesis that, on average, men are paid less or the same as women, construct a 95% confidence interval of the expected average difference.

If either average level of *responsibility* or average years of *experience* differs, construct *95% confidence intervals* for the expected average difference.

Briefly summarize your results using language that a businessperson (who may not remember quantitative analysis) could understand.

Illustrate your results with Column charts. Choose bottom line titles that help your audience see the results.
Be sure to round your statistics to two or three significant digits.

CASE 3-3 Polaski Vodka: Can a Polish Vodka Stand Up to the Russians?

Seagrams management decided to enter the premium vodka market with a Polish vodka suspecting that it would be difficult to compete with Stolichnaya, a Russian vodka and the leading premium brand. The product formulation and the package/brand impact on perceived taste were explored with experiments to decide whether the new brand was ready to launch.

The taste: First, Seagrams managers asked, "Could consumers distinguish between Stolichnaya and Seagrams' Polish vodka in a *blind* taste test, where the impact of packaging and brand name were absent?"

Consultants designed an experiment to test the null and alternative hypotheses:

H_0: The taste rating of Seagram's Polish vodka is at least as high as the taste rating of Stolichnaya. The average difference between taste ratings of Stolichnaya and Seagrams' Polish vodka does not exceed zero:

$$\mu_{STOLICHNAYA} - \mu_{POLISH} \leq 0.$$

H_1: The taste rating of Seagram's Polish vodka is lower than the taste rating of Stolichnaya. The average difference between taste ratings of Stolichnaya and Seagram's Polish vodka is positive:

$$\mu_{STOLICHNAYA} - \mu_{POLISH} > 0.$$

In this first experiment, each participant tasted two unidentified vodka samples and rated the taste of each on a 10 point scale. Between tastes, participants cleansed palates with water. Experimenters flipped a coin to determine which product would be served first: if heads, Seagrams' Polish vodka was poured first; if tails, Stolichnaya was poured first. Both samples were poured from plain clear beakers. The only difference between the two samples was the actual vodka.

These experimental data in **Case 3-3 Polaski Taste.xls** are repeated measures. From each participant, we have two measures whose difference is the difference in taste between the Russian and Polish vodkas.

Test the difference between taste ratings of the two vodkas.

Construct a *95% confidence interval* of the difference in taste ratings.

Illustrate your results with a PivotChart and interpret your results for management.

The brand and package. Seagrams management proceeded to test the packaging and name, Polaski. The null hypothesis was

> H₀: The taste rating of Polaski vodka poured from a Polaski bottle is at least as high as the taste rating of Polaski vodka poured from a Stolichnaya bottle. The mean difference between taste ratings of Polaski vodka poured from a Stolichnaya bottle and Polaski vodka poured from the Seagrams bottle bearing the Polaski brand name does not exceed zero.

Alternatively, if the leading brand name and distinctive bottle of the Russian vodka affected taste perceptions, the following could be true:

> H₁: The mean difference between taste ratings of Polaski vodka poured from Stolichnaya bottle and Polaski vodka poured from the Seagrams bottle bearing the Polaski brand name is positive.

In this second experiment, Polaski samples were presented to participants twice, once poured from a Stolichnaya bottle and once poured from the Seagrams bottle bearing the Polaski name. Any minute differences in the actual products were controlled for by using Polaski vodka in both samples. Differences in taste ratings would be attributable to the difference in packaging and brand name.

Thirty new participants again tasted two vodka samples, cleansing their palates with water between tastes. As before, a coin toss decided from which bottle the first sample would be poured: Stolichnaya if heads, Polaski if tails. Each participant rated the taste of the two samples on a 10 point scale.

These data are in **Case 3-3 Polaski Package.xls.**

Test the difference in ratings due to packaging.

Construct a *95% confidence interval* of the difference in ratings due to the packaging.

Illustrate your results with a PivotChart.

Interpret your results for management.

CASE 3-4 American Girl in Starbucks

Mattel and Warner Brothers are considering a partnership with Starbucks to promote their new American Girl movie. Starbucks previously backed Lionsgate's *Akeelah and the Bee*, which earned $19 million. In exchange for $8 million, Starbucks would install signage and stickers in 6,800 of its stores, print American Girl branded cup sleeves, and sell the picture's soundtrack. Materials for the movie would also appear on the company's website. Starbucks claims 44 million customers in the 6,800 stores.

In a pretest of the promotion during 1 week in one Starbucks store, **184 of the 924, or 20%** of Fast Card customers served that week agreed that they had heard of the movie when surveyed by phone the following week.

Mattel managers believe that roughly **35%** of those who are aware of the movie will buy tickets.

There is only a 2.5% chance that the percentage buying tickets would be less than 5%.

There is only a 2.5% chance that the percentage buying tickets will be more than 9%.

95% of moviegoers are expected to bring **1** to **3** family members or friends, on average.

Mattel would earn **$1** royalty from each ticket.

To justify the promotion, Mattel management wants to be sure that royalties are likely to **exceed $8 million**.

1. What are the chances that *royalties* from ticket sales to Starbucks customers would **exceed $8 million**?

   Mattel and Warner Brothers are also considering McDonalds as a potential promoter of the new movie. Mattel management suspects that Starbucks customers are wealthier than McDonalds customers. (Because wealthier families have the resources to buy American Girl products, this is the target market for the new movie audience, and Mattel would favor the sponsor with wealthier customers.)
   Household income data from intercept interviews of 30 McDonalds customers and 30 Starbucks customers are in **Case 3-4 StarbucksvMcD.xls**.

2. Can Mattel managers conclude that Starbucks customers are wealthier than McDonalds customers?
   What evidence allows you to reach this conclusion?

3. Estimate the average *income* difference between Starbucks and McDonalds customers using a 95% *confidence interval*.

Chapter 4
Quantifying the Influence of Performance Drivers and Forecasting: Regression

Regression analysis is a powerful tool for quantifying the influence of continuous, *independent, drivers* X on a continuous *dependent, performance* variable Y. Often we are interested in both explaining how an independent decision variable X drives a dependent performance variable Y and also in predicting performance Y to compare the impact of alternate decision variable X values. X is also called a *predictor* because from X we can predict Y. Regression allows us to do both: quantify the nature and extent of influence of a performance driver and predict performance or response Y from knowledge of the driver X.

With regression analysis, we can statistically address the following questions:

- Is variation in a dependent performance response variable Y influenced by variation in an independent variable X?

If X is a driver of Y, with regression, we can answer these questions:

- What percent of variation in performance Y can be accounted for with variation in driver X?
- If driver X changes by one unit, what range of response can we expect in performance Y?
- At a specified level of the driver X, what range of performance levels Y is expected?

In this chapter, simple linear regression with one independent variable is introduced, and we explore ways to address each of these questions linking a continuous driver, which may be a decision variable, to a continuous performance variable. We also explore the link between correlation and simple linear regression, because the two are closely related.

4.1 The Simple Linear Regression Equation Describes the Line Relating a Decision Variable to Performance

Regression produces an equation for the line that best relates changes or differences in a continuous, dependent performance variable Y to changes or differences in a continuous, independent driver X. This line comes closest to each of the points in a scatterplot of Y and X:

$$\hat{Y} = b_0 + b_1 X,$$

where $\hat{Y}$ is the expected value of the dependent performance, or response, variable, called "y hat"; X is the value of an independent variable, decision variable, or driver; b_0 is the *intercept* estimate, which is the expected value of $\hat{Y}$ when X is zero; b_1 is the estimated slope of the regression line,

C. Fraser, *Business Statistics for Competitive Advantage with Excel 2010: Basics, Model Building, and Cases,* DOI 10.1007/978-1-4419-9857-6_4, © Springer Science+Business Media, LLC 2012

which indicates the expected change in performance $\hat{Y}$ in response to a unit change from the driver's average $\overline{X}$.

Example 4.1 HitFlix Movie Rentals

An owner of a movie rental vending business is planning to add new vending units and needs to decide how many titles to offer in each. The plan is to design the new units to stock and offer 100 titles, but the owner thinks a larger number of titles might generate more revenue. Titles offered may drive revenues. The null and alternate hypotheses that the owner would like to test are

H_0: Titles offered X has no effect on movie rental revenues Y.

H_1: Titles offered X drives movie rental kiosk revenues Y.

Scatterplots of titles offered, X, and annual revenues, Y, for a random sample of 52 vending units from the chain are shown in Fig. 4.1 from Excel:

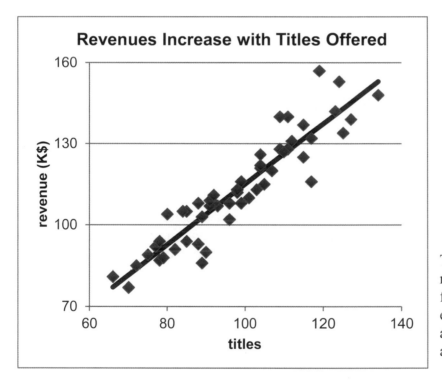

The scatterplot indicates that revenues may be a linear function of titles offered. For each additional title, average annual revenues increase by about $1.1K or $1100.

Fig. 4.1 Vending unit revenues by titles

The average difference in revenues between vending units with 70 and 80 titles, $11,000 [=(80 − 70) × $1,100], is identical to the average difference in revenues between vending units with 120 and 130 square feet, $11,000 [=(130 − 120) × $1,100]. Expected revenues $\hat{Y}$ increase at a constant rate of $1,100 with each increase of one title offered. Because variation in revenues Y is related linearly to variation in titles X, the linear regression line is a good summary of the data:

$$rev\hat{e}nues(K\$) = 3.4(K\$) + 1.1(K\$ per\_title)title.$$

In this example, the intercept estimate b_0 is 3.4(K$). Were a vending unit to offer zero titles (which isn't likely), the expected revenue would be $3,400. The estimated slope b_1 is 1.1(K$ per title), indicating that we expect an average increase in revenue of $1,100 in response to an increase in titles offered of one.

4.2 *F* Tests Significance of the Hypothesized Linear Relationship *R Square* Summarizes Its Strength and Standard Error Reflects Forecasting Precision

Using the regression formula, we can predict the expected revenue $\hat{Y}$ for any given vending unit offering a number of titles X. Table 4.1 contains predictions for five vending units offering different numbers of titles.

Table 4.1 Expected revenue

Titles	Expected revenue($K)		
X	b_0	$+b_1X$	$= \hat{Y}$
70	3.4	1.1(70)	80
80	3.4	1.1(80)	91
90	3.4	1.1(90)	102
110	3.4	1.1(110)	124

The differences between expected and actual revenue are the *residuals* or *errors*. Errors from these four vending units are shown in Table 4.2 and Fig. 4.2.

Table 4.2 Residuals from the regression line

Titles	Actual revenue($K)	Expected revenue($K)	Residual ($K)
X	Y	$\hat{Y}$	$e = Y\text{-}\hat{Y}$
80	87	91	−4
100	102	113	−11
110	140	124	−16
120	142	135	7

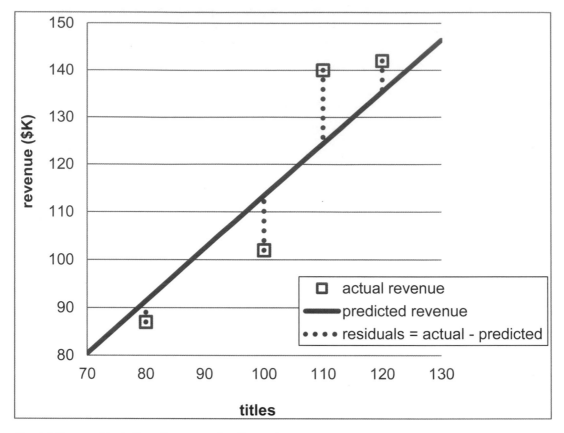

Fig. 4.2 Four residuals from the regression line

The *sum of squared errors* in a sample,

$$\text{SSE} = \sum e_i^2 = \sum (Y_i - \hat{Y})^2 = \sum (Y_i - b_0 - b_1 X_i)^2,$$

is the portion of total variation in the dependent variable SST, which remains unexplained after accounting for the impact of variation in X. The *least squares* regression line is the line with the smallest *SSE* of all possible lines relating X to Y.

The regression *standard error*, equal to the square root of *mean square error*, MSE (=SSE/(N–2)),

$$\text{standard error} = \sqrt{MSE}$$

reflects the precision of the regression equation. We expect forecasts to be within approximately two standard errors of actual performance 95% of the time.

The difference, SST–SSE, called the *regression sum of squares*, SSR, or *model sum of squares*, is the portion of total variation in Y influenced by variation in X. To test the hypothesis that the independent variable influences the dependent variable in the population,

H_0: Variation in X does not drive variation in y

versus H_1: Variation in X does drive variation in y,

we use our sample data to calculate the ratio of the explained to unexplained variation for a given size model and sample size. Adding independent variables to a model adds explanatory power. Dividing explained variation *SSR* by the number of independent variables focuses the hypothesis test on the variation explained per independent variable. Dividing unexplained variation *SSE* by the sample size, less the number of variables in the model, makes the relevant comparison of the variation explained per independent variable *MSR* to unexplained variation for a model of given size and sample size *MSE*. This ratio of mean squares is distributed as an *F*, and the particular *F* distribution is indexed by model size and sample size. The numerator degrees of freedom is the number of predictors and the denominator degrees of freedom is the sample size less the number of variables in the model:

$$F_{1,(N-2)} = \frac{SSR/1}{SSE/(N-2)} = \frac{MSR}{MSE}.$$

F distributions are skewed with a minimum value of zero. Several *F* distributions with one numerator degree of freedom (corresponding to one independent variable) and various sample sizes are shown in Fig. 4.3.

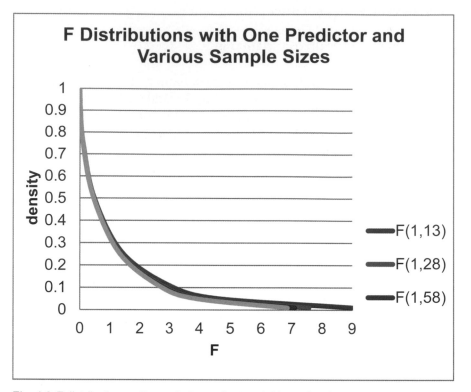

Fig. 4.3 *F* distributions with one independent variable and various sample sizes

The percentage of total variation in the dependent performance variable Y that can be accounted for by variation in the independent decision variable X is *R Square* given by

$$R\ Square = SSR\ /\ SST.$$

R Square ranges between zero and one, or zero and 100%. The greater the influence of X on Y, the closer *R Square* is to 100%, and the larger F is, for a given sample size, N. Our model hypotheses can also be stated as follows:

H_0: *R Square* is zero percent.

versus H_1: *R Square* is greater than zero percent.

The percent of total variation in the dependent performance variable Y, which can be accounted for by variation in the independent decision variable X, is *R Square* given by

$$R\ Square = SSR\ /\ SST$$

R Square and the standard error appear in SUMMARY OUTPUT, which is followed by the ANOVA table in regression output. The SUMMARY OUTPUT and ANOVA tables from Excel for the **HitFlix Movie Rental** regression are shown in Table 4.3.

Table 4.3 Model summary of fit and ANOVA table

SUMMARY OUTPUT					
Regression Statistics					
Multiple R	.93				
R Square	.86				
Adjusted R Square	.86				
Standard Error	7.44				
Observations	52				
ANOVA	*df*	*SS*	*MS*	*F*	*Significance F*
Regression	1	16,800	16763	303	.0000
Residual	50	2,800	55		
Total	51	19,500			

R Square, the ratio of regression sum of squares (16,800) to total sum of squares (19,500), is .86, or 86%:

$$R\,Square = \frac{Regression\,Sum\,of\,Squares}{Total\,Sum\,of\,Squares} = \frac{16,800}{19,500} = .86.$$

Variation in titles offered X accounts for 86% of the variation in revenues Y. Other factors account for the remaining 14%.

The regression standard error is 7.4($K). We can expect 95% of revenue forecasts for vending units offering a certain number of titles to be no further than $t_{.05,(N-2)}$ times this standard error, 14.8($K) (= 2.01 × 7.4($K)), or $14,800, from average revenues of all vending units offering the same number of titles.

The $F_{1,50}$ statistic is 303. With a sample of size 52 and one independent variable, the significance of F is a very small number, less than .0001. There is little chance that we would observe the sample data patterns were titles offered not driving revenues, that is, were the null hypothesis true.

Based on regression analysis of this sample, we have sufficient evidence to reject the null hypothesis:

H_0: Variation in titles offered X does not drive variation in movie rental revenues Y.

Sample evidence suggests, instead

H_1: Variation in titles offered X drives variation in movie rental revenues Y.

4.3 Test and Infer the Slope

Because the true impact β_1 of a driver X on performance Y is unknown, this slope, or *coefficient*, is estimated from a sample. This estimate b_1 and its sample standard error s_{b_1} are also used to test the hypothesis that variation in X drives variation in Y:

H_0: Variation in the independent variable X does not drive variation in the dependent variable Y.
or
H_0: The regression slope is zero: $\beta_1 = 0$.

Alternatively,

H_1: Variation in the independent variable X drives variation in the dependent variable Y.
or
H_1: The regression slope is not zero: $\beta_1 \neq 0$.

In many instances, from experience or logic, we know the likely direction of influence. In those instances, the alternate hypothesis requires a one tail test:

H_1: The independent variable X positively influences the dependent variable Y.
or
H_1: The regression slope is greater than zero: $\beta_1 > 0$.

This one sided alternate hypothesis describes an upward slope. A similar alternate hypothesis could be used when logic or experience suggests a downward slope. In the **Movie Rentals** example, if revenue did not depend on titles offered, the scatterplot would resemble a spherical cloud and the regression line would be flat at the dependent variable mean $\overline{Y}$, as in Fig. 4.4.

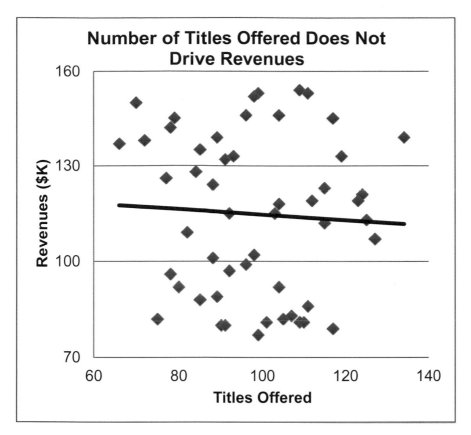

Fig. 4.4 *X* does not drive *Y*, and the regression line slope is flat ($b_1 = 0$)

To form a conclusion about the significance of the slope, calculate the number of standard errors that separate the estimate b_1 from zero:

$$t_{(N-2)} = b_1 / s_{b_1}$$

In **Movie Rentals**, the standard error of the slope estimate s_{b_1} is .064. The slope is more than 17 standard errors from zero:

$$t_{50} = 1.12 / .064 = 17.4$$

At this *t* value, a two tail test has a *p Value* of .0001. From both experience and logic, the movie rental business owner had a good idea that the titles offered had a positive impact on revenues, so the alternate hypothesis is that the slope is positive. Dividing the two tail *p Value* by 2, the one tail *p Value* is .00005. There is a very small chance that we would observe the sample data were titles offered not driving revenues. From our sample evidence, we reject the null hypothesis of a flat slope and accept the alternate hypothesis of a positive slope. Sample evidence suggests that the number of titles offered has a positive impact on revenues.

Excel does these calculations for us. The slope and intercept estimates are labeled *Coefficients* in Excel, shown in Table 4.4.

Table 4.4 Coefficient estimates, standard errors, and *t tests*

	Coefficients	*Standard error*	*t Stat*	*p Value*	*Lower 95%*	*Upper 95%*
Intercept	3.43	6.38	.5	.5931	−9.39	16.25
Titles	1.12	.064	17.4	.0000	.99	1.24

There is a 95% chance that the true population slope will fall within $t_{.05,(N-2)}$ standard errors of the slope estimate:

$$b_1 - t_{.05,50}s_{b_1} < \beta_1 < b_1 + t_{.05,50}s_{b_1}$$
$$1.12 - 2.01(.064) < \beta_1 < 1.12 + 2.01(.064)$$
$$.99 < \beta_1 < 1.24$$

The impact of one additional title on vending unit revenue is within the range of .99 ($K) to 1.24 ($K) or $990 to $1,240.

4.4 Analyze Residuals to Learn Whether Assumptions Are Met

When we use linear regression, we assume that the errors are uncorrelated with the independent variable. Explanation and prediction of revenues should be as good for vending units with a limited number of titles as for units offering many titles. To confirm that this assumption is met, look at a plot of the residuals by predicted values. There should be no pattern.

A plot of the residuals by predicted values, Fig. 4.5, is not pattern free. The residuals show more variation for units with many titles. Within the range of existing titles offered, predictions for units with a limited number of titles are likely to be more accurate than predictions for units offering many titles.

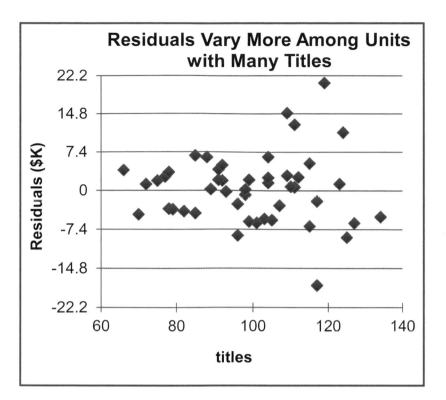

Fig. 4.5 Residuals by titles

This situation, in which residual variation is nonconstant, is termed *heteroskedasticity*. A remedy may be rescaling the dependent variable, the independent variable, or both, perhaps to natural logarithms.

Linear regression assumes that the residuals are Normally distributed. The distribution of residuals, shown in Fig. 4.6, is bell shaped and has a skewness of .1. Roughly, 95% (actually 92% = (4 + 17 + 23 + 4)/52) of predictions are within two standard errors, $14.8K, of actual revenues. Eight percent are more than two standard errors, $14,900, from actual, which is slightly more than the 5% we expect from Normally distributed residuals.

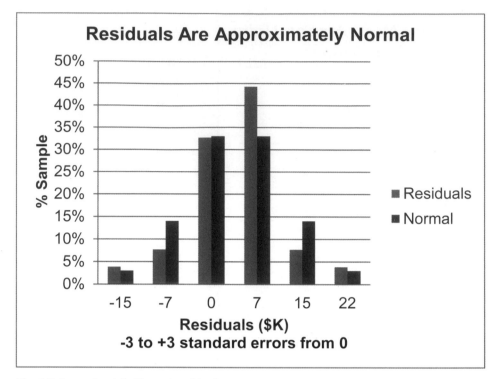

Fig. 4.6 Approximately Normal residuals

4.5 Prediction Intervals Estimate Average Response

Regression analysis can be used to forecast a 95% confidence interval for the value of the dependent variable Y given a specific value for the independent variable X. The standard error for this prediction s_Y depends on how much X influences Y, the sample size N, the standard deviation of X, and how far the particular specific value of X is from the average $\overline{X}$. However, if the sample size is large, the standard prediction errors will be close to the regression standard error or root mean square error, s. As its name suggests, *root mean square error s* is the square root of *SSE*.

In **HitFlix Movie Rentals**, s is 7.4($K). Forecasts for individual units can be expected to be within $15K [$=t_{.05,(N-2)} \times 7.4$ ($K) = 2.01 \times 7.4($K)]$ of actual revenues. The prediction margin of error is $15K. Table 4.5 and Fig. 4.7 show 95% prediction intervals for vending units offering various numbers of titles.

Table 4.5 Individual 95% Prediction Intervals

Titles	Expected revenue ($K) $\hat{Y}$	Standard error s	Margin of error $t_{.05,50}s$	95% Prediction interval $\hat{Y} \pm t_{.05,50}s$	
70	82	7.4	15	67	96
100	115	7.4	15	100	130
130	149	7.4	15	134	163

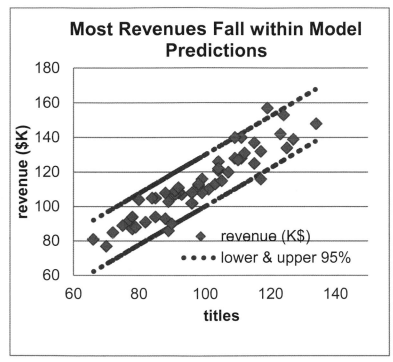

Fig. 4.7 95% prediction intervals for individual vending units

4.6 Use Sensitivity Analysis to Explore Alternative Scenarios

Comparing possible revenues from the planned number of titles offered, 100, with a larger 130 title option, the HitFlix owner learns that the additional 30 titles is expected to produce $34K (= $149K – $115K) additional revenue.

Titles	Predicted revenue ($K)
100	115
130	149

4.7 Explanation and Prediction Create a Complete Picture

From the regression analysis, the **HitFlix Movie Rental** owner can

- Conclude that titles offered drives revenues
- Estimate the extent that titles offered drives revenues
- Compare predicted revenues at alternate levels of titles offered

In the presentation of results to management, the owner would conclude the following:

Sample evidence suggests that the number of titles offered drives vending unit revenues. Variation in titles offered accounts for 86% of the variation in revenues among a random sample of 52 vending units.

With knowledge of titles offered, revenue can be estimated with a margin of error of $15K.

For each title offered over the average of 100, we can expect an average increase in revenue of $990 to $1,240.

Comparing expected revenue from vending units offering 100 titles and 130 titles, the additional titles are expected to generate $34K more revenue.

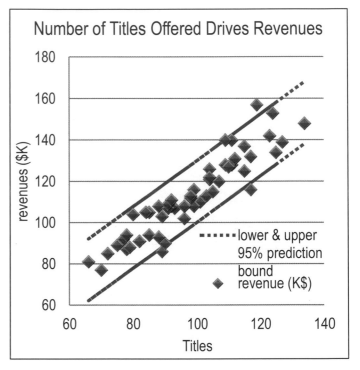

$$\hat{revenue}(K\$) = 3.4(K\$)^a + 1.1(\$K)^a\, titles$$
$$R\ Square:\ .86^a$$
$$^a Significant\ at\ .0001.$$

Titles offered	Expected revenue
100	$115,000
130	$149,000

The HitFlix owner presented results of his regression analysis by illustrating the regression line with 95% confidence prediction intervals on top of the actual data. This demonstrates how well the model fits the data.

4.8 Present Regression Results in Concise Format

The regression equation is included, in standard format, with the dependent variable on the left, *R Square* below the equation, and significance levels of the model and parameter estimates indicated with superscripts:

$$\hat{Y} = b_0{}^a + b_1{}^a X$$
$$R\ Square = \underline{\quad}{}^a$$
$${}^a Significant\ at\ \underline{\quad}.$$

Not everyone who reads this memo will understand these three lines. For the general business audience, the verbal description with graphical illustration conveys all of the important information. The three additional lines provide the information that statistically savvy readers will want in order to assess how well the model fits and which parameter estimates are significant.

4.9 Assumptions We Make When We Use Linear Regression

If we attempt to explain or predict a dependent variable with an independent variable, but omit a third (or fourth) important influence, results will be misleading. It will seem that the independent variable chosen is more important than it actually is. Often a group of independent variables together jointly influence a dependent variable. If just one from the group is included in a regression, it may seem to be responsible for the joint impact of the group. Chapters 8 and 9 introduce diagnosis of *multicollinearity*, the situation in which predictors are correlated and jointly influence a dependent variable.

Linear regression of time series data assumes that the unexplained portions of a model, the residuals, are stable over time, and that predictions do not get better or worse with time. Patterns uncovered in the data are stable over time. Chapter 9 introduces diagnosis of and remedies for *autocorrelated* errors that break this assumption and vary with time.

Linear regression assumes that the dependent variable, which is often a performance variable, is related linearly to the independent variable, often a decision variable. In reality, few relationships are linear. More often, performance increases or decreases in response to increases in a decision variable, but at a diminishing rate. The dependent variable is often limited. Revenues, for example, are never negative and are limited (probably at some very high number) by the number of customers in a market. In these cases, linear regression doesn't fit the data perfectly. Extrapolation beyond the range of values within a sample can be risky if we assume constant response when the response is actually diminishing or increasing. Although often imperfect reflections of reality, linear relationships can be useful approximations. In Chap. 11, we explore simple remedies to improve linear models of nonlinear relationships by simply rescaling to square roots, logarithms, or squares.

4.10 Correlation Reflects Linear Association

A correlation coefficient ρ_{XY} is a simple measure of the strength of the linear relationship between two continuous variables X and Y. The sample estimate of the population correlation coefficient ρ_{XY} is calculated by summing the product of differences from the sample means $\overline{X}$ and $\overline{Y}$, standardized by the standard deviations s_X and s_Y:

$$r_{XY} = \frac{1}{(N-1)} \sum_i \frac{(x_i - \overline{X})}{s_X} \frac{(y_i - \overline{Y})}{s_Y},$$

where x_i is the value of X for the ith sample element and y_i is the value of Y for the ith sample element. When X and Y move together, they are positively correlated. When they move in opposite directions, they are negatively correlated.

Example 4.2 HitFlix Movie Rentals
Table 4.6 contains titles offered X and revenues Y from a sample of eight vending units, and a scatterplot in Fig. 4.8 reveals that units that stock more titles also have greater revenues.

Table 4.6 Titles stocked and revenues ($K) for eight vending units

Vending unit	Titles offered X	Revenues ($K) Y	Vending unit	Titles offered X	Revenues ($K) Y
1	110	75	5	150	115
2	110	80	6	160	135
3	120	85	7	170	140
4	130	105	8	170	145
			Sample mean	140	$110

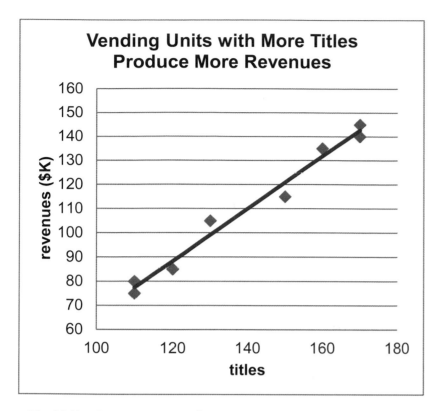

Fig. 4.8 Vending unit revenues ($K) by titles stocked

Differences from the sample means and their products are shown in Table 4.7.

Table 4.7 Differences from sample means and crossproducts

Unit$_i$	Titles stocked			Revenues ($K)			
	x_i	$\overline{X}$	$x_i - \overline{X}$	y_i	$\overline{Y}$	$y_i - \overline{Y}$	$(x_i - \overline{X})(y_i - \overline{Y})$
1	110	140	−30	$75	$110	$−35	1,050
2	110	140	−30	80	110	−30	900
3	120	140	−20	85	110	−25	500
4	130	140	−10	105	110	−5	50
5	150	140	10	115	110	5	50
6	160	140	20	135	110	25	500
7	170	140	30	140	110	30	900
8	170	140	30	145	110	35	1,050

The sample standard deviations are $s_X = 25.6$ titles and $s_Y = 28.2$ ($K). The correlation coefficient is

$$r_{Xy} = \frac{1}{(8-1)} \left[\frac{1,050 + 900 + 500 + 50 + 50 + 500 + 900 + 1,050}{(25.6)(28.2)} \right]$$

$$= \frac{1}{7} [5,000 / 722]$$

$$= .990.$$

A correlation coefficient can be as large in absolute value as 1.00, if two variables are perfectly correlated. All of the points in the scatterplot would lie on the regression line in that case. *R Square*, which is the squared correlation in a simple regression, would be 1.00, whether the correlation coefficient were −1.00 or +1.00.

In the **HitFlix Movie Rentals** example above, *R Square* is

$$R\,Square = r_{Xy}^2 = .99^2 = .98.$$

If two variables are strongly negatively correlated, their scatterplot would look like the top panel in Fig. 4.9. Two scatterplots of uncorrelated variables are shown in the middle and lower panels of the figure.

Notice that although X and Y are not related linearly in the third panel, they are strongly related. There are situations, for example, where more is better up to a point and improves performance, then *saturation* occurs and, beyond this point, response deteriorates:

- Without enough advertising, customers will not be aware of a new product. Spending more increases awareness and improves performance. Beyond some saturation point, customers grow weary of an advertisement, decide that the company must be desperate to advertise so much, and switch to another brand, reducing performance.
- A factory with too few employees X to work all of the assembly positions would benefit from hiring. Adding employees increases productivity Y up to a point. Beyond some point, too many employees would crowd the facility and interfere with each other, reducing performance.

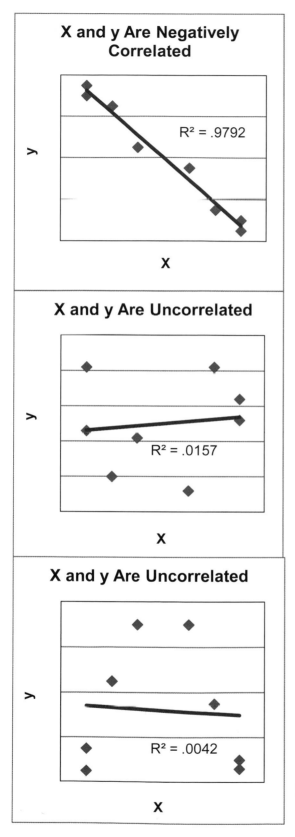

Fig. 4.9 Negatively correlated and uncorrelated variables

4.11 Correlation Coefficients Are Key Components of Regression Slopes

Correlation coefficients are closely related to regression slopes. From the correlation between X and y, as well as their sample standard deviations s_X and s_y, the regression slope estimate can be calculated:

$$b_1 = r_{Xy} \frac{s_y}{s_X}.$$

Similarly, from the regression slope estimate and sample standard deviations s_X and s_Y, the correlation coefficient can be calculated:

$$r_{Xy} = b_1 \frac{s_X}{s_y}.$$

In the **HitFlix Movie Rentals** example, with the correlation coefficient $r_{titles,revenues} = .99$, the sample standard errors are $s_{titles} = 26.5$ and $s_{revenues} = 28.2$, and the regression slope estimate can be calculated:

$$b_{titles} = .99 \frac{28.2}{26.5} = 1.09$$

Based on sample evidence, there is little chance that titles stocked and vending unit revenues are uncorrelated.

Corresponding simple regression results are shown in Table 4.8.

Table 4.8 Regression of revenue by titles

SUMMARY OUTPUT

Regression statistics

Multiple R	.99				
R Square	.98				
Standard error	4.38				
Observations	8				

ANOVA

	df	SS	MS	F	Significance F
Regression	1	5,435	5435	283.0	.0000
Residual	6	115	19		
Total	7	5,550			

	Coefficients	Standard error	t Stat	p Value	Lower 95%	Upper 95%
Intercept	−42.2	9.2	−4.6	.004	−64.6	−19.8
Titles stocked	1.09	.065	16.8	.0000	.93	1.25

Example 4.3 Pampers

Procter & Gamble hoped that targeted customers who value fit in a preemie diaper would use price as a quality of fit cue and prefer a higher priced diaper. Ideally, fit importance would be negatively correlated with price responsiveness. In the concept test of the new preemie diaper using a sample of 97 preemie mothers, price responsiveness was measured as the difference between trial intentions at competitive and premium prices, each measured on a 5 point scale (1 = Definitely Will Not Try to 5 = Definitely Will Try). Fit importance was measured on a 9 point scale (1 = Unimportant to 9 = Very Important). The correlation between price responsiveness and fit importance from Excel are shown in Fig. 4.10.

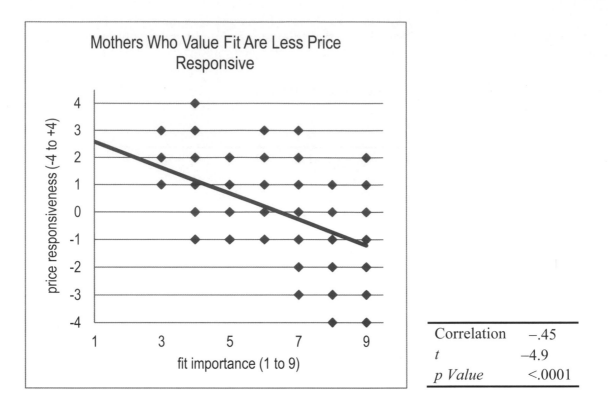

Fig. 4.10 Correlation between price responsiveness and fit importance

The correlation between price responsiveness y and fit importance X is moderately large and negative:

$$r_{Xy} = -.45.$$

The lower the importance of fit to a preemie mom, the greater her responsiveness to a price reduction. Regression analysis from Excel, shown in Table 4.9, quantifies this negative, linear relationship.

Table 4.9 Regression of price responsiveness by fit importance

SUMMARY OUTPUT						
Regression statistics						
R Square	.20					
Standard error	1.66					
Observations	97					
ANOVA						
	df	*SS*	*MS*	*F*	*Significance F*	
Regression	1	67	67	24.1	.00001	
Residual	95	262	2.8			
Total	96	329				
	Coefficients	*Standard error*	*t Stat*	*p Value*	*Lower 95%*	*Upper 95%*
Intercept	3.1	.65	4.7	.0000	1.8	4.4
Fit importance	−.47	.10	−4.9	.0000	−.67	−.28

From the results of correlation and regression analysis, Procter & Gamble management concluded the following:

> Price responsiveness is negatively correlated with fit importance of diapers to preemie mothers. Variation in fit importance accounts for 20% of the variation in price responsiveness. Though not a large influence on price responsiveness, fit importance does drive responsiveness, along with other factors. A difference between "Moderately Important" and "Important," which is a two point difference on the 9 point importance scale, reduces price responsiveness by about one (.5–1.3) scale point on a 9 point responsiveness scale.
> It is likely that preemie mothers seeking a high quality diaper with superior fit find claims of superior fit at a lower price unbelievable. A higher price supports the higher quality, superior fit image.

4.12 Correlation Complements Regression

The correlation coefficient summarizes direction and strength of linear association between two continuous variables. Because it is a standardized measure, taking on values between −1 and +1, it is readily interpretable. Unlike regression analysis, it is not necessary to designate a dependent and an independent variable to summarize association with correlation analysis. Later, in the context of multiple regression analysis, the correlations between independent variables are an important focus in our diagnosis of multicollinearity, introduced in Chaps. 8 and 9.

Correlation analysis should be supplemented with visual inspection of data. It would be possible to overlook strong nonlinear associations with small correlations. Inspection of a scatterplot will reveal whether association between two variables is linear.

Correlation is closely related to simple linear regression analysis:

- The squared correlation coefficient is *R Square*, our measure of percentage of variation in a dependent variable accounted for by an independent variable.
- The regression slope estimate is a product of the correlation coefficient and the ratio of the sample standard deviation of the dependent variable to sample standard deviation of the independent variable:
    - Slope estimates from simple linear regression are unstandardized correlation coefficients.
    - Correlation coefficients are standardized simple linear regression slope estimates.

4.13 Linear Regression Is Doubly Useful

Linear regression handles two modeling jobs, quantification of a driver's influence and forecasting. Regression models quantify the direction and nature of influence of a driver on a response or performance variable. Regression models also enable forecasts and the comparison of decision alternatives. This latter use of regression to answer "what if" questions, *sensitivity analysis*, is an important tool for decision making.

Excel 4.1 Build a Simple Linear Regression Model: Impact of Titles Offered on HitFlix Movie Rental Revenues

Use regression analysis to explore the linear influence of differences in *titles* offered on *revenue ($K)* differences across a random sample of 52 movie rental vending units.

Open **Excel 4.1 HitFlix Movie Rental Revenues.xls**.

Use shortcuts to run regression: **Alt AY11, Regression**:

For **Input Y Range**, enter label and observations on the dependent variable, *revenues ($K)*, and for **Input X Range**, enter label and observations on the independent variable, *titles*. Specify **Labels**, **Residuals**, and **Residual Plots**:

titles	revenue (K$)
78	87
89	86
70	77
79	88
90	90.0
77	92
66	81
72	85
93	107
82	91
85	105
96	102
85	94
78	94

Regression

Input
Input Y Range: $A$1:$A$53
Input X Range: $B$1:$B$53
☑ Labels   ☐ Constant is Zero
☐ Confidence Level: 95 %

Output options
○ Output Range:
◉ New Worksheet Ply:
○ New Workbook

Residuals
☑ Residuals   ☑ Residual Plots
☐ Standardized Residuals   ☐ Line Fit Plots

Normal Probability
☐ Normal Probability Plots

OK   Cancel   Help

▲	A	B	C	D	E	F	G
1	SUMMARY OUTPUT						
2							
3	Regression Statistics						
4	Multiple R	0.92643					
5	R Square	0.85827					
6	Adjusted I	0.85543					
7	Standard	7.44072					
8	Observati	52					
9							
10	ANOVA						
11		df	SS	MS	F	gnificance F	
12	Regressic	1	16762.8	16762.8	302.772	7.4E-23	
13	Residual	50	2768.21	55.3643			
14	Total	51	19531				
15							
16		Coefficient	andard En	t Stat	P-value	Lower 95%	Jpper 95%c
17	Intercept	3.43228	6.38194	0.53781	0.59309	-9.3862	16.2508
18	titles	1.11604	0.06414	17.4004	7.4E-23	0.98722	1.24487

To see the distribution of residuals, find the residual standard deviation and skew.

C78		▼	f_x	=SKEW(C25:C76)	
▲	A	B	C	D	E
74	50	145.17	-6.1699		
75	51	152.982	-4.9822		
76	52	140.706	1.29429		
77			7.37	SD	
78			0.10	skew	

Copy and paste in the formulas in **histogram bins.xls**, replace the standard deviation with the residual standard deviation, and make a histogram of the residuals:

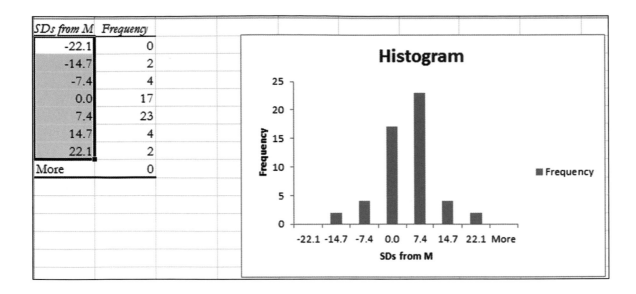

SDs from M	Frequency
-22.1	0
-14.7	2
-7.4	4
0.0	17
7.4	23
14.7	4
22.1	2
More	0

Excel 4.2 Construct Prediction Intervals

To see 95% prediction intervals for a vending unit offering a specific number of *titles*, copy *Predicted Revenues* from the regression sheet and paste into the Hitflix Sheet.

	A	B	C
21			
22	RESIDUAL OUTPUT		
23			
24	Observatio	ed revenu	Residuals
25	1	90.4837	-3.4837
26	2	102.76	-16.76
27	3	81.5554	-4.5554
28	4	91.5998	-3.5998
29	5	103.876	-13.876
30	6	89.3677	2.63232

	A	B	C
	titles	revenue (K$)	Predicted revenue (K$)
1			
2	78	87	90
3	89	86	103
4	70	77	82
5	79	88	92
6	90	90.0	104
7	77	92	89

fx	=T.INV.2T(0.05,50)

C	D
Predicted revenue (K$)	critical t
90	2.00856

To make prediction intervals, find the *critical t* value that corresponds to a 95% confidence level and 50 (=*N* – 2) degrees of freedom with the Excel function **T.INV.2T**(*probability, df*). For *probability*, enter .05 for a 95% level of confidence, and for *df* enter 50 (= *N* – 2), the residual degrees of freedom (from **B13** of the regression sheet).

=D2*E2		
D	**E**	**F**
ted (K$)	critical t	s	margin of error
90	2.01	7.44	14.9

To find the *margin of error* for 95% *prediction intervals*, select and copy the *standard error* in **B7** of the regression sheet and paste into sheet 1, then multiply the *standard error* by the *critical t*:

Find the lower 95% and upper 95% prediction interval bounds by subtracting and adding the margin of error from *predicted revenues*. Lock the margin of error cell reference with **fn4** so that Excel returns to row 2 for this part of the formula.

fx =C2+$F$2					
C	**D**	**E**	**F**	**G**	**H**
Predicted revenue (K$)	critical t	s	margin of error	lower 95%	upper 95%
90	2.01	7.44	14.9	76	105

Select the two new cells, **Shift+dn** through the last data row, then **Cntl+D** to fill in the columns.

fx =C2-$F$2					
C	**D**	**E**	**F**	**G**	**H**
Predicted revenue (K$)	critical t	s	margin of error	lower 95%	upper 95%
90	2.01	7.44	14.9	76	105
103				88	118
82				67	97
92				77	107

To see the model fit and prediction intervals, rearrange columns so that *titles* is followed by actual *Revenues* and 95% lower and upper prediction intervals.

Select columns **F** and **G**, use shortcuts to cut those columns, **Cntl+X**, and paste into columns **C** and **D** by selecting column **C,** then **Alt HIE. (Cntl+X** cuts selected cells. **Alt HIE** selects the <u>H</u>ome menu and <u>I</u>nsert function and inserts cut or copied c<u>E</u>lls to the left of the selected column or cell.)

Select filled cells in columns **A** through **D**, *titles*, actual *Revenues*, and 95% lower and upper prediction intervals and make a scatterplot:

Adjust the axes to make better use of white space:

Alt JAAIIM and **Alt JAAVM. (JA** opens the L<u>A</u>yout menu, the <u>A</u>xis menu, the <u>H</u>orizontal axis or the <u>V</u>ertical axis, and <u>M</u>ore options.)

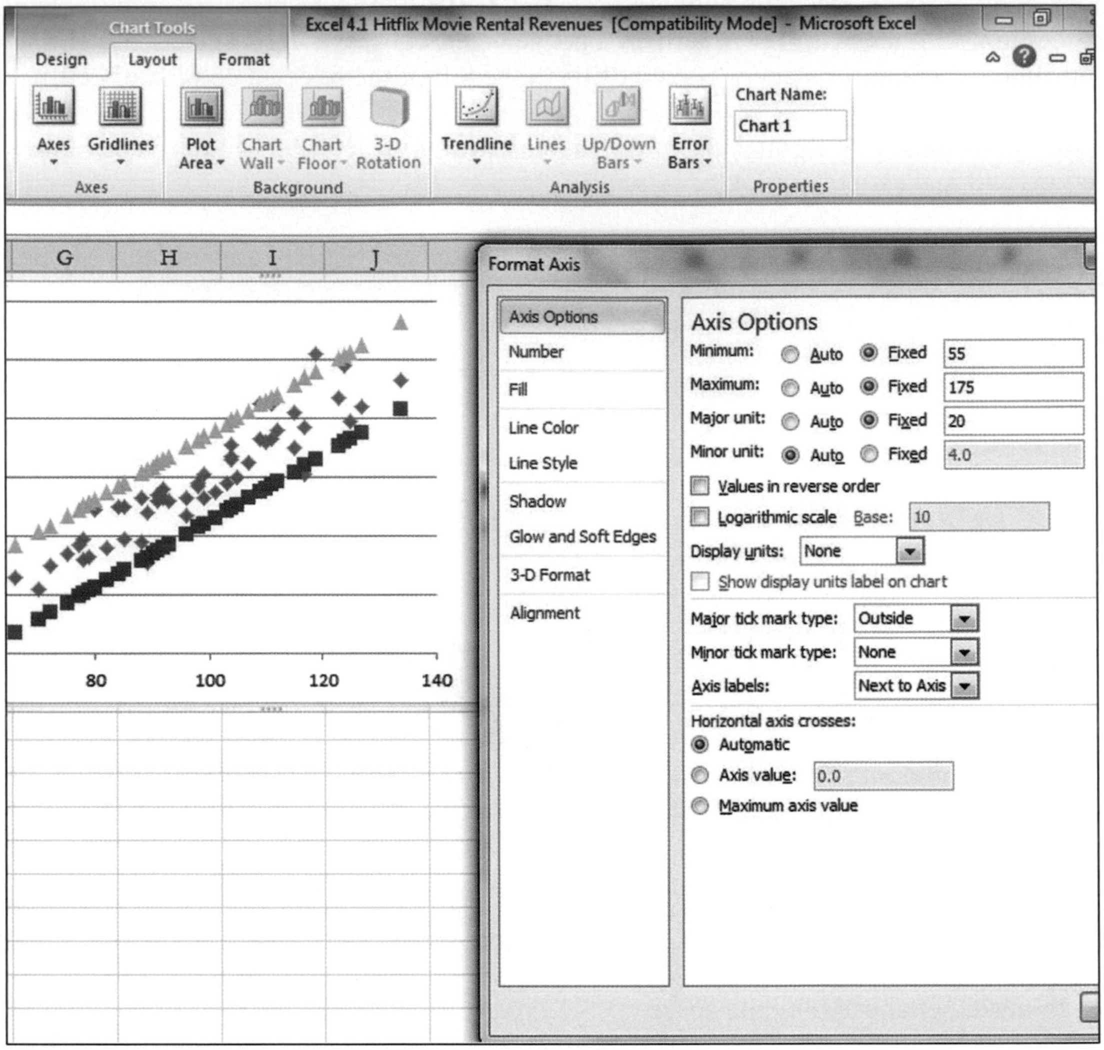
Chart Tools
Excel 4.1 Hitflix Movie Rental Revenues [Compatibility Mode] - Microsoft Excel
Design
Layout
Format
Axes
Gridlines
Plot Area
Chart Wall
Chart Floor
3-D Rotation
Trendline
Lines
Up/Down Bars
Error Bars
Chart Name:
Chart 1
Axes
Background
Analysis
Properties
G
H
I
J
80
100
120
140
Format Axis
Axis Options
Number
Fill
Line Color
Line Style
Shadow
Glow and Soft Edges
3-D Format
Alignment
Axis Options
Minimum: Auto Fixed 55
Maximum: Auto Fixed 175
Major unit: Auto Fixed 20
Minor unit: Auto Fixed 4.0
Values in reverse order
Logarithmic scale Base: 10
Display units: None
Show display units label on chart
Major tick mark type: Outside
Minor tick mark type: None
Axis labels: Next to Axis
Horizontal axis crosses:
Automatic
Axis value: 0.0
Maximum axis value

Add axes titles and a chart title that conveys the primary regression result: **Alt JCL.** (**JC** opens the Design menu and the **L**ayout choices.)

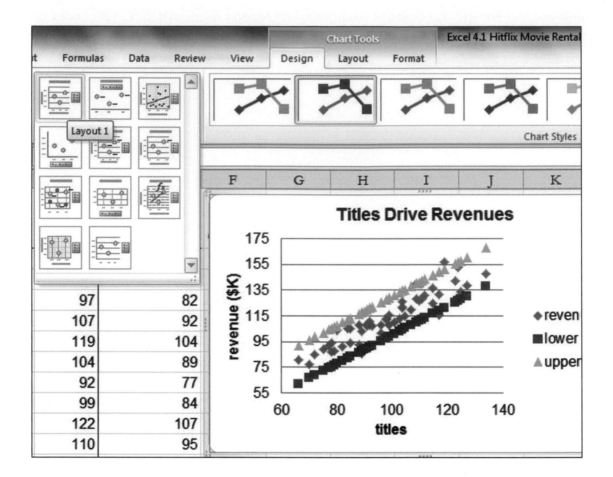

Choose a style, **Alt JCS** (**JC** opens the Design menu and **S** opens **S**tyles.)

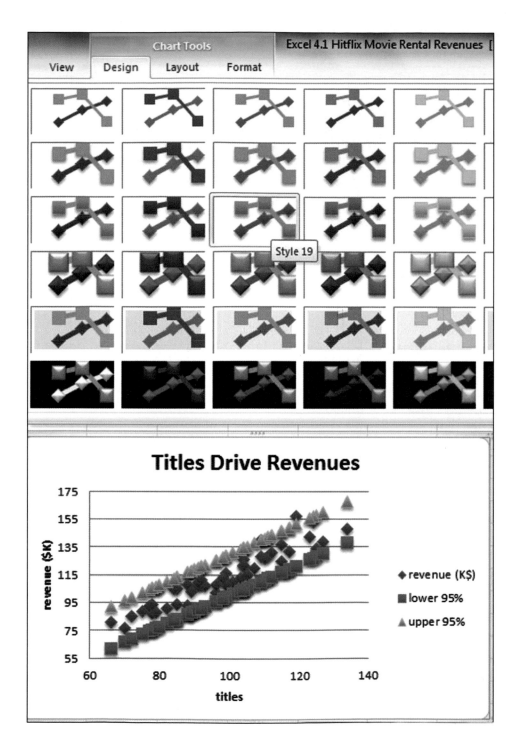

Change the lower and upper 95% bounds to lines in a single color:

Alt JAE dn to *Series lower 95%* or *Series upper 95%*, **Alt JAM (JA** invokes the L**A**yout menu and **M** opens for**M**at.)

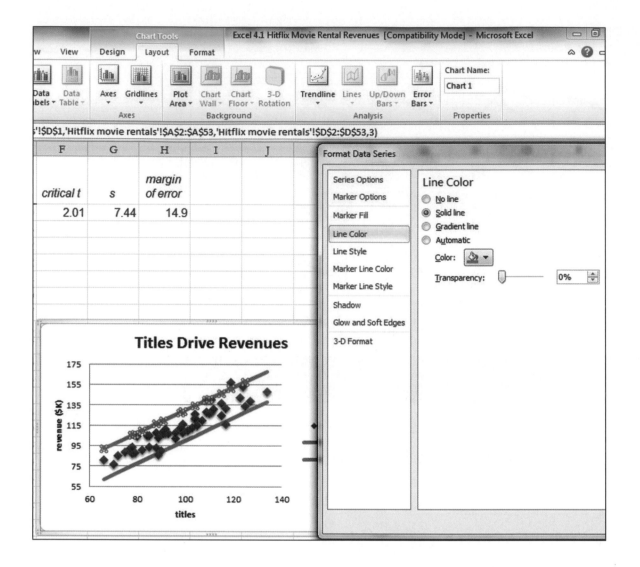

Excel 4.3 Find Correlations Between Variable Pairs

Management would like to know whether there is an association between the perceived importance of diaper fit and price responsiveness among preemie mothers.

Fit importance ratings and *price responsiveness* from a concept test sample of 97 preemie mothers are in **Excel 4.3 Pampers Price Responsiveness.xls.**

At the bottom of the dataset, use the Excel function **CORREL(***array1,array2***)** to find the correlation between *fit importance* rating and *price responsiveness*.

B99	▾	f_x	=CORREL(A2:A98,B2:B98)	
	A	**B**	**C**	**D**
1	fit importance	price responsiveness		
96	4	4		
97	3	1		
98	3	2		
99	correlation	-0.45		

Lab Practice 4 Oil Price Forecast

Rolls-Royce is concerned that demand for their jet engines may decline if the price of oil rises. Management believes that the growing Indian and Chinese economies, declining US crude oil production and world supply, and rising US crude oil imports may be driving oil prices.

Lab Practice 4 Oil Price Forecast.xls contains 19 years of annual data on
> *spot price* of Texas crude oil$_t$,
> *China GDP($B)$_{t-4}$*
> *India GDP($B)$_{t-4}$*
> *US production(K barrels/day)$_{t-4}$*
> *World supply(M barrels/day)$_{t-4}$*
> *US imports(K barrels/day)$_{t-4}$*

Build a simple regression model to estimate the impact of one of these hypothetical drivers on crude oil *spot price*. Round all of your numerical answers to two or three significant digits.

1. Present your regression equation in standard format.

2. What percentage of variation in *spot prices* can be accounted for by variation in the driver that you chose? _____

3. How close to actual *spot prices* could you expect a forecast to be 95% of the time? _____

4. What range in *spot prices* could Management be 95% certain to expect in 2015, given
> An increase in *China GDP* to $3,000 billion in 2011
> An increase in *India GDP* to $2,000 billion in 2011
> An increase in *US production* to 8,500 thousand barrels/day
> An increase in *world supply* to 87 million barrels/day
> or
> A decrease in *US imports* of 3,200 thousand barrels/day
> _____(Choose the one driver that you chose for your model.) _____ to _____

5. Make a scatterplot of *95% individual prediction intervals* and *actual spot prices* by *year* and attach. Insert a chart title that summarizes your conclusion.

Lab 4 Simple Regression Dell Slimmer PDA

Dell is considering the introduction of an ultraslim PDA that would fit in a shirt pocket, come in an array of colors, and be sold in Wal-Mart. Dell withdrew its Axim PDA after its share fell to 3%. Developers want to be sure that the new PDA will offer the features most desired by the target segments of young, lower income high school students, and service workers. Managers believe from past research that there are three PDA lifestyle segments.

- *Younger Players.* The youngest segment, high school students, who are fashion conscious and technically savvy. Some PDAs in this segment are provided by **higher** income parents. PDAs are primarily used for text messaging and playing music and video games. Penetration in this segment is low.
- *Older Players.* High school graduates employed in service jobs. These users are the least technically savvy. PDAs are a luxury used to play music and video games. Penetration in this segment is the lowest.
- *Professionals and Soon to Be.* College students and college graduates. This segment is technically savvy and uses PDA software in classes or on the job. PC connectivity is important, although text messaging and music are also important. This market is saturated and most purchases are upgrades.

Palm and HP cater to the *Professionals* and *Soon to Be* segments. **Dell** is targeting *Younger* and *Older Players*, hoping to avoid competition. The new PDA would be ultraslim and also fit in a shirt pocket (unlike the withdrawn Axim).

Data from a concept test of 14- to 34-year olds in **Lab 4 Dell Slimmer.xls** include
- Measures of the importance of thinness and ability to fit in a shirt pocket, on a 1 to 9 point scale (1 = unimportant to 9 = extremely important)
- Key demographics: age, household income (in thousands), and years of education

1. *Importance of thinness:* Use a *t* test to determine whether *thinness* is an important attribute to customers like those surveyed. An attribute is considered important if the average customer rating is **greater than 5** on the 9 point scale.

   A ___*one tail* ____*two tail* *t test* is required.

   The null hypothesis is _____

   The alternate hypothesis is _____

   Management can conclude that 14- to 34-year olds rate *thinness* important (**at least 5** on a 9 point scale): _____Y _____N

2. Construct a *95% confidence interval* for the average *importance* of *thinness* in the population and illustrate your result with a clustered Column chart.

   Margin of error: _____

   Average importance of *thinness:* _____ to _____ on a 9 point scale.

3. Demographics that drive *thinness importance*: Use simple regression to identify demographics that drive the *importance of thinness* and the variation in *thinness importance* explained by variation in each demographic.

Demographic	*p Value*	Drives *thinness importance*	% Var in *thinness importance* explained
Age		Y or N	
Education		Y or N	
Income		Y or N	

4. Find the expected difference in *thinness importance* associated with each demographic difference in the sample. (If a potential driver is not significant, leave its row blank.)

Demographic	Expected difference in *thinness importance* due to demographic difference
Age (years)	
Education (years)	
Income ($k)	

5. Illustrate one of the significant driver's influence with a scatterplot showing population average response to driver differences by adding the line of fit with *95% prediction intervals*.

Case 4-1 GenderPay (B)

The human resources manager of Slam's Club was shocked by the recent revelations of gender discrimination by Wal-Mart ("How Corporate America Is Betraying Women," *Fortune*, January 10, 2005), but believes that the employee salaries in his company reflect levels of responsibility (and not gender). You have been asked to analyze this hypothetical link between level of responsibility and salary.

   Case 4-1 GenderPay.xls contains employee *salaries* and levels of *responsibility* from a random sample of employees.

1. Determine whether *responsibility* drives *salaries*. If level of *responsibility* drives *salaries*, determine
   (a) The percent of variation in *salaries* that can be accounted for by variation in level of *responsibility*
   (b) The margin of error in forecasts of *salaries* from level of *responsibility* with 95% certainty
   (c) How much *expected salary* in the population changes with each additional *responsibility* level

2. The human resources manager noticed that many employees are working at *responsibility* level 5. Determine how much payroll might be reduced, on average, if a level 5 employee were replaced with a new level 1 employee with similar experience.

3. Present the model that you built, including
   (a) The regression equation in the standard format
   (b) A scatterplot of *salaries* by level of *responsibility* with *95% prediction intervals*
   (c) A chart title that helps your audience see your conclusion
   Be sure to round your results to two or three significant digits.

Case 4-2 GM Revenue Forecast[2]

The General Motors Management would estimate the percentage of customers who will return again to choose a GM car. GM's award winning customer Loyalty has been widely publicized, although in 2009, for the first time in 9 years, Toyota and Honda overtook GM, claiming the annual Automotive Loyalty Awards.

---

[2] This case is a hypothetical scenario using actual data.

Toyota and Honda Win Top Honors in 14th Annual Event

SOUTHFIELD, Mich. (January 12, 2010) – Toyota and Honda took top honors in R. L. Polk & Co.'s 14th Annual Automotive Loyalty Awards, which were presented this evening at the 2010 Automotive News World Congress in Detroit.

Edging out General Motors for the first time in nine years, Toyota ranked number one in Overall Loyalty to Manufacturer, indicative of the manufacturer's ability to retain previous customers.

Honda also was a big winner, taking top honors in the Overall Loyalty to Make category.

 "Maintaining a solid loyal customer base is not easy, but it is essential to survive in today's competitive environment," said Stephen Polk, chairman, president and CEO of R. L. Polk & Co. "Tonight's winners are all excellent examples of what customer retention can do for your brand and your bottom line."

About The Polk Automotive Loyalty Awards.

The Polk Automotive Loyalty Awards recognize manufacturers for superior owner loyalty performance. Loyalty is determined when a household that owns a new vehicle returns to market and purchases or leases another new vehicle of the same model or make. For a complete list of current and past Polk Automotive Loyalty Award winners, please visit http://usa.polk.com/Company/Loyalty/.

Table
Polk Automotive Loyalty Award Winners – 2009 Model Year

Categories	Winners	Loyalty %
Overall Awards		
Overall Loyalty to Manufacturer	Toyota	58.60%
Overall Loyalty to Make	Honda	54.86%

Case 4-2 General Motors Revenue.xls contains quarterly data of 5 years, including
Quarter
Revenues,
Revenues q-4, lagged revenues from four quarters ago

Build a simple regression model to estimate the impact of past year *revenues* on current *revenues*. Round your numerical results to two or three significant digits.

1. Present your regression equation in standard format.

2. What percentage of variation in *revenues* can be accounted for by past *revenues*?

3. How close to actual *revenues* could you expect a forecast to be 95% of the time?

4. What range in percentages of this quarter's GM *revenues* could management be 95% certain will repeat next year?

5. Present a scatterplot of *95% individual prediction intervals* with *actual revenues* by quarter. Insert a chart title that summarizes your conclusion.

Assignment 4-1 Impact of Defense Spending on Economic Growth

Some experts have suggested that the US economy thrives when the nation is involved in global conflict. **Assignment 4-1 Defense.xls** contains quarterly *GDP* and past quarter *Defense* spending in billion dollars.

Create a scatterplot and calculate the correlation coefficient to see whether *GDP* and *defense spending* are related linearly.

Fit a simple linear regression to estimate the impact on quarter *GDP* of changes in past quarter *defense spending*.

Analyze the residuals. Are they
  Pattern free?
  Approximately Normally distributed?

Summarize your results, in a single spaced report, 12 pt font, with one embedded figure and your regression equation in standard format.
Round to two or three significant digits.
Choose a chart title that summarizes your conclusions.
Use language that policy makers could easily understand, whether or not they have recently taken statistics.
Include in your report
  Whether past quarter *defense spending* is correlated with *GDP*
  The percentage of variation in *GDP* that can be explained by variation in past quarter *defense spending*
  The margin of error in forecasts of *GDP* from past quarter *defense spending*
  The expected range of possible impacts on *GDP* of a $1 billion increase in past quarter *defense spending*

In a technical footnote, include your conclusions from your residual analysis.

Chapter 5
Market Simulation and Segmentation with Descriptive Statistics, Inference, Hypothesis Tests, and Regression

5.1 Case 5-1 Simulation and Segmentation of the Market for Preemie Diapers

Deb Henretta is about to commit substantial resources to launch Pampers Preemies. The following article from the *Wall Street Journal* describes Procter & Gamble's involvement in the preemie diaper market:

New York, N.Y.
May 5, 2003
P&G Targets the 'Very Pre-Term' Market
Wall Street Journal

Copyright Dow Jones & Company Inc May 5, 2003
THE TARGET MARKET for Procter & Gamble Co.'s newest diaper is small. Very small.
Of the nearly half a million infants born prematurely in the U.S. each year, roughly one in eight are deemed "very pre-term," and usually weigh between 500 grams and 1,500 grams (one to three pounds). Their skin is tissue-paper-thin, so any sharp edge or sticky surface can damage it, increasing the chance of infection. Their muscles are weak, and unlike full-term newborns, excessive handling can add more stress that in turn could endanger their health.
Tiny as they are, the number of premature infants is increasing—partly because of improved neonatal care: From 1985 to 2000, infant mortality rates for premature babies fell 45%, says the National Center for Health Statistics. Increasingly, such babies are being born to older or more affluent women, often users of fertility drugs, which have stimulated multiple births.
It's a testament to the competitiveness of the $19 billion global diaper market that a behemoth like Procter & Gamble, a $40 billion consumer-products company, now is focusing on a niche that brought in slightly more than $1 million last year; just 1.6% of all births are very pre-term. But P&G sees birth as a "change point," at which consumers are more likely to try new brands and products. Introducing the brand in hospitals at an important time for parents could bring more Pampers customers, the company reasons.
P&G's Pampers, which is gaining ground on rival Kimberly-Clark, but still trails its Huggies brand, has made diapers for premature infants for years. (P&G introduced its first diaper for "preemies" in 1973; Kimberly-Clark in 1988), but neither group had come up with anything that worked well for the very smallest of these preemies. The company that currently dominates the very-premature market is Children's Medical Ventures, Norwell, Mass., which typically sells about four million diapers a year for about 27 cents each. The unit of Respironics Inc., Murrysville, Pa., has been making its "WeePee" product for more than a decade. But the company, which also makes incubator covers, feeding tubes and extra small bathtubs for preemies, hadn't developed certain features common in mass-market diapers, such as softer fabric coverings.
By contrast, P&G's preemie diapers, which it started distributing to hospitals in August, sell for about 36 cents each; about four cents more than P&G's conventional diapers. P&G's "Preemie Swaddler" fits in the palm of an adult's hand and has no adhesives or hard corners. It closes with mild velcro-like strips and is made of breathable fabric, not plastic. It has an extra layer of fabric close to the infant's skin to avoid irritation.
Children's Medical Ventures is coming out with another size of the WeePee, and plans to introduce velcro-like closures, a development the company says was in the works before P&G came out with a rival diaper. The new diapers won't cost any more, Children's Medical Ventures says.

P&G says the new diaper is the natural extension of its Baby Stages initiative, which took effect in February 2002 when P&G revamped its Pampers brand in the U.S. to cater to various stages of a baby's development. Working with very small preemies helps the company better understand infant development and become "more attuned to new products they might need," says Deb Henretta, president of P&G's global baby-care division. But the marketing director for Children's Medical Ventures believes the increasing affluence of preemie parents is a greater inducement for big companies to enter the market. In the past, the typical mother of a preemie was poorer, often a teenager, but today more preemie "parents tend to be older, well-educated, and have money for things like fertility treatments," says Cathy Bush, marketing director for Children's Medical Ventures.

The competition may raise the bar for the quality of diapers for these smallest of preemies. P&G says the parents of premature babies are demanding better products. "They have much higher expectations than they did years ago," Ms. Henretta says.

Neonatal nurses have all sorts of opinions about the relative merits of Preemie Swaddlers and WeePees. Pat Hiniker, a nurse at the Carilion Roanoke Community Hospital in Virginia, says the new Pampers diaper, while absorbent, is too bulky for small infants. Allison Brooks of Alta Bates Hospital in Berkeley, Calif., says P&G's better absorbency made the babies less fidgety when they needed to be changed. "That sounds small, but you don't want them wasting their energy on squirming around," she says. "They need all their energy to grow."

In any case, if health professionals have their way, the very-premature market will shrink, or at least stop growing. The March of Dimes recently launched a $75 million ad campaign aimed at stemming the rise of premature births. P&G is donating 50,000 diapers to the nonprofit organization.

Before resources are dedicated, Deb wants to confirm that preemie parents are attracted to the *Pampers Preemies* concept of superior comfort and fit. She has commissioned a concept test to assess consumers' intentions to try the product.

The Market for Preemie Diapers

The market for preemie diapers is unusual in that the first diapers that a preemie baby wears are chosen by the hospital. Procter & Gamble is banking on positive experiences with Pampers Preemies in the hospital and consumer brand loyalty once baby goes home. If parents see Pampers Preemies in the hospital, are satisfied with their performance, and find them widely available at the right price, parents may adopt the Pampers Preemies brand after the infant comes home. Satisfaction and brand loyalty to Pampers Preemies could then lead to choice of other Pampers products as the baby grows. If the concept test indicates that consumers' intentions to try are high, then the results will be included in promotional materials and selling efforts to hospital buyers.

Preemie Parent Segments

Based on focus group interviews and market research, Deb's team has learned that there are five broad segments of preterm parents:

1. *Younger* (14–19), unemployed mothers who live with their parents. These young mothers are inexperienced and their pregnancies are unplanned. They tend to differ widely in their attitudes and preferences, and so a further breakdown is necessary:

1.a) *Younger, Single, Detached.* These young mothers are relatively unattached to their babies and relatively indifferent about the particular diapers they use. Their means are limited and they are highly responsive to low prices and price promotions.

1.b) *Younger, Single, Committed.* These young mothers are attached to their babies and want the best diapers. They are inexperienced consumers and could be attracted by a premium diaper, although resources may limit their buying power. Brand name appears to be very important to these young women, and they believe that better mothers choose name brands seen on television.

2. *Young* (20–35) mothers tend to be married and have adequate resources. Their pregnancies tend to be planned and this segment is virtually indistinguishable from the larger segment of disposable diaper users for full term babies. This group has the fewest preterm births.

3.a) and b) *Older Victorious over Biological Clocks* (35–39) and *Oldest* (40+) mothers tend to be wealthier, more highly educated professionals with higher incomes. A large proportion has no other children and has undergone fertility treatment. Multiple preemie births are more likely in this segment. Some of these mothers are single parents. This group is particularly concerned about functional diaper features and wants the best diaper their dollars can buy. They are willing to pay for a premium diaper perceived as the highest quality, offering superior fit and comfort.

The Concept Test

A market research agency has conducted a concept test of *Pampers Preemies* to gauge interest among consumers in a variety of potential target markets. The 97 mothers with preemies who had been born at two local hospitals were asked to fill out a survey about purchase intentions after trying the product on their babies. If those data support the launch, Deb will need to know which functional feature(s) to stress in advertising and the type of mother and family to feature in the ads. Therefore, questions regarding attribute importance and demographic information were also collected in the survey.

Data from the concept test are contained in **Case 5-1 Pampers Concept test.xls**. Below is an overview of the questions asked in the survey, the manner in which they were coded, and the variable names contained in the dataset (which are in italics).

Trial Likelihood

Participants were asked, "How likely would you be to try Pampers Preemies if they were available in the store where you normally buy diapers and were sold at a price of $X.XX per diaper?"

The question was asked twice at two different price points: a "premium" price of $.36 (*premium intent*) and a "value" price of $.27 (*value intent*).

Responses were coded as follows.

Definitely Would Not Try = .05
Probably Would Not Try = .25
Maybe Would Try   = .50
Probably Would Try   = .75
Definitely Would Try   = .95

Attribute Importance

Participants were asked, "How important is each of the following attributes to you when choosing a diaper?" for the attributes

"Brand name" (*brand importance*)
"Comfort/fit" (*fit importance*)
"Keeps baby dry/doesn't leak" (*staysdry importance*)
"Natural composition" (*natural importance*)

Responses were given on a scale from 1 to 5 where 1 = "Not Important at All" and 5 = "Extremely Important."

Demographic Information

Consumers were asked to report their age (*age*), annual household income (*income*), family size including the new baby (*family size*), and the number of other children in the home (*other children*).

Data Recoding

Some of the original variables were recoded to make new variables for analysis.

Likely and Unlikely Triers

Two new variables, *premium trier* and *value trier*, were created from the intention to try questions (*premium intent* and *value intent*) to identify "Likely Triers" of the product at both price points tested. "Likely Triers" were identified using a "Top two box rule" (i.e., those who indicated that they "Probably" or "Definitely" would try the product). Therefore, for *premium intent* $\geq$ .75, *premium trier* = 1; otherwise *premium trier* = 0. Likewise, for *value intent* $\geq$ .75, *value trier* = 1; otherwise *value trier* = 0.

I. Information Needed

Deb's team needs an estimate of revenue potential plus additional information in four areas.

Revenue Potential: Deb's team has devised a method to estimate potential revenues, which link demographic factors. Their logic is explained below.
The potential market for Pampers Preemies depends on several key demographic factors.
Births: Births in a year is a product of *women* 15–44 who could have babies and the *birthrate* among those women:

$$births_t = women\ 15\ 44_t \times birthrate_t$$

The number of *women* 15–44 and the *birthrate* vary from year to year, and their future values are uncertain.

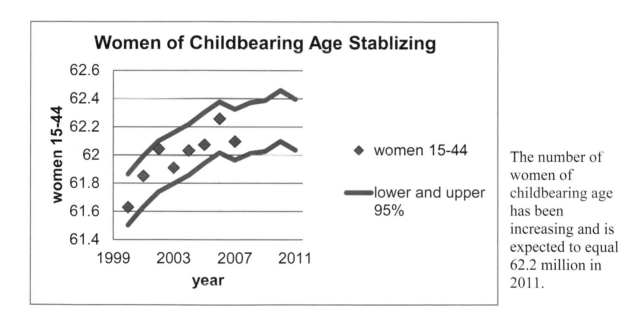

The number of women of childbearing age has been increasing and is expected to equal 62.2 million in 2011.

Forecasters agree that there is only about a 5% chance that the number of women of childbearing age will fall below 62.0 million or exceed 62.4 million in 2011.

Medical advances and changing demographics, including immigration, have led to an increasing birthrate among women of childbearing age.

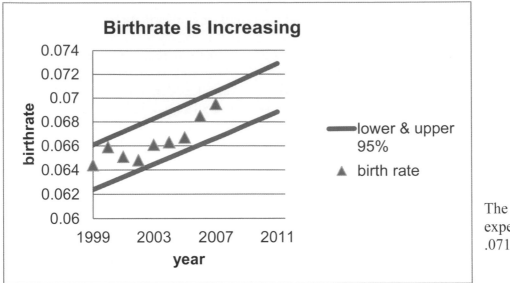

The *birthrate* is expected to equal .071 in 2011.

Forecasters agree that there is only about a 5% chance that the birthrate will fall below .069 or exceed .073 in 2011.

Given the fairly stable number of *women* 15–44 and an increasing *birthrate*, the number of births in future years are likely to rise.

Preterm Births: The number of *very preterm births* in a year is the product of number of *births* and the chance that a newborn will be very preterm, that is, the *very preterm birthrate*:

$$\text{very preterm births}_t = \text{births}_t \times \text{very preterm birthrate}_t.$$

Advances in infertility treatments have led to more births by older, high risk mothers. Immigration has led to more births by the youngest mothers, many with little information about prenatal care.

The percentage of babies born very preterm has been increasing and is expected to fall within the range .0211 to .0215 in 2011, with any value within this range equally likely.

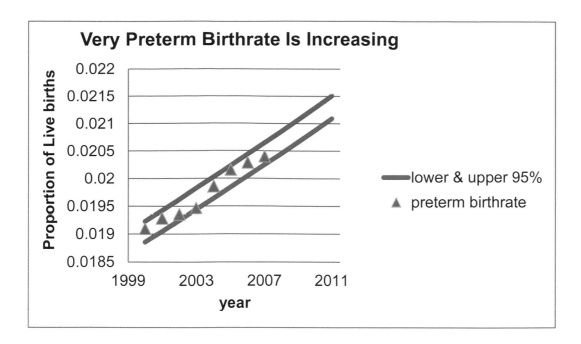

Surviving Preterm Babies. The number of *surviving very preterm babies* is the product of the *very preterm rate*, less *preterm mortality rate*, and *births*:

$$surviving\ very\ preterm\ babies_t = (very\ preterm\ rate_t - preterm\ mortality\ rate_t) \times births_t.$$

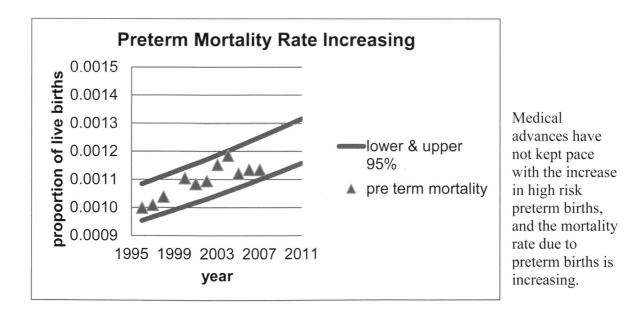

Medical advances have not kept pace with the increase in high risk preterm births, and the mortality rate due to preterm births is increasing.

In 2011, the preterm mortality rate is expected to lie within the range .00116 to .00132 of live births, with all values within this range equally likely.

Births and the *very preterm* percentage of births are expected to increase in future years, although the *preterm mortality rate* is expected to increase, as well. There will be more babies and more very preterm babies, although the number of *surviving very preterm* babies could be either greater or smaller, given the increasing *preterm mortality rate*.

Preterm Diaper Market. The preemie diaper *market* is a product of *surviving very preterm babies*, the average number of days a very preterm baby remains very preterm, approximately 30, and the average number of diapers used per day, approximately 9:

> *Very preterm diapers sold$_t$* = 30 days per very preterm baby
> × 9 diapers per very preterm baby per day   × *surviving very preterm babies$_t$*.

Procter & Gamble's Preemie Business: Procter & Gamble *sales* from Pampers Preemies would depend on *market* size, above, and their *market share* (which is expected to vary with *price*):

> *P&G very preterm diapers sold$_t$* = *market share$_t$* × *very preterm diapers sold$_t$*.

From past experience, Procter & Gamble managers have learned that 75% of the proportion of *Likely Triers*, the *trial rate*, become loyal customers in the first year:

> *market share$_t$* = .75 *trial rate$_t$*.

Procter & Gamble Preemie Diaper Revenue: Revenue($)$_t$ is the product of demographic variables, the trial rate$_t$, all uncertain, and *price($)*:

> Revenue($)$_t$ = price ($) × *P&G very preterm diapers sold$_t$*.

To be a viable investment, revenue in the first year following commercialization of Pampers Preemies must be greater than $3MM (million).

To effectively evaluate revenue potential, managers believe the uncertainties in each of the driving demographics and in market response must be incorporated in analyses. Use Monte Carlo simulation to account for uncertainties.

1. Find *distribution of possible* 2011 *revenues at the premium price* and the chance that *revenues* will exceed $3MM at the premium price in 2011.

   (a) *Births:* Use Procter & Gamble managers' assumptions for 2011 to simulate samples of 1,000 possible values for *women 15–44$_{2011}$* and *birthrate$_{2011}$*, and then use those samples to find a sample of 1,000 possible values for *births$_{2011}$*.

   (b) *Preterm Diaper Market:* Use Procter & Gamble managers' assumptions for 2011 to simulate samples of 1,000 possible values for *very preterm birthrate$_{2011}$*, and *very preterm mortality rate$_{2011}$*, and then use those samples with your sample of 1,000 possible values for $births_{2011}$ to find samples of 1,000 possible values for *surviving very preterm$_{2011}$* babies and *very preterm diapers sold$_{2011}$*.

(c) *P&G Preemie Diaper Sales:* Use P&G assumptions regarding trial intent and market share with a *conservative standard error* to simulate samples of 1,000 possible values for *market share$_{2011}$* proportion, and then use this sample and your sample of *very preterm diapers sold$_{2011}$* to find a sample of 1,000 possible values for *P&G very preterm diapers sold$_{2011}$*.

(d) *P&G Revenues:* From your sample of 1,000 possible *P&G very preterm diapers sold$_{2011}$*, find the distribution of possible Pampers Preemies revenues at the premium price, $.36.

(e) From the cumulative distribution of possible revenues, find the chance that revenues will exceed $3MM at the *premium price*.

2. Find the distribution of possible Pampers Preemies revenues and chance that revenues will exceed $3MM *at the value price*.

II. Additional Information Needed

1. Demographic differences between Likely and Unlikely Triers and identification of lifestyle segments most likely to try.

(a) Test suspected population differences between Likely and Unlikely Triers (premium trier) using a *two sample t test* along each of the demographics, *age, income, family size*, and *number of other children*.

(b) For each significant demographic difference between Trier segments, estimate the extent of difference between Likely and Unlikely Trier segments in the population. (The **Likely v Unlikely** worksheet in **Case 5-1 pampers concept test.xls** has been sorted by Trier segment for these tests.)

(c) Illustrate significant differences with a Column chart.

(d) From differences in (a), identify the lifestyle segments you believe will be most attracted to the concept (*younger detached, younger committed, young, older victorious over biological clock, oldest*).

2. Identification of attributes likely to be considered important by Likely Triers
   The worksheet page, **Likely Triers Only**, of **Case 5-1 pampers concept test.xls** contains importance ratings from the segment of *Likely Triers* only.

   (a) Determine which attributes are likely to be considered important to the segment of *Likely Triers* (*premium trier* = 1) from sample ratings of *brand importance, fit importance, staysdry importance,* and *natural importance.* To qualify as an important attribute, the average importance rating for that attribute by the *Likely Trier* segment would exceed 5 on a 9 point scale.

   (b) Illustrate your results with a Column chart of the 95% lower and upper confidence interval bounds.

Team Assignment

To prepare for the case discussion, your team should estimate revenue potential and find the additional information needed by Deb Henretta, listed earlier.

1. Each team is responsible for the presentation of *revenue* forecasts at the two alternate prices OR information regarding attribute importances, demographics, and target segments. To facilitate your presentation, construct PowerPoint slides that illustrate your key results, using the guidelines from class:

   (a) Slide 1 introducing your team
   (b) Slides presenting your results
   (c) One concluding slide with TakeAways

   Use graphs, rather than tables. Round to no more than three significant digits. Use fonts no smaller than 24 pt, including text in graphs. Adjust axes.

2. Each team is also responsible for creating a single page, single spaced memo, using a 12 pt font, presenting your analysis of prices, market shares, revenues, demographics, and attributes to P&G management. Include one embedded figure that illustrates a key result. Follow the format suggested in Chap. 5. Include attachments with other graphics that are referred to in your memo. Attach only graphs, and only those that are referred to in the memo. Round to no more than three significant digits.

Each team's memo should be accompanied by a printout showing that the correct analysis was used and identifying the relevant statistics which led to the results and conclusions. Do not attach printouts of data; attach only analyses.

5.2 Use PowerPoints to Present Statistical Results for Competitive Advantage

PowerPoint presentations are a powerful tool that can greatly enhance your presentation of the results of your analysis. They are your powerful sidekick, Tonto to your Lone Ranger. PowerPoints help your audience remember key points and statistics and make available graphics to illustrate and enhance the story you are telling.

The key to effective use of PowerPoints for presenting your results for competitive advantage is to be sure that they are not competing with you. PowerPoints with too much text draw audience attention away from you. Cliff Atkinson, in his 2008 book, *beyond bullet points* (Microsoft Press), explains clearly how audience members process information during PowerPoint presentations and why you should move beyond bullet points in the design of your PowerPoints. Much of the material that follows reflects Mr. Atkinson's wisdom, and his book is a recommended investment.

Audience Brains Are Designed to Process and Remember Information

Our brains are ingeniously designed to filter and process large amounts of information, selecting the most relevant to be stored in long term memory. Only a small portion of incoming information gains admission into working memory, and only some portion of information processing in working memory survives and is stored in long term memory.

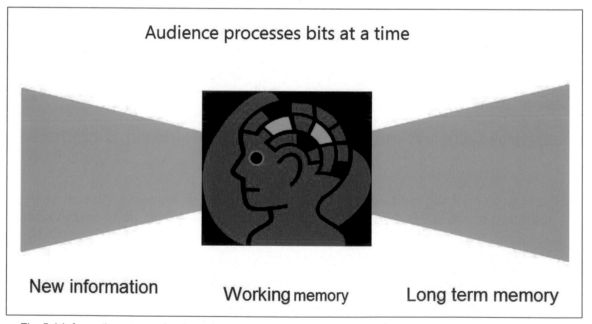

Fig. 5.1 Information processing in a given moment

The goal is to help your audience filter information and direct their attention to your key results and interpretation so that your message will be remembered.

Limit Text to a Single Complete Sentence per Slide

Inasmuch as brains process only a few select bits of information in a given moment, increase the chance that the key points in your presentation become those select elements. If your slides are loaded with text, the critical point has a small chance of being processed in working memory.

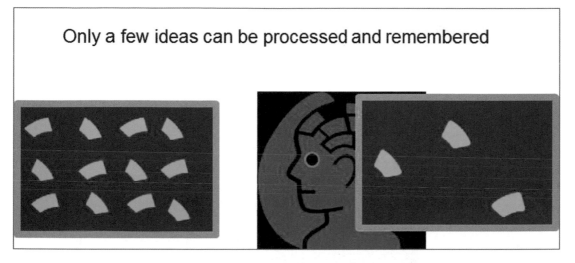

Fig. 5.2 Limited processing in working memory

Work to design your slides so that each slide presents a single idea. Use only one complete sentence per slide.

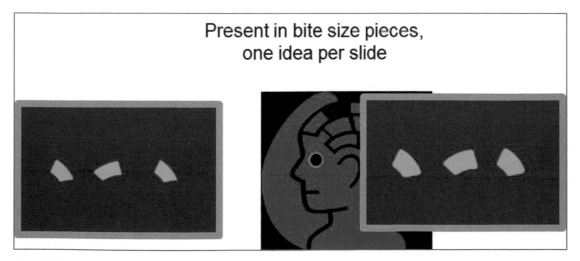

Fig. 5.3 Present one idea at a time

Pause to Avoid Competing with Your Slides

Brains process a single channel at a time. Attention is directed toward either visuals or audio in any given moment.

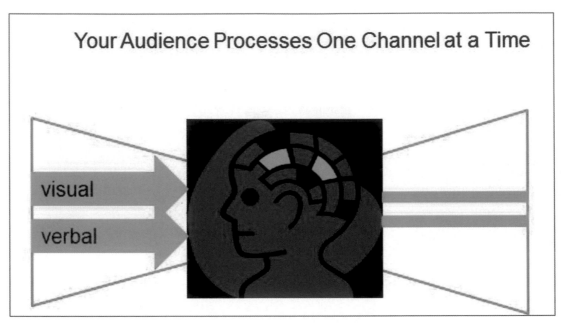

Fig. 5.4 Processing of a single channel in a given moment

Your PowerPoints should complement the story that you are delivering. They should not compete with you for attention. You are the star and the focus of attention. Your PowerPoints should play a supporting role. Pause to allow time for the audience to process that single idea, and then elaborate and explain. This will avoid competition between your slides and you.

Illustrate Results with Graphs Instead of Tables

Tables are effective elements in reports that convey a lot of information for readers to refer to and ponder. Tables are not processed in seconds, which is the time available to process each of your PowerPoint slides.

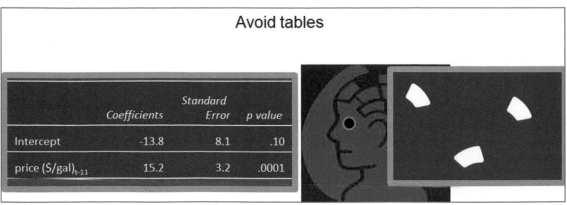

Fig. 5.5 Information overload from tables

Synthesize the results in your tables into graphs. Graphs organize your results and illustrate key takeaways. Well designed graphs can be processed in seconds, allowing audience attention to flow from a slide back to you, the speaker, and, ideally, the focus of attention. An effective slide contains a single complete sentence, the *headline,* and a graph to illustrate it.

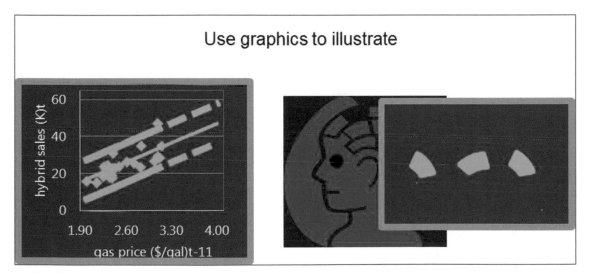

Fig. 5.6 Effective presentation of results with graphs

Start PowerPoint Design in Slide Sorter

Ensure that your PowerPoints are organized effectively. Build your deck by beginning in Slide Sorter view. Choose the main points that you want the audience to remember.

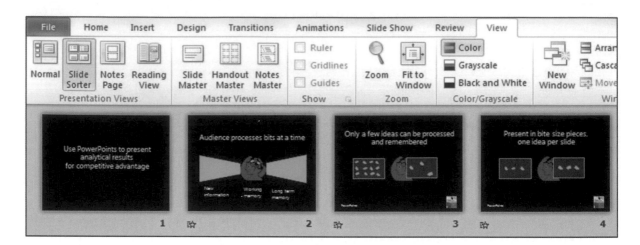

Fig. 5.7 Slide sorter view

Next, add slides with supporting information for the main point slides.

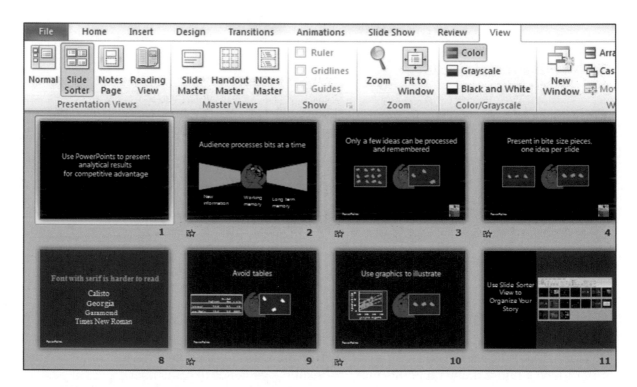

Fig. 5.8 Slide sorter view of main and supporting ideas

Put Supporting Text in Slide Notes

Presenters sometimes worry that they will forget the story. For insurance, they include all of the text to be delivered in their slides. You can guess the consequence. Audience members attempt to read and process all the text in the slides. To do this, they must ignore the presenter. In the few seconds that a slide appears, there is much too little time to read and process all the text. As a result, the audience processes only remnants of the story. Audience members are frustrated, because at the end of the presentation, they have incomplete information that doesn't make sense.

In addition to supporting your presentation, your PowerPoint slides deliver an impression. Slides filled with text deliver the impression that the presenter lacks confidence. When audience members fail to process slides laden with text and tables, the natural conclusion is that the speaker is ineffective. "She spoke for fifteen minutes, but I can't remember what she said. Made no sense."

Audience members can reach a second, unfortunate conclusion in cases where a presenter has simply converted report pages into slides. Slides converted from reports are crammed with text and tables and too often look like report pages, with white backgrounds and black text. This sort of unimaginative PowerPoint deck delivers the impression that the speaker is lazy.

In contrast, slides with a single, complete sentence headline and graph deliver the impression that the presenter is confident. After easily processing the slides and then focusing on the explanation and elaboration delivered by the presenter, audience members understand and remember the story.

If you present a single idea in each slide, you will remember what you want to say to explain the idea and add elaboration. The audience will focus on your presentation, because you will provide the missing links.

You can have the best of both worlds. You can include your explanation and elaboration of the main points in the slide Notes. The Notes are not seen during your presentation, but they are available later. Provide handouts at the end of your presentation from the Notes view of your slides.

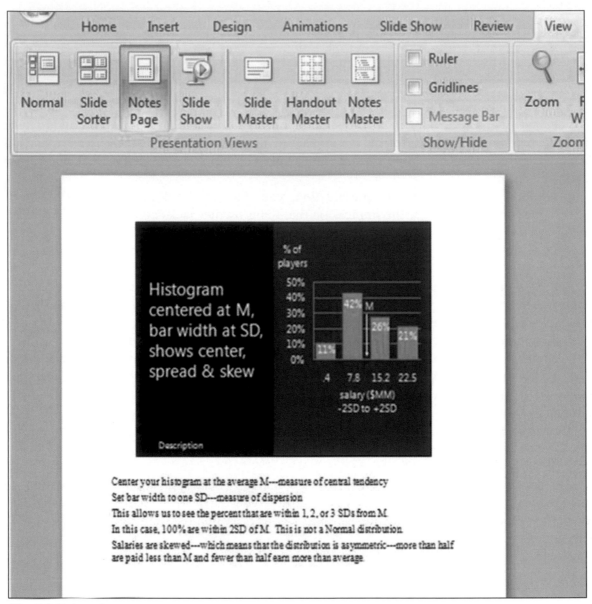

Fig. 5.9 Notes view

Choose a Slide Design That Reduces Distraction

Design your slides so that elements are minimally distracting. You want the audience to be able to quickly and easily see the idea in each slide and then focus on you for explanation and elaboration.

Use a Font That Can Be Easily Read

Use at least a 24 pt font so that audience members can easily read your headline, numbers, and labels. If you include numbers in your graphs, in the axes or as data labels, they must be easily read. (Be sure to round your numbers to two or three significant digits.) Axes labels and other text must also be easily read. Any font smaller than 24 pt will challenge easy reading.

Choose a Sans serif font (Ariel, Lucida, or Garamond). Sans serif fonts (without "feet") are easier to read in PowerPoints. San serif characters, without extra lines, are clearer in slides. (The opposite is true for reports, where the serif enhances reading ease.) If you have any doubts about readability, test your slides in a room similar in size and shape to the presentation location.

Choose Complementary Colors and Limit the Number

In cases where the slides will be presented in a darkened room, the background should be darker than the title and key words. Choose a medium or darker background, with complementary, contrasting, lighter text color. PowerPoints in this setting are more like television, movies, and Internet media, and less like books or reports, and should feature darker backgrounds like those you see in movie credits.

When presentations are in well lit rooms, backgrounds can be lighter than title and key words. In a light setting, PowerPoints resemble text pages, with lighter backgrounds and darker text colors.

If we see more than five colors on a slide (including text), our brains overload and we have difficulty processing the message and remembering it. Limit the number of distinct colors in each slide.

5.3 Write Memos That Encourage Your Audience to Read and Use Results

Memos are the standard for communication in business. They are short and concise, which encourages the intended audience to read them right away. Memos that present statistical analysis to decision makers

- Feature the bottom line in the subject line
- Quantify how the bottom line result influences decisions
- Are ideally confined to one single spaced page
- Include an attractive embedded graphic that illustrates the key result

Many novice analysts copy and paste pages of output. The output is for consumption by analysts, whose job it is to condense and translate output into general business language for decision makers. Decision makers need to be able to find the bottom line results easily without referring to a statistics textbook to interpret them. It is our job to explain in easily understood language how the bottom line result influences decisions. For the quantitative members of the audience, key statistics are included.

An example of a memo that might have been written by the quantitative analysis team at Procter & Gamble to present a key result of a concept test of Pampers Preemies to brand management is presented below.

Notice that

- The subject line contains the bottom line result.
- The regression analysis tables are omitted.

- Results are illustrated with a scatterplot of the fit.
- Results are described in general business English.
- The regression equation is visible for the quantitatively adept, who are assumed to be a minority proportion of the readers.

Description of the concept test and results are condensed and translated. Brand management learns from reading the memo what was done, who was involved, what the results were, and what the implications are for decision making.

MEMO

Re: Importance of Fit Drives Trial Intention
To: Pampers Preemies Management
From: Procter & Gamble Quantitative Analysis Team
Date: October 2010

Summary: what was done & learned

Results of a concept test of the Pampers Preemies suggest that the *Importance of fit* drives trial intentions, supporting the expected market salience of superior diaper fit.

Data source & method

The Concept Test Sample: The Preemies concept with premium price was described to a convenience sample of 97 preemie mothers in three hospitals in Cincinnati during the week of August 10–14, 2010. Demographics of this sample mirror national demographics of preemie mothers and are representative of all preemie mothers.

Data & scales

Concept Test Measures: The mothers indicated intent to purchase on a nine point scale (1 = "Definitely Won't Try" . . . 9 = "Definitely Will Try") and rated the importances of diaper attributes, including fit, brand, capability to protect from insults, and natural composition on balanced 9 point scales (1 = "Unimportant" . . .9 = "Very Important"). 51 mothers who were either "Probably" or "Definitely" likely to try were classified as Likely Triers. The remaining 46 mothers who were "Maybe," "Probably Not," or "Definitely Not" likely to try were classified as Unlikely Triers.

Results in English

Concept Test Results: Likely Triers rate fit more important than Unlikely Triers by 1 to 3 scale points on the 9 point scale.

Conclusion

Conclusions: The importance of fit distinguishes Likely and Unlikely Triers. Offering exceptional fit promises to deliver a salient feature to mothers who would try.

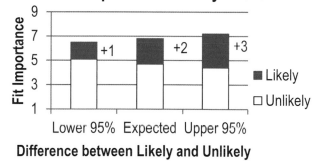

Fit More Important to Likely Triers

What else might matter

Other Potential Drivers: Other attributes, including brand, composition, capability to keep baby dry, and price, and demographics may also distinguish mothers who are more likely to try Pampers Preemies.

Chapter 6
Finance Application: Portfolio Analysis with a Market Index as a Leading Indicator in Simple Linear Regression

Simple linear regression of stock rates of return with a Market index provides an estimate of *beta*, a measure of risk, which is central to finance investment theory.

Investors are interested in both the mean and the variability in stock price growth rates. Preferred stocks have higher expected growth – expected *rates of return* – shown by larger percentage price increases over time. Preferred stocks also show predictable growth, low variation, which makes them less risky to own. A portfolio of stocks is assembled to diversify risk, and we can use our estimates of portfolio beta to estimate risk.

6.1 Rates of Return Reflect Expected Growth of Stock Prices

Example 6.1 General Electric and Apple Returns
Figure 6.1 contains plots of share prices of two well known companies, General Electric and Apple, over a 58 month period, January 2004 to December 2008.

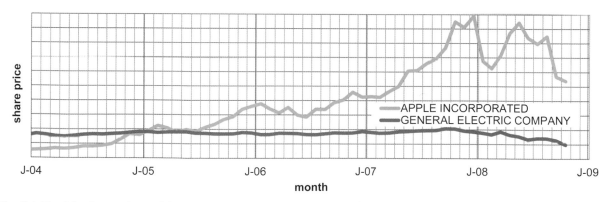

Fig. 6.1 Monthly share prices of General Electric and Apple, January 2004 to December 2008

It is important to note that although prices in some months were statistical outliers, those unusual months were not excluded. A potential investor would be misled were unusually high or low prices ignored. Extreme values are expected and included, because they influence conclusions about the appeal of each potential investment. The larger the number of unusual months, the greater the dispersion in a stock price, and the riskier the investment.

C. Fraser, *Business Statistics for Competitive Advantage with Excel 2010: Basics, Model Building, and Cases,*
DOI 10.1007/978-1-4419-9857-6_6, © Springer Science+Business Media, LLC 2012

To find the growth rate in each of the stock investments, calculate the monthly percent change in price, or *rate of return, RR*:

$$RR_{stock,t} = \frac{(price_{stock,t} - price_{stock,t-1})}{price_{stock,t-1}}$$

where t is time period (month).

Investors seek stocks with higher average rates of return and lower standard deviations. They would prefer to invest in stocks that exhibit higher expected, average growth, and less volatility or risk. The standard deviation in the rate of return captures risk. If a stock price shows little variability, it is a less risky investment.

Figure 6.2 illustrates monthly rates of return in GE and Apple stocks and a Market index, the S&P 500, over the 5 year period:

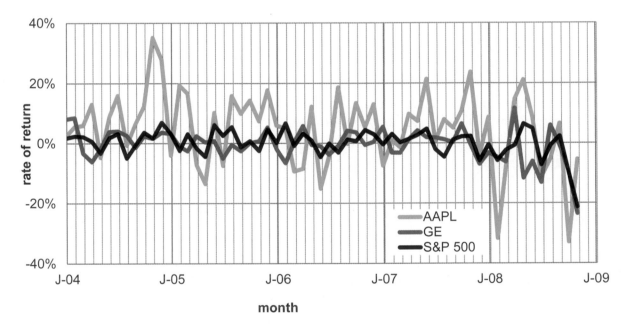

Fig. 6.2 Monthly rates of return of General Electric and Apple, January 2004 to December 2008

Table 6.1 Monthly rates of return of General Electric and Apple Stock[a]

Monthly rate of return					
General Electric		Apple		S&P 500	
M	−.005	M	.048	M	.022
SD	.056	SD	.127	SD	.046
Minimum	−.235	Minimum	−.330	Minimum	−.213
Maximum	.117	Maximum	.352	Maximum	.069

[a]January 2004 to December 2008

From Table 6.1, notice that Apple's mean monthly rate of return of 4.8% exceeds the Market mean monthly rate of return of 2.2%, although Apple stock prices are more volatile: the standard deviation in monthly rates of return is .127, compared with the Market's standard deviation of .046. The greater expected return from Apple comes at the cost of added risk. Over this 5 year period, General Electric stock had a lower monthly rate of return, –.05%, than the Market average, and its prices were more volatile, with a standard deviation of .056, than the Market.

We would report to a potential investor:

- Over the 58 months examined, Apple offers a greater expected monthly rate of return of 4.8%, relative to the S&P 500 Index of the Market, with expected monthly return of 2.2%, but at higher risk with standard deviation in return .13 versus .05.
- In this 5 year period, General Electric offered a smaller expected rate of return, –.05%, than the Market index, and was also riskier, with standard deviation in return .06.

6.2 Investors Trade Off Risk and Return

Investors seek stocks that offer higher expected rates of return RR and lower risk. Relative to a Market index, such as the S&P 500, which is a composite of 500 individual stocks, many individual stocks offer higher expected returns, but at greater risk. Market indices are weighted averages of individual stocks. Like other weighted averages, a Market index has an expected rate of return in the middle of the expected returns of the individual stocks making up the index. An investor attempts to choose stocks with higher than average expected returns and lower risk.

6.3 Beta Measures Risk

A Market index reflects the state of the economy. When a time series of an individual stock's rates of return is regressed against a Market index, the simple linear regression slope β_1 indicates the expected percent change in a stock's rate of return in response to a percent change in the Market rate of return. β_1 is estimated with b_1 using a sample of stock prices:

$$\hat{RR}_{stocki,t} = b_0 + b_1 RR_{Market,t},$$

where $RR_{stocki,t}$ is the estimated rate of return of a stock i in month t, and $RR_{Market,t}$ is the rate of return of a Market index in month t.

In this specific case, the simple linear regression slope estimate b is called *beta*. Beta captures Market specific risk. If, in response to a percent change in the Market rate of return, the expected change in a stock's rate of return b is greater than 1%, the stock is more volatile and exaggerates Market movements. A 1% increase in the Market value is associated with an expected change in the stock's price of more than 1% change. Conversely, if the expected change in a stock's rate of return b is less than 1%, the stock dampens Market fluctuations and is less risky. A 1% change in the Market's value is associated with an expected change in the stock's price of less than 1%. Beta reflects the amount of risk a stock contributes to a well diversified portfolio.

Recall from Chap. 4 that the sample correlation coefficient between two variables r_{xy} is closely related to the simple regression slope estimate b_1:

$$b_1 = r_{xy} \frac{s_y}{s_X}.$$

In a leading indicator model of an individual stock's rate of return against a Market index, our estimate of beta is directly related to the sample correlation between the individual stock's rate of return and the Market rate of return:

$$beta_{stock_i} = b_{stock_i} = r_{stock_i, Market} \frac{s_{stock_i}}{s_{Market}}.$$

The estimate of beta is a direct function of the sample correlation between an individual stock's rate of return and the Market rate of return, as well as Market sample variance. Stocks with rates of return that are more strongly correlated with the Market rate of return and those with larger standard deviations have larger betas.

Notice in Fig. 6.2 that General Electric stock has a smaller variance than Apple stock or the Market. General Electric is a less risky investment. Notice also that both stocks tend to move with the Market, although Apple moves more and General Electric moves less.

It would not be surprising to find that Apple stock returns are riskier than General Electric returns, since iPhones and iPads are relatively expensive luxury items. In boom cycles, companies that sell luxuries do more business. General Electric sells many necessities, including appliances and light bulbs. The demand for these products is affected less by economic swings, making GE stock relatively less correlated with Market swings, and, hence, less risky.

Table 6.2 Correlations, standard deviations, and betas[a]

	Correlation with the Market, $r_{stock, Market}$	SD	beta, b_{stock}
SP500 RR		.046	
GE RR	.50	.056	.61[a,b]
Apple RR	.54	.127	1.51[a]

[a]Significant at .01
[b]Significantly less than 1.0 at a 95% confidence level

General Electric's and Apple's returns are both strongly correlated with the Market index returns ($r_{GeneralElectric, Market} \approx r_{Apple, Market} \approx .5$). However, Apple returns are considerably more volatile ($s_{Apple} = .127 > s_{GeneralElectric} = .056$). Because Apple rates of return are more volatile than General Electric, Apple will also have a larger beta than General Electric.

Betas b_{stock_i} are shown in the last column of Table 6.3. A percent increase in the Market produces

- Less than 1% expected increase in General Electric's price
- A 1% expected increase in Apple's price

Beta estimates are shown in Table 6.3 and Fig. 6.3.

Table 6.3 Estimates of betas

General Electric
SUMMARY OUTPUT

Regression statistics

R Square	.246				
Standard error	.049				
Observations	59				

ANOVA	*df*	*SS*	*MS*	*F*	*Significance F*
Regression	1	.046	.0455	18.6	.0001
Residual	57	.139	.0024		
Total	58	.185			

	Coefficients	*Standard error*	*t Stat*	*p Value*	*Lower 95%*	*Upper 95%*
Intercept	− .006	.006	− .9	.37	− .019	.007
S&P RR	.611	.142	4.3	.0001	.328	.894

Apple
SUMMARY OUTPUT

Regression statistics

R Square	.296				
Standard error	.108				
Observations	59				

ANOVA	*df*	*SS*	*MS*	*F*	*Significance F*
Regression	1	.278	.278	23.9	.0000
Residual	57	.663	.012		
Total	58	.941			

	Coefficients	*Standard error*	*t Stat*	*p Value*	*Lower 95%*	*Upper 95%*
Intercept	.046	.014	3.3	.002	.018	.074
S&P RR	1.510	.309	4.9	.0000	.892	2.128

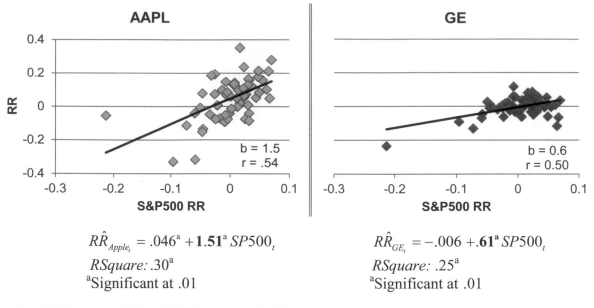

$$RR\hat{}_{Apple_t} = .046^a + \mathbf{1.51}^a\,SP500_t$$
$$RSquare: .30^a$$
$$^a Significant\ at\ .01$$

$$RR\hat{}_{GE_t} = -.006 + \mathbf{.61}^a\,SP500_t$$
$$RSquare: .25^a$$
$$^a Significant\ at\ .01$$

Fig. 6.3 Response of GE and AAPL stocks to the Market

Both stocks have similar, moderately high correlations with the Market, yet their betas differ. The difference in risk, in this case, is due to the difference in the standard deviations in their rates of return. Comparing betas, a potential investor would conclude the following.

General Electric, with an estimated beta less than one (b_{GE}=.61), is a low risk investment. GE returns dampen Market swings. With a percent increase in the Market, we expect to see an average increase of .61% in GE's price.

Apple stock, with an estimated beta of one (b_{Apple}=1.51) is riskier than GE, and mirrors Market movement. With a percent increase in the Market, we expect to see an average increase of about 2%, 1.51%, in Apple's price.

6.4 A Portfolio Expected Return, Risk, and Beta Are Weighted Averages of Individual Stocks

An investor is really interested in the expected return and risk of her portfolio of stocks. These are weighted averages of the expected returns and betas of the individual stocks in a portfolio:

$$E(RR_P) = \sum_i w_i E(RR_i)$$
$$b_P = \sum_i w_i b_i$$

where $E(RR_P)$ is the expected portfolio rate of return, w_i is the percent of investment in the ith stock, $E(RR_i)$ is the expected rate of return of the ith stock, b_P is the portfolio beta estimate, and b_i is the beta estimate of the ith stock.

Example 6.2 Three Alternate Portfolios

An investment manager has been asked to suggest a portfolio of three stocks from four being considered by a client: Exxon Mobil, Apple, and IBM. The prospective investor wanted to include computer stock in his portfolio and had heard that IBM was a desirable "Blue Chip." She suspected that holding both Apple and IBM stocks might be risky, were the computer industry to falter.

To confidently advise her client, the investment manager compared four portfolios of three equally weighted stocks from the three requested options. Individual stock weights in each portfolio equal one half. Table 6.4 contains the expected portfolio rates of return and betas for the three possible combinations.

Table 6.4 Expected portfolio returns and beta estimates

Portfolio	Expected portfolio return $\sum E(RR_i)/2$	$E(RR_P)$	Portfolio beta estimate $\sum b_i/2$	b_P
Exxon Mobil+Apple	(.014 +.048)/2	.031	(.50 +1.51)/2	1.01
Exxon Mobil+IBM	(.014 +.002)/2	.008	(.50 + .83)/2	.64
Apple+IBM	(.048 +.002)/2	.025	(1.51 + .83)/2	1.17

Alternatively, she could find expected portfolio returns and betas with software, and this would be the practical way to compare more than a few portfolios. Figure 6.5 shows expected (mean) rates of return and regression beta estimates for the three portfolios from Excel.

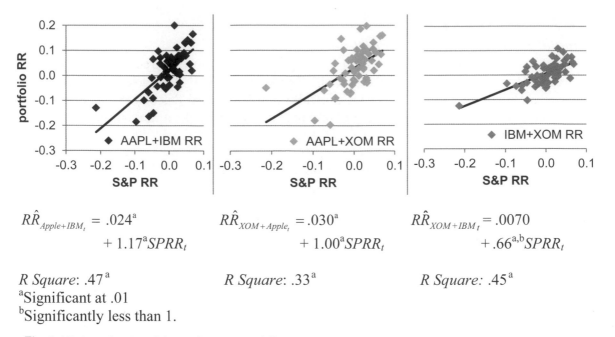

$$R\hat{R}_{Apple+IBM_t} = .024^a$$
$$+ 1.17^a SPRR_t$$

$$R\hat{R}_{XOM+Apple_t} = .030^a$$
$$+ 1.00^a SPRR_t$$

$$R\hat{R}_{XOM+IBM_t} = .0070$$
$$+ .66^{a,b} SPRR_t$$

R Square: .47[a]   *R Square*: .33[a]   *R Square:* .45[a]

[a]Significant at .01
[b]Significantly less than 1.

Fig. 6.4 Beta estimates of three alternate portfolios

6.5 Better Portfolios Define the Efficient Frontier

In the comparison of alternative portfolios, the investment manager wanted to identify alternatives that promised greater expected return without greater risk or, alternatively, those that reduced risk without reducing return. Better portfolios, which promise the highest return for a given level of risk, define the *Efficient Frontier*. To see the Efficient Frontier, she made a scatterplot of portfolio expected rate of return by portfolio risk. Those relatively efficient portfolios lie in the upper left.

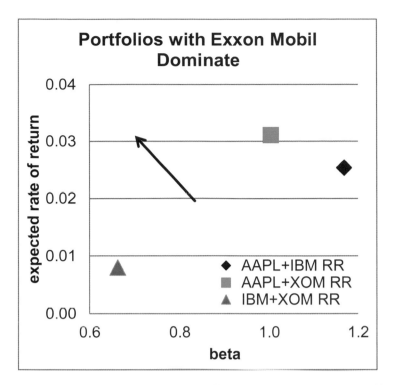

Fig. 6.5 Relatively efficient portfolios offer greater expected return and lower risk

Comparing portfolios in Fig. 6.6, the investment manager found that the more diversified portfolios that contain Exxon Mobil and one of the computer stocks offer either higher expected rate of return (with Apple) or lower risk (with IBM) than the portfolio with both computer stocks.

$$E(RR_{XOM+AAPL}) = .031 > E(RR_{AAPL+IBM}) = .025 > E(RR_{XOM+IBM}) = .008$$

$$b_{XOM+IBM} = .66 < b_{XOM+AAPL} = 1.00 < b_{AAPL+IBM} = 1.17$$

However, the choice between the two diversified portfolios will depend upon the prospective investor's risk preference.

The investment manager presented results of her analysis with recommendations in the following memo to her client.

MEMO

Re: Recommended Portfolios are Diversified
To: Mr. Rich N. Vest
From: Madison Monroe, Investment Advisor, Stellar Investments
Date: October 2007

Portfolios that contain Exxon and either Apple or IBM stocks outperform the Apple + IBM combination and promise expected monthly returns of 2.5–3.1%.

Assessment and Comparison of Alternate Portfolios: Portfolios containing two from the candidate set of three stocks, Exxon Mobil, Apple, and IBM, have been compared to assess their expected returns and risk levels. Assessments were based on 5 years of monthly prices, December 2003 through November 2008, and movement relative to the S&P1000 Market index during this period.

Three Portfolios Compared: Expected monthly rates of return range from .8 to 3.1%. Portfolios with Apple stock yield higher expected returns. That with Apple and Exxon Mobil yields the highest expected return, 3.1%.

In response to a 1% change in the S&P1000, the Exxon Mobil + IBM combination is expected to move less, .7%, dampening Market movement. This is the conservative choice. Other combinations mirror the Market and are expected to move 1.0–1.2%.

Conclusions: The choice of Exxon Mobil with Apple promises the highest expected return, although riskier than the similar portfolio with IBM instead of Apple, which offers a lower expected return with substantially less risk. We suggest purchase of Exxon Mobil, and *either* Apple or IBM.

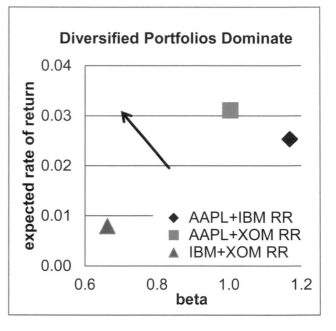

Other Options: You may wish to consider a portfolio with a larger number of stocks to increase your diversification and reduce your risk.

6.6 Portfolio Risk Depends on Correlations with the Market and Stock Variability

Both the expected rate of return of a portfolio and its risk, measured by its beta, depend on the expected rates of return and betas of the individual stocks in the portfolio. Individual stock betas are direct functions of

- The correlation between a stock's rate of return and the Market index rate of return
- The standard deviation of a stock's rate of return

Beta for a stock or a portfolio is estimated by regressing the stock or portfolio monthly rates of return against monthly Market rates of return. The resulting simple linear regression slopes are estimates of the stock or portfolio beta.

Excel 6.1 Estimate Portfolio Expected Rate of Return and Risk

Three Portfolios with Exxon Mobil, IBM, and Apple
Monthly rates of return for each of the three stocks and the S&P1000 index of the Market, adjusted for inflation, are in **Excel 6.1 Three Portfolios.xls**.

Correlations between stocks and the Market: Correlations between rates of return of pairs of stocks and the Market sometimes suggest combinations that might reduce risk through diversification.

To see the pairwise correlations, **Alt AY11, Correlation**.

For **Input Range, s**elect the rates of return of the three stocks and the S&P1000**.**

◢	A	B	C	D	E
1		*S&P RR*	*AAPL RR*	*IBM RR*	*XOM RR*
2	S&P RR	1			
3	AAPL RR	0.544	1		
4	IBM RR	0.639	0.318	1	
5	XOM RR	0.405	0.445	0.224	1

Monthly portfolio returns formula: Make three new columns, one for each of the three portfolios… monthly rate of return, which will be the average of rates of return of each of the two stocks in the portfolio.

Insert three new columns. Select column **C, D,** and **E,** and **Alt Home Insert Columns.**

In the first row of each new column enter a formula for the average of two stocks. For the *AAPL+IBM* portfolio, for example, enter

=AVERAGE(F4,G4).

Select the three new cells and double click the lower right corner to fill in the monthly rates of return for the three portfolios.

f_x	=AVERAGE(F4,G4)					
	C	D	E	F	G	H
	AAPL+IBM	AAPL+XOM	IBM+XOM	AAPL RR	IBM RR	XOM RR
	0.023	0.077	0.078	0.022	0.024	0.133
	0.063	0.025	0.033	0.056	0.071	-0.005
	0.016	0.047	0.003	0.060	-0.028	0.034
	0.041	0.058	-0.031	0.130	-0.048	-0.014
	-0.043	-0.012	-0.008	-0.047	-0.040	0.023

Expected monthly rates of return: Find the expected monthly return for the three portfolios, **Alt MUA.**

f_x	=AVERAGE(C4:C62)		
B	C	D	E
S&P RR	AAPL+IBM	AAPL+XOM	IBM+XOM
-0.008	0.015	-0.069	-0.004
0.023	0.009	0.031	-0.027
-0.096	-0.184	-0.179	-0.034
-0.213	-0.129	-0.049	-0.125
ERR	0.025	0.031	0.008

Estimate betas from simple regression: To find the Market specific risk, *beta*, find the simple regression slope of each portfolio rate of return with *S&P RR*.

For the first portfolio, *AAPL+IBM*, run regression with *AAPL+IBM* in the **Input Y Range**, and *S&P RR* in the **Input X Range**.

	A	B	C	D	E	F	G
16		*Coefficients*	*Standard Error*	*t Stat*	*P-value*	*Lower 95%*	*Upper 95%*
17	Intercept	0.024	0.008	3.1	0.0027	0.009	0.039
18	S&P RR	1.17	0.17	7.1	2E-09	0.84	1.50

	A	B	C	D	E	F	G
16		*Coefficients*	*Standard Error*	*t Stat*	*P-value*	*Lower 95%*	*Upper 95%*
17	Intercept	0.0296	0.0086	3.4	0.0012	0.012	0.047
18	S&P RR	1.004	0.190	5.3	2E-06	0.624	1.384

	A	B	C	D	E	F	G
16		*Coefficients*	*Standard Error*	*t Stat*	*P-value*	*Lower 95%*	*Upper 95%*
17	Intercept	0.0070	0.0044	1.6	0.1194	-0.0019	0.0158
18	S&P RR	0.66	0.10	6.8	6E-09	0.47	0.86

Excel 6.2 Plot Return by Risk to Identify Dominant Portfolios and the Efficient Frontier

To compare the expected rates of return and estimated risk of the three portfolios, plot the portfolio rates of return against their betas to identify the Efficient Frontier.

Create a summary of the portfolio betas and expected returns below the data:

Copy the row containing portfolio labels and insert the copy below the data rows. (**Cntl+C**opy, **Alt H**ome **I**nsert copied c**E**lls.)

Add a row following the new label row, **Alt H**ome **I**nsert **R**ow.

Use the Excel function **slope**(*y array*, *x array*) to add betas below each of the three portfolio labels.

f_x	=SLOPE(C4:C62,$B$4:$B$62)		
B	**C**	**D**	**E**
S&P RR	AAPL+IBM	AAPL+XOM	IBM+XOM
-0.008	0.015	-0.069	-0.004
0.023	0.009	0.031	-0.027
-0.096	-0.184	-0.179	-0.034
-0.213	-0.129	-0.049	-0.125
S&P RR	AAPL+IBM	AAPL+XOM	IBM+XOM
beta	1.17	1.00	0.66
ERR	0.025	0.031	0.008

Copy the three sets of labels, betas, and expected rates of return and paste without formulas (**Cntl+C**opy; **Alt H**ome **V S**pecial val**U**es transpos**E**).

S&P RR	AAPL+IBM	AAPL+XOM	IBM+XOM
beta	1.17	1.00	0.66
ERR	0.025	0.031	0.008
S&P RR	beta	ERR	
AAPL+IBM	1.17	0.025	
AAPL+XOM	1.00	0.031	
IBM+XOM	0.66	0.008	

Select and plot the beta and expected return for the first portfolio, **Alt iN**sert **D.**

-0.008	0.015	-0.069	
0.023	0.009	0.031	
-0.096	-0.184	-0.179	
-0.213	-0.129	-0.049	
S&P RR	*AAPL+IBM*	*AAPL+XOM*	
beta	1.17	1.00	
ERR	0.025	0.031	
S&P RR	beta	ERR	
AAPL+IBM	1.17	0.025	
AAPL+XOM	1.00	0.031	
IBM+XOM	0.66	0.008	

Because we are plotting a single point, Excel will read both *beta* and the *ERR* as a single series, plotting two points. Edit the series, **Alt JC** sElect, to correct this.

Add the other two portfolio series of betas and expected returns, **Alt JC** s<u>E</u>lect.

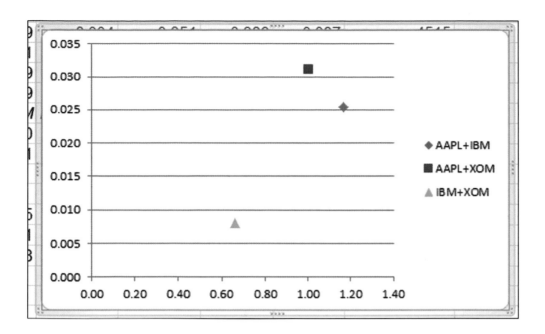

Add chart and axes titles; adjust axes and style.

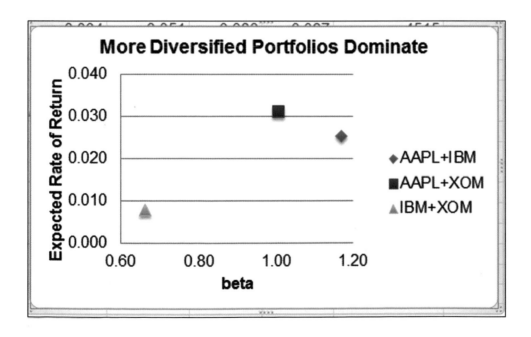

Assignment 6-1 Individual Stocks' Beta Estimates

Use logic to choose two stocks to analyze from **Assignment 6-1 Stock RR.xls.**
Choose a stock that you would expect to have a beta less than 1 and a stock that you expect to have a beta more than 1.
Be prepared to explain the logic of your choices.

The **Assignment 6-1 Stock RR.xls** dataset contains 5 years of monthly prices from December 2004 to October 2009, for individual stocks, as well as monthly value of a Market index, the S&P500.
Stock rates of return included in the dataset are as follows:

COSTCO	GOLDMAN SACHS	NUCOR
DELL	HEWLETT-PACKARD	STARBUCKS
DISNEY WALT	MCDONALDS	TOYOTA
GENERAL ELECTRIC	MICROSOFT	WALMART

1. Plot rates of return for both stocks and the S&P1000 return by month in a scatterplot:
   (a) Do the stocks track the Market?
   (b) Do they dampen or exaggerate Market swings?

2. Conduct two simple linear regressions to estimate the betas of the two stocks you chose.
   (The two dependent variables will be the monthly *rates of return* of the two stocks and the independent variable will be monthly *S&P 500 rates of return*, S&P500RR.)

Assignment 6-2 Expected Returns and Beta Estimates of Alternate Portfolios

A potential investor has asked you to recommend two stocks that together would produce a desirable portfolio. He expects to invest half in each stock.

Choose three stocks from the set in **Assignment 6-2 Stock RR.xls** to potentially combine.
Compare the *expected return* and *risk (beta)* of the three portfolios from all possible pairs and make a recommendation to the investor.

1. Make three new portfolio variables equal to averages of each of the stock pairs' *rates of return*, and then find the average sample portfolio return, which is the *expected portfolio return*.

2. Run simple regressions of the portfolio monthly *rates of return* against the *Market rate of return* to find portfolio *betas*.

3. Construct a chart of the Efficient Frontier and offer your investment recommendation to the potential investor, based on a comparison of *expected rates of return* and estimated *risk*.

Chapter 7
Association Between Two Categorical Variables: Contingency Analysis with Chi Square

Categorical variables, including nominal and ordinal variables, are described by tabulating their frequencies or probability. If two categorical variables are associated, the frequencies of values of one will depend on the frequencies of values of the other. Chi square tests the hypothesized association between two categorical variables, and contingency analysis quantifies their association.

7.1 Evidence of Association When Conditional Probabilities Differ from Joint Probabilities

Contingency analysis begins with the crosstabulation of frequencies of two categorical variables. Figure 7.1 shows a crosstabulation of sandwich spreads and topping combinations chosen by 40 students.

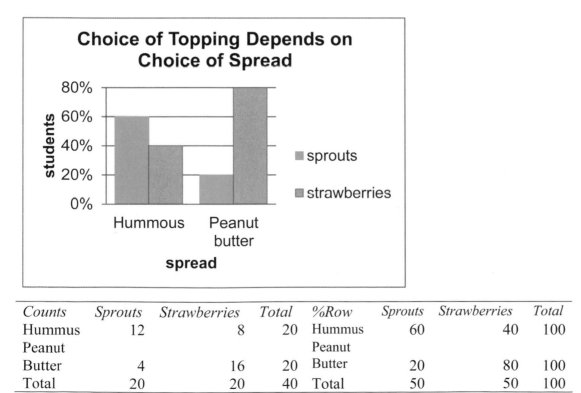

Counts	Sprouts	Strawberries	Total	%Row	Sprouts	Strawberries	Total
Hummus	12	8	20	Hummus	60	40	100
Peanut Butter	4	16	20	Peanut Butter	20	80	100
Total	20	20	40	Total	50	50	100

Fig. 7.1 Crosstabulation: sandwich topping depends on spread

C. Fraser, *Business Statistics for Competitive Advantage with Excel 2010: Basics, Model Building, and Cases*,
DOI 10.1007/978-1-4419-9857-6_7, © Springer Science+Business Media, LLC 2012

If the unconditional probabilities of category levels, such as sprouts versus strawberry topping, differ from the probabilities, conditional on levels of another category, such as hummus or peanut butter spread, we have evidence of association. In this sandwich example, sprouts were chosen by half the students, making its unconditional probability .5. If a student chose hummus spread, the conditional probability of sprouts topping was higher (.60). If a student chose peanut butter spread, sprouts was the less likely topping choice (.40).

Example 7.1 Recruiting Stars

The Human Resource managers are hoping to improve the odds of hiring outstanding performers and to reduce the odds of hiring poor performers by targeting recruiting efforts. Management believes that recruiting at the schools closer to firm headquarters may improve the odds of hiring stars. Students familiar with local customs may feel more confident at the firm. Removing schools far from headquarters may reduce the odds of hiring poor performers. Management's hypotheses are

H_0: Job performance is not associated with undergraduate program location.
H_1: Job performance is associated with undergraduate program location.

To test these hypotheses, department supervisors throughout the firm sorted a sample of 40 recent hires into three categories based on job performance: poor, average, and outstanding. The sample employees were also categorized by the proximity to headquarters: Home State, Same Region, and Outside Region. These crosstabulations are shown in the PivotTable in Table 7.1 and the PivotChart in Fig. 7.2.

Table 7.1 Job performance depends on program location

Count	Performance			
Location	Poor	Average	Outstanding	Total
Outside region	5	2	3	10
Same region	2	10	3	15
Home state	3	3	9	15
Total	10	15	15	40
% of Row	Performance			
Location	Poor	Average	Outstanding	Total
Outside region	50	20	30	100
Same region	13	67	20	100
Home state	20	20	60	100
Total	25	38	38	100

χ_4^2 12.5   *p Value*   .02

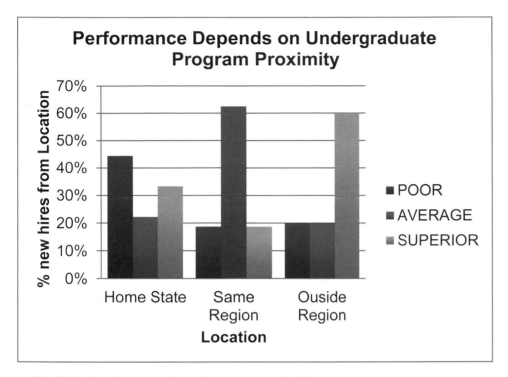

Fig. 7.2 Job performance depends on program location

The crosstabs indicate that a quarter of the firm's new employees are *Poor* performers, about 40% are *Average* performers, and about 40% are *Outstanding* performers. From the PivotChart we see that more than a quarter of employees from programs *Outside Region* are *Poor* performers, and more than 40% of employees from *Home State* programs are *Outstanding* performers. Were program location and performance not associated, a quarter of the recruits from each location would be *Poor* performers. We would, for example, expect a quarter of ten employees recruited from *Outside Region* to be Poor performers, that is, 2.5 (=.25(10)). Instead, there are actually five (*Outside Region, Poor*) employees. There is a greater chance, 50%, of *Poor* performance, given *Outside Region*, relative to *Same Region* or *Home State*. Ignoring program location, the probability of poor performance is .25; acknowledging program location, this probability of poor performance varies from .13 (*Same Region*) to .50 (*Outside Region*). These differences in row percentages suggest an association between program rank and performance.

7.2 Chi Square Tests Association Between Two Categorical Variables

The chi square (χ^2) statistic tests the significance of the association between performance and program location by comparing expected cell counts with actual cell counts, squaring the differences, and weighting each cell by the inverse of expected cell frequency:

$$\chi^2_{(R-1),(C-1)} = \sum_{ij}^{RC} (e_{ij} - n_{ij})^2 / e_{ij}$$

,

where R is the number of row categories, C is the number of column categories, n is the number in the ith row and jth column, and e is the number expected in the ith row and jth column.

χ^2 gives more weight to the least likely cells. χ^2 distributions are skewed and with means equal to the number of degrees of freedom. Several χ^2 distributions with a range of degrees of freedom are shown in Fig. 7.3.

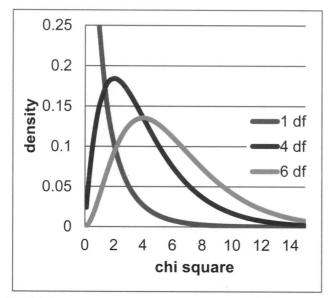

Fig. 7.3 Chi square distributions for a range of degrees of freedom

In the **Recruiting Stars** example, Table 7.1, chi square, χ^2_4, is 12.5, which can be verified using the formula

$$
\begin{aligned}
\chi^2 = \ & (2.5-5)^2/2.5 + (3.8 - 2)^2/3.8 + (3.8 - 3)^2/3.8 \\
& + (3.8 - 2)^2/3.8 + (5.6 - 10)^2/5.6 + (5.6 - 3)^2/5.6 \\
& + (3.8 - 3)^2/3.8 + (5.6 - 3)^2/5.6 + (5.6 - 9)^2/5.6 \\
= \ & \qquad\quad 2.5 + \qquad .9 \qquad + .2 \\
& + \qquad\quad .9 + \qquad 3.5 \qquad + 1.2 \\
& + \qquad\quad .2 + \qquad 1.2 \qquad + 2.0 = 12.5.
\end{aligned}
$$

From a table of χ^2_4 distributions, we find that for a crosstabulation of this size, with three rows and three columns ($df =$ (rows $-$ 1) $\times$ (columns $-$ 1) $= 2 \times 2 = 4$), $\chi^2_4 = 12.5$ indicates that the p *Value* is .02. Two percent of the distribution lies to the right of 12.5. There is little chance of observing this sample data were performance and program tier not associated. The null hypothesis of lack of association is rejected.

Those cells that contribute more to chi square indicate the nature of the association. In this example, we see in Table 7.2 that these are the (*Outside Region, Poor*), (*Same Region, Average*), and (*Home State, Outstanding*) cells:

Table 7.2 Contribution to chi square by cell

	Poor	Average	Outstanding
Outside region	**2.5**	.9	.2
Same region	.9	**3.5**	1.2
Home state	.2	1.2	**2.0**

$$\chi^2 = \mathbf{2.5} + .9 + .2$$
$$+ .9 + \mathbf{3.5} + 1.2$$
$$+ .2 + 1.2 + \mathbf{2.0} = 12.5.$$

Poor performance is more likely if a new employee comes from a program *Outside Region*, *Average* performance is more likely if a new employee comes from a program in the *Same Region*, and *Outstanding* performance is more likely if a new employee comes from a *Home State* program. Thus, job performance is associated with program location.

7.3 Chi Square Is Unreliable If Cell Counts Are Sparse

There are two possible reasons why the chi square statistic is large and apparently significant. The first reason is the likely actual association between program location and performance. The second reason is that there are few (less than five) expected employees in five of the nine cells, shown in Table 7.3.

Table 7.3 Expected counts by cell

	Poor	Average	Outstanding
Outside region	2.5	3.8	3.8
Same region	3.8	5.6	5.6
Home state	3.8	5.6	5.6

Inasmuch as the chi square components include expected cell counts in the denominator, *sparse* (with expected counts less than five) cells inflate chi square. When sparse cells exist, we must either combine categories or collect more data.

In the **Recruiting Stars** example, management was most interested in increasing the chances of hiring *Outstanding* performers. Some believed that *Outstanding* performers were recruited from programs in the *Home State*; therefore, these categories were preserved. *Same Region* and *Outside Region* program locations were combined. *Poor* and *Average* performance categories were combined. We are left with a 2 × 2 contingency analysis (Fig. 7.4).

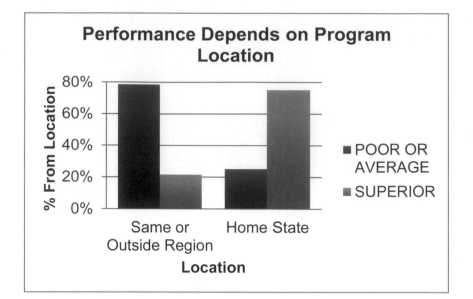

Count	Performance poor/			% Row	Performance poor/		
Location	average	outstanding	Total	Location	average	outstanding	Total
Same or Outside Region	19	6	25	Same or Outside Region	76	24	100
Home State	6	9	15	Home State	40	60	100
Total	25	15	40	Total	63	38	100

Chi Square	5.2
df	1
p Value	.02

Fig. 7.4 PivotChart of performance by program location with fewer categories

With fewer categories, all expected cell counts are now greater than five, providing a reliable $\chi^2_1 = 5.2$, which remains significant (*p Value* = .02). The PivotChart continues to suggest that the incidence of *Outstanding* performance is greater among employees recruited from *Home State* programs. The impact of program location on *Poor* performance is unknown, because *Poor* and *Average* categories were combined. Also unknown is the difference between employees from *Same* and *Outside Regions* programs, because these categories were likewise combined.

Recruiters would conclude the following.

Job performance of newly hired employees is associated with undergraduate program location. Twenty-four percent of our new employees recruited from Same or Outside Region undergraduate programs have been identified as Outstanding performers. Within the group recruited from Home State undergraduate programs, more than twice this percentage, 60%, are Outstanding performers, a significant difference. Results suggest that in order to achieve a larger percentage of Outstanding performers, recruiting should be focused on Home State programs.

7.4 Simpson's Paradox Can Mislead

Using contingency analysis to study the association between two variables can be potentially misleading, because all other related variables are ignored. If a third variable is related to the two being analyzed, contingency analysis may indicate that they are associated, when they may not actually be. Two variables may appear to be associated because they are both related to a third, ignored variable.

Example 7.2 American Cars

The CEO of American Car Company was concerned that the oldest segments of car buyers were avoiding cars that his firm assembles in Mexico. Production and labor costs are much cheaper in Mexico, and his long term plan is to shift production of all models to Mexico. If older, more educated, and more experienced buyers avoid cars produced in Mexico, American Car could lose a major market segment unless production remained in the United States.

The CEO's hypotheses were

H_0: Choice between cars assembled in the United States and cars assembled in Mexico is not associated with age category.

H_1: Choice between cars assembled in the United States and cars assembled in Mexico is associated with age category.

He asked Travis Henderson, Director of Quantitative Analysis, to analyze the association between age category and choice of US made versus Mexican made cars. The research staff drew a random sample of 263 recent car buyers, identified by age category. After preliminary analysis, age categories were combined to ensure that all expected cell counts in a [age category × origin choice] crosstabulation were each at least five. Contingency analysis is shown in the PivotChart and PivotTables in Fig. 7.5.

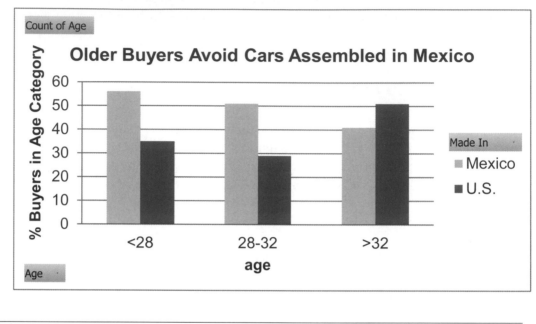

Count	Assembled in			% Rows	Assembled in		
Age	U.S.	Mexico	Total	Age	U.S.	Mexico	Total
Under 28	35	56	91	Under 28	38	62	100
28 to 32	29	51	80	28 to 32	36	64	100
33 Plus	51	41	92	33 Plus	55	45	100
Total	115	148	263	Total	44	56	100

Chi Square	8.0
df	2
p Value	.02

Fig. 7.5 Contingency analysis of US made versus Mexican made car choices by age

A glimpse of the PivotChart confirmed suspicions that older buyers seemed to be rejecting cars assembled in Mexico. The *p Value* for chi square was .02, indicating that the null hypothesis, lack of association, ought to be rejected. Choice between US- and Mexican-made cars seemed to be associated with age category. Fifty-six percent of the entire sample across all ages chose cars assembled in Mexico. Within the oldest segment, however, the Mexican assembled car share was lower, 45%. Although nearly two thirds of the younger segments chose cars assembled in Mexico, less than half of the oldest buyers chose Mexican made cars.

The CEO was alarmed at these results. His company could lose the business of older, more experienced buyers if production were shifted south of the border. Brand managers were about to begin planning "Made in the USA" promotional campaigns targeted at the oldest car buyers. Emily Ernst, the Director of Strategy and Planning, suggested that age was probably not the correct basis for segmentation. She explained that the older buyers shop for a particular *type* of car – a family sedan or station wagon – and few family sedans or wagons were being assembled in Mexico. Models assembled at home in the United States tended to be large sedans and station wagons, styles sought by older buyers. She proposed that it was *style* that influenced the US assembled versus Mexican assembled choice, and not age, and that it was style that was dependent on age. Her hypotheses were

H_0: Choice of car style is not associated with age category.
H_1: Choice of car style is associated with age category.

To explore these alternate hypotheses, the research team ran contingency analysis of style choice (SUV, sedan/wagon, and coupe) by age category, Fig. 7.6.

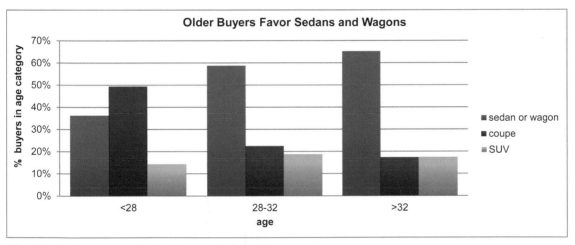

Fig. 7.6 Contingency analysis of car style choice by age category

Count	Style sedan/				Row%	Style sedan/			
Age	wagon	coupe	SUV	Total	Age	wagon	coupe	SUV	Total
< 28	33	45	13	91	< 28	36	49	14	100
28 to 32	47	18	15	80	28 to 32	59	23	19	100
33+	60	16	16	92	33+	65	17	17	100
Total	140	79	44	263	Total	53	30	17	100

χ_4^2	26.2	*p Value*	.0000

Contingency analysis of this sample indicates that choice of style is associated with age category. More than half (53%) of the car buyers chose a sedan or wagon, although only about a third (36%) of the younger buyers chose a sedan or wagon, and nearly twice as many (65%) older buyers chose a sedan or wagon. Thirty percent of the sample bought a coupe, and just nearly half (49%) of the younger buyers chose a coupe. Only 17% of the oldest buyers bought a coupe. These are significant differences supporting the conclusion that style of car chosen is associated with age category.

This is the news the CEO was looking for. If older car buyers are choosing US made cars because they desire family styles, sedans and wagons, which tend to be assembled in the United States, then perhaps these older buyers aren't shunning Mexican made cars. His hypotheses were

H_0: Given choice of a sedan or wagon, choice of US assembled versus Mexican assembled is not associated with age category.
H_1: Given choice of a sedan or wagon, choice of US assembled versus Mexican assembled is associated with age category.

To test these hypotheses, the analysis team conducted three contingency analyses of origin choice (US versus Mexican assembled) by age category, looking at each style separately in Table 7.4 and Fig. 7.7.

Table 7.4 Contingency analysis: origin choice by age given syle

% Age by style		Assembled in					
Style	Age	USA	Mexico	Total	χ^2	df	p Value
Sedan or	Under 28	52	48	100			
wagon	28–32	60	40	100			
	33 plus	45	55	100			
Total		53	47	100	2.5	2	.29
coupe	Under 28	29	71	100			
	28–32	44	56	100			
	33 plus	17	83	100			
Total		29	71	100	3.0	2	.22
SUV	Under 28	38	62	100			
	28–32	50	50	100			
	33 plus	33	67	100			
Total		41	59	100	.9	2	.63
Grand total		44	56	100			

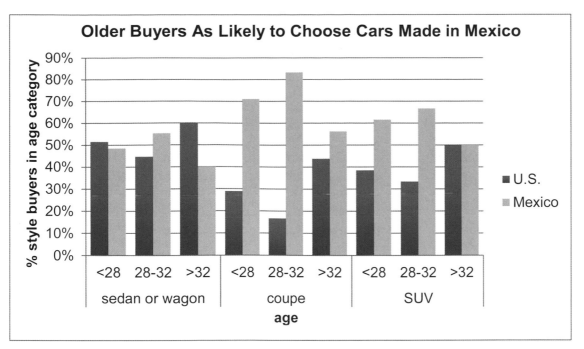

Fig. 7.7 Contingency analysis: origin choice given age by style

Controlling for style of car by looking at each style separately reveals a lack of association between origin preference for US made versus Mexican made cars and age category. Across all three car styles, *p Values* are greater than .05. There is not sufficient evidence in this sample to reject the null hypothesis. We conclude from this sample that the US assembled versus Mexican assembled choice is not associated with age category. The domestic automobile manufacturer should therefore not alter plans to move production South.

Simpson's paradox describes the situation where two variables appear to be associated only because of their mutual association with a third variable. If the third variable is ignored, results are misleading. Because contingency analysis focuses upon just two variables at a time, analysts should be aware that apparent associations may come from confounding variables, as the American Cars example illustrates.

The research team summarized these results in the following memo.

MEMO

Re.: Country of Assembly Does Not Affect Older Buyers' Choices
To: CEO, American Car Company
 Emily Ernst, Director of Planning and Strategy
 Brand Management
From: Travis Hendershott, Director of Quantitative Analysis

Analysis of a sample of new car buyers reveals that styles of car drive the choices of distinct age segments. Choices of all ages of buyers are independent of country of manufacture.

Contingency Analysis: Choices of 263 new car buyers were analyzed to assess the dependence of choice on country of manufacture, United States or Mexico, and age category.

Results: Choice between US and Mexican assembled cars is not associated with age category.

Style of car chosen is associated with age category. Younger buyers are more likely to choose a sporty coupe.
Older buyers are more likely to buy a sedan or wagon.

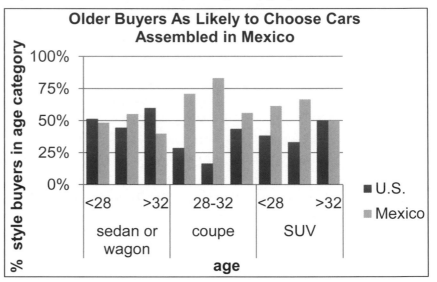

$$\chi_2^2 = 2.5, ns \, ; \chi_2^2 = 3.0, ns \, ; \chi_2^2 = .9, ns$$

Conclusions: Production in Mexico is not expected to affect car buyer choices, providing the opportunity to shift assembly south to take advantage of cheaper labor.

Limitations: A larger sample would enable examination of more representative age categories, and specifically, a broader middle segment and older oldest segment.

7.5 Contingency Analysis Is Demanding

Contingency analysis requires a large and balanced dataset to ensure a stable chi square. Even large samples may contain small proportions of particular categories, forcing combinations that aren't ideal. In the **American Cars** example, a broad category was used for the oldest age segment, combining fairly different ages, 33–60, and a narrow category was defined for the middle age segment, ages 28–32. The sample, although large, was not balanced and contained a large proportion of car buyers aged 3039. This group was split and combined with sparse younger and older age categories to allow expected cell counts greater than five. With smaller samples, just two categories for a variable may remain, which may limit hypothesis testing. In the **Recruiting Stars** example, final results could not be used to assess the association between recruiting and poor employee performance after Poor and Average performing employees were combined.

7.6 Contingency Analysis Is Quick, Easy, and Readily Understood

Despite the fairly demanding data requirements, contingency analysis is appealing because it is simple and results are easily understood. For very large samples, sparse cells are not a problem and many categories may be used, increasing the specificity of results and allowing a range of hypothesis tests.

For smaller samples, other alternatives, such as logit analysis (discussed in detail in Chap. 13), exist for analyzing categorical variable associations. These carry fewer data demands and allow incorporation of multiple variables. Multivariate analysis helps us avoid drawing incorrect conclusions in cases where Simpson's paradox might mislead.

Excel 7.1 Construct Crosstabulations and Assess Association Between Categorical Variables with PivotTables and PivotCharts

American Cars
In order to explore the possible association between choice of US and Mexican assembled cars by age, begin by making a *PivotTable* to see the crosstabulation.

Open Excel 7.1 **American Cars.xls**.

Select filled cells in the *Age* and *Made In* columns and then insert a PivotTable.

Drag *Age* to **ROW**, *Made In* to **COLUMN**, and *Age* to **DATA**.

	A	B	C	D
1	Drop Report Filter Fields Here			
2				
3	Count of Made In	Made In ▾		
4	Age ▾	Mexico	U.S.	Grand Total
5	<28	56	35	91
6	>32	41	51	92
7	28-32	51	29	80
8	Grand Total	148	115	263

PivotTable Field List

Choose fields to add to report:
- ☑ Age
- ☑ Made In

Drag fields between areas below:

▽ Report Filter	⊞ Column Labels
	Made In ▾

⊞ Row Labels	Σ Values
Age ▾	Count of Mad... ▾

Interest is in the conditional probabilities of choice of cars *Made In* the United States and Mexico given *age* category.

The **% of row** are the conditional probabilities. To see these, use shortcuts:

Alt JT field settin**G**s to **Show Values As % of Row Total**.

	A	B	C	D	E	F	G	H	I	J
1	Drop Report Filter Fields Here									
2										
3	Count of Made In	Made In								
4	Age	Mexico	U.S.	Grand Total						
5	<28	56	35	91						
6	>32	41	51	92						
7	28-32	51	29	80						
8	Grand Total	148	115	263						

Value Field Settings

Source Name: Made In

Custom Name: Count of Made In

Summarize Values By | Show Values As

Show values as

No Calculation

- No Calculation
- % of Grand Total
- % of Column Total
- % of Row Total
- % Of
- % of Parent Row Total

Number Format   OK   Cancel

To put the age categories in order, select and right click the >32 cell, **M**ove to **E**nd.

Drop Report Filter Fields Here

Count of Ma...

Age

<28 ... .00%

>32 ... 44.57% 55.43% 100.00%

28-32 ... 25% 100.00%

Grand Total ... 73% 100.00%

- Copy
- Format Cells...
- Refresh
- Sort ▶
- Filter ▶
- ✓ Subtotal "Age"
- Expand/Collapse ▶
- Group...
- Ungroup...
- Move ▶
  - Move ">32" to Beginning
  - Move ">32" Up
  - Move ">32" Down
  - Move ">32" to End
- ✕ Remove "Age"
- Field Settings...
- PivotTable Options...

Make a **PivotChart** of *Made In* by *Age:*

Alt JT **C**hart

Choose a design and style, and add a chart title that reflects your conclusion.

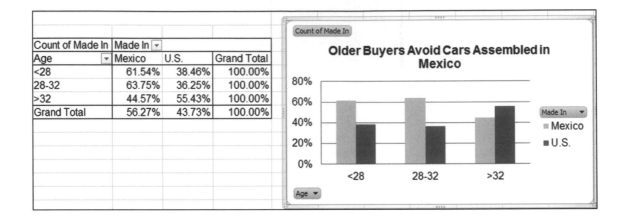

Excel 7.2 Use Chi Square to Test Association

To find the chi square statistic, change the PivotTable cells back to counts.

Select a cell in the PivotTable.

Alt JT field settin**G**s

To **Show Values As No Calculation**

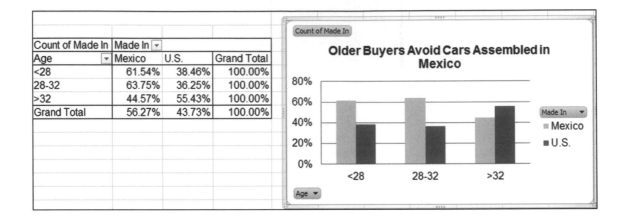

For chi square, make a table of *expected* cell counts and a table of cell contributions to chi square.

Select the two empty rows above the PivotTable, plus the PivotTable, copy, and paste right of the PivotTable with values and formats, but not formulas, **Alt H**ome **V S**pecial val**U**es.

Repeat to paste in a second copy.

A	B	C	D	E	F	G	H	I	J	K	L
Count of Made In	Made In			Count of Made In	Made In			Count of Made In	Made In		
Age	Mexico	U.S.	Grand Total	Age	Mexico	U.S.	Grand To	Age	Mexico	U.S.	Grand To
<28	56	35	91	<28	56	35	91	<28	56	35	91
28-32	51	29	80	28-32	51	29	80	28-32	51	29	80
>32	41	51	92	>32	41	51	92	>32	41	51	92
Grand Total	148	115	263	Grand Total	148	115	263	Grand Total	148	115	263

In the first cell of the second, *expected* table, enter the formula for the expected count, multiplying cells containing the Grand Total of youngest buyers and the Grand Total of cars assembled in Mexico, and then dividing by the Grand Total.

=$D5 × B$8/D8 fn4.

A dollar sign preceding **D** locks the column and a dollar sign preceding **8** locks the row, so that we can fill in the remaining cells in the table with this formula.

	F5			f_x	=$D5*B$8/$D$8		
	A	B	C	D	E	F	G	H
1								
2								
3	Count of Made In	Made In			Count of Made In	Made In		
4	Age	Mexico	U.S.	Grand Total	Age	Mexico	U.S.	Grand Tot
5	<28	56	35	91	<28	51.2091	35	91
6	28-32	51	29	80	28-32	51	29	80
7	>32	41	51	92	>32	41	51	92
8	Grand Total	148	115	263	Grand Total	148	115	263

Select the new cell, fill in the column:

Cntl+dn, Cntl+Down

and then fill in the adjacent row:

Shift+->, Cntl+Right

	F5			f_x	=$D5*B$8/$D$8		
	A	B	C	D	E	F	G
1							
2							
3	Count of Made In	Made In ▾			Count of Made In	Made In	
4	Age ▾	Mexico	U.S.	Grand Total	Age	Mexico	U.S.
5	<28	56	35	91	<28	51	40
6	28-32	51	29	80	28-32	45	35
7	>32	41	51	92	>32	52	40
8	Grand Total	148	115	263	Grand Total	148	115

In the third table, find each cell's contribution to chi square, the squared difference between expected counts, in the second table, and actual counts, in the first table, divided by expected counts in the second table.

In the first cell of the third, chi square, table enter:

=(F5-B5)^2/F5.

Select then new cell and fill in the column and the rows.

	J5				f_x	=(F5-B5)^2/F5					
	A	B	C	D	E	F	G	H	I	J	K
1											
2											
3	Count	Made In ▾			Count	Made In			Count	Made In	
4	Age	Mexico	U.S.	Grand Total	Age	Mexico	U.S.	Grand Tot	Age	Mexico	U.S.
5	<28	56	35	91	<28	51	40	91	<28	0.4	0.6
6	28-32	51	29	80	28-32	45	35	80	28-32	0.8	1.0
7	>32	41	51	92	>32	52	40	92	>32	2.2	2.9

Use the Excel function **SUM(*array1,array2*)** to add the cell contributions in the three rows.

	L5		f_x	=SUM(J5:K5)
	I	J	K	L
1				
2				
3	Count of Made In	Made In		
4	Age	Mexico	U.S.	Grand Tot
5	<28	0.4	0.6	1.0
6	28-32	0.8	1.0	1.8
7	>32	2.2	2.9	5.1

In the Grand Total row, find the Mexico sum:

Alt for**M**ula a**U**to **S**um.

Then select the Mexico Grand Total and fill in the row to find chi square in the last cell.

J8		f_x =SUM(J5:J7)	
	I	J	K	L
1				
2				
3	Count of Made In	Made In		
4	Age	Mexico	U.S.	Grand Total
5	<28	0.4	0.6	1.0
6	28-32	0.8	1.0	1.8
7	>32	2.2	2.9	5.1
8	Grand Total	3.5	4.5	8.0

Find the *p Value* for this chi square using the Excel function **CHISQ.DIST.RT**(*chisquare,df*) with degrees of freedom *df* of 2.

M8		f_x =CHISQ.DIST.RT(L8,2)			
	I	J	K	L	M	N
1						
2						
3	Count of Made In	Made In				
4	Age	Mexico	U.S.	Grand Total		
5	<28	0.4	0.6	1.0		
6	28-32	0.8	1.0	1.8		
7	>32	2.2	2.9	5.1		
8	Grand Total	3.5	4.5	8.0	0.018612	p value

Excel 7.3 Conduct Contingency Analysis with Summary Data

Sometimes data are in summary form. That is, we know the sample size and we know the percentage of the sample in each category.

Marketing Cereal to Children

Kooldogg expects that many Saturday morning cartoon viewers would be attracted to their sugared cereals. A heavy advertising budget for sugared cereals is allocated to Saturday morning television. We use contingency analysis to analyze the association between Saturday morning cartoon viewing and frequent consumption of Kooldogg cereal with sugar added. From a survey of 300 households, researchers know whether children aged 2–5 *Watch Saturday Morning Cartoons* on a regular basis (at least twice a month) and whether those children *Eat Kooldogg Cereal with Added Sugar* (at least once a week).

Open **Excel 7.3 Kooldogg Kids Ads.xls**.

Select the summary data and make a PivotTable, with *Watches Saturday Morning Cartoons* in **Rows**, *Eats Kooldogg Sugary Cereal* in **Columns**, and *Number of Children* in Σ **Values**:

Copy rows **1** and **2** with the table and paste twice with formats and values, **Alt HVSU**.

In the second table, find the expected cell counts under the assumption that Kooldog cereal consumption is independent of Saturday morning TV viewing.

In the third table, find cell contributions to chi square with squared differences between expected cell counts and actual cell counts, divided by expected cell counts.

	J5			fx =(F5-B5)^2/F5						
	B	C	D	E	F	G	H	I	J	K
1										
2										
3	Eats Kooldog Sugary Cereal			Sum of Number of children	Eats Kooldog Sugary Cereal			Sum of Number of children	Eats Kooldog Sugary Cereal	
4	0	1	Grand Total	Watches Saturday morning cartoons	0	1	Grand Total	Watches Saturday morning cartoons	0	1
5	36	4	40	0	5.3	34.7	40	0	176.3	27.1
6	4	256	260	1	34.7	225.3	260	1	27.1	4.2

Sum the cell contributions to chi square to find chi square.

	I	J	K	L
	L7			fx =SUM(L4:L6)
1				
2				
3	Sum of Number of children	Eats Kooldog Sugary Cereal		
4	Watches Saturday morning cartoons	0	1	Grand Total
5	0	176.3	27.1	203.5
6	1	27.1	4.2	31.3
7	Grand Total	203.5	32.3	235

Use **CHISQ.DIST.RT()** to find the *p Value* of chi square.

	I	J	K	L	M	N
	M7				fx =CHISQ.DIST.RT(L7,1)	
1						
2						
3	Sum of Number of children	Eats Kooldog Sugary Cereal				
4	Watches Saturday morning cartoons	0	1	Grand Total		
5	0	176.3	27.1	203.5		
6	1	27.1	4.2	31.3		
7	Grand Total	203.5	32.3	235	5.45241E-53	p value

Based on sample evidence, the null hypothesis of independence is rejected. Eating cereal with added sugar is associated with Saturday morning cartoon viewing.

To see the association, copy rows **1** and **2** with the PivotTable and paste below the original, this time with formulas, using **Cntl+V**.

Change the cell counts to percents of row: right click a cell in the copied table, **Alt JTG Show Values As % Row Total**.

Type in the labels *No* and *Yes* and *Don't Eat Kooldogs* and *Eat Kooldog*.

	A	B	C	D
1				
2				
3	Sum of Number of children	Eats Kooldog Sugary Cereal ▾		
4	Watches Saturday morning cartoons ▾	0	1	Grand Total
5	0	36	4	40
6	1	4	256	260
7	Grand Total	40	260	300
8		Drop Report Filter Fields Here		
9				
10	Sum of Number of children	Eats Kooldog Sugary Cereal ▾		
11	Watches Saturday morning cartoons ▾	Don't Eat Kooldoggs	Eat Kooldoggs	Grand Total
12	No	90%	10%	100%
13	Yes	2%	98%	100%
14	Grand Total	13%	87%	100%

Make a **PivotChart** with shortcuts **Alt JT C**hart to see the association.

Choose a design and style, add axes titles, and a chart title that summarizes your conclusion.

Lab 7 Skype Appeal

Following the launch of Google's Android phone, rumors surfaced that Google is considering a joint venture with Skype. Skype boasts 330 million users worldwide. Google management believes that Skype appeals most to younger consumers, who make long distance calls with Skype, instead of cell phones or land lines.

Google conducted a survey of 101 randomly chosen consumers, from ages 14–65. Consumers were asked which they relied on most for long distance: (1) cell phone, (2) Skype, or (3) landline.

A crosstabulation of the responses is in **Lab 7 Skype Appeal.xls**:

1. Make a PivotTable of long distance users by age and type of phone.

   Copy the PivotTable and paste in next to the original, and then change cell counts to percentages of row.

   What percentage of long distance users surveyed rely on Skype? _____%

   What percentage of 18- to 21-year-olds rely on Skype? _____%

2. Make a table next to the second with the *expected* number of long distance users by age and type of phone *given no association* between age and type of phone:

   expected count in row i column j = number in row i × percent in column j

   $$e_{ij} = n_i p_j$$

   or

   expected users in age segment i of phone j = number in age segment i × percent who use phone j.

   How many of the fifteen 18- to 21-year-olds would you expect to rely on Skype if long distance phone choice is *not* associated with age? _____

3. Mark (X) cells that are *sparse*.

   Group (circle) age segments so that no cells are sparse.

   Update your *expected* cell counts.

Age	Cell phone	Landline	Skype
14–17			
18–21			
22–29			
30–39			
40–49			
50–59			
60–65			

4. Make a table of cell contributions to chi square: $(e_{ij} - n_{ij})^2/e_{ij}5.$

5. Find the *p Value* for your chi square with __ ($=$(rows $-$ 1) $\times$ (columns $-$ 1)) *df*: _____

6. Is choice of long distance phone type dependent on age? Y or N

7. *Type* of phone most dependent on age: _____

8. Phone *type* choices depend most on *age segment*: _____

9. Make a PivotChart to illustrate your results.

10. Which *age segment*(s) is more likely than average to rely on *Skype*? _____

Assignment 7-1 747s and Jets[3]

Boeing Aircraft Company management believes that demand for particular types of aircraft is associated with a particular global region across their three largest markets, North America, Europe, and China. To better plan and set strategy, they have asked you to identify region(s) where demand is uniquely strong for 747s and for regional jets.

Assignment 7-1 JETS747.xls contains Boeing's actual and projected deliveries 2005–2024 of each type of aircraft in each of the three regions.

Use contingency analysis to test the hypothesis that *demand* for a particular aircraft is associated with *global region*.

If the association is significant, explain the nature of the association.

Include a PivotChart and explain what it illustrates.

Assignment 7-2 Fit Matters

Procter & Gamble management would like to know whether intent to try their new preemie diaper concept is associated with the importance of fit. If Likely Triers value fit more than Unlikely Triers, fit could be emphasized in advertisements.

Assignment 7-2 Fit Matters.xls contains data from a concept test of 97 mothers of preemie diapers, including trial *Intention* and *Fit Importance*, measured on a 9 point scale. You may decide to combine categories.

Use contingency analysis to test the hypothesis that *intent* to try is associated with the *importance of fit*.

If the association is significant, explain the nature of the association.

Include a PivotChart and explain what it illustrates.

---

[3]This case is a hypothetical scenario using actual data.

Assignment 7-3 Allied Airlines

Rolls-Royce management has observed the growth in commercial airline alliances. Airline companies that are allied tend to purchase the same aircraft. Management would like to know whether alliance is associated with global region.

Data including the number of allied airline companies, *Allied*, and *Global Region* are contained in **Assignment 7-3 Allied Airlines.xls**. You may decide to combine global regions.

Use contingency analysis to test the hypothesis of association between *alliance* and *global region*.

If the association is significant, describe the nature of the association.

Include a PivotChart and explain what it illustrates.

Assignment 7-4 Netbooks in Color

Dell managers want to know whether college students' preferences for light weight netbooks and wide color selection are associated with major. Dell's netbook is lighter than many competing netbooks and comes in more colors than any other netbook. Managers believe that light weight and wide choice of colors may appeal to Arts & Sciences and Commerce students more than to Engineering students, which would give Dell an advantage to be promoted in those segments.

A sample of netbook and iPad owners was drawn from each of three schools on the UVA campus, Commerce, Arts & Sciences, and Engineering. Netbook or iPad brands owned by students were recorded. Those data, with number of *colors* available and *weight* are in **netbooks.xls**.

Determine whether preference for light weight and variety of colors are associated with college major.

1. State the hypotheses that you are testing.

2. What are your conclusions? (Include the statistical tests that you used to form your conclusions.)

3. What is the probability that
    a. A netbook or iPad owner will own a light weight brand?
    b. An Arts & Science student will own a light weight brand?
    c. A Commerce student will own a light weight brand?
    d. An Engineering student will own a light weight brand?

4. What is the probability that
   a. A netbook or iPad buyer will own a brand available in at least six colors?
   b. An Arts & Science student will own a brand available in at least six colors?
   c. A Commerce student will own a brand available in at least six colors?
   d. An Engineering student will own a brand available in at least six colors?

A Dell Intern believes that conclusions may differ if only netbooks owners are considered, excluding the unique segment of iPad owners.

5. Repeat your analyses excluding iPad owners. Summarize your conclusions, including the statistics that you used.

6. Illustrate your netbook results (excluding iPad owners) with PivotCharts.

Case 7-1 Hybrids for American Car

Rising gas prices and environmental concerns have led some customers to switch to hybrid cars. American Car (AC) offers two hybrids, AC sapphire and AC durado, a full size SUV and a pickup. AC executives believe that with their hybrid SUV and pickup, they will be able to attract loyal AC customers who desire a hybrid.

AC offers no hybrid sedans or coupes. Major competitors, Ford, Toyota, and Honda, offer hybrid sedans and coupes. Shawn Green, AC division head, is worried that customers who were driving sedans, coupes, or wagons may not want a truck or an SUV. They might switch from AC to Ford, Toyota, or Honda in order to purchase a hybrid car.

To investigate further, Mr. Green commissioned a survey of car buyers. The new car purchases of a representative random sample of 4,000 buyers were sorted into eight groups, based on the type of car they had owned and *Traded* (Prestige Sport, Compact SUV, Large and Full size SUV) and whether they bought *Hybrid* or Conventional.

These data are in **Case 7-1 Hybrid.xls**. The number of *Buyers* indicates popularity of each *Traded*, *Hybrid* combination.

Conduct contingency analysis with these data to determine whether *choice of hybrid vehicles* depends on *type of vehicle owned previously*.
Specifically,

1. Is there an association between the *type of car owned and Traded* and *choice of a Hybrid* instead of a Conventional car?

2. What is the probability that a new car buyer will choose a *hybrid*?

3. How likely is each of the segments to switch to hybrids?

4. Illustrate your results with a PivotChart. Include a bottom line title.

5. What are the implications of results for American Car Division and your advice to Mr. Green?

Case 7-2 Tony's GREAT Advertising

Kellogg spends a hefty proportion of its advertising budget to expose children to ads for sweetened cereal on Saturday mornings. Kellogg brand ads feature cartoon hero characters similar to the cartoon hero characters that children watch on Saturday morning shows. This following press release is an example.

Advertising Age, Dec 6, 2004 v75 i49 p1
Kellogg Pounces on Toddlers; Tiger Power to Wrest Tot Monopoly Away from General Mills' $500M Cheerios brand. (News) *Stephanie Thompson.*

Byline: STEPHANIE THOMPSON

In the first serious challenge to General Mills' $500 million Cheerios juggernaut, Kellogg is launching a toddler cereal dubbed Tiger Power.

The cereal, to arrive on shelves in January, will be endorsed by none other than Frosted Flakes icon Tony the Tiger and will be "one of our biggest launches next year," according to Kellogg spokeswoman Jenny Enochson. Kellogg will position the cereal-high in calcium, fiber and protein-as "food to grow" for the 2-to-5 set in a mom-targeted roughly $20 million TV and print campaign that begins in March from Publicis Groupe's Leo Burnett, Chicago.

Cereal category leader Kellogg is banking on Tiger Power's nutritional profile as well as the friendly face of its tiger icon, a new shape and a supposed "great taste with or without milk" to make a big showing in take-along treats for tots.

Kellogg spent $7.3 million on Frosted Flakes in 2003 and $7 million on the brand for January through July of this year.

Tony Grate, the brand manager for Frosted Flakes would like to know whether there is an association between Saturday morning cartoon viewing and consumption of his brand.

The Saturday morning TV viewing behaviors, *Saturday Morning Cartoons*, and consumption of Frosted Flakes, *Frosted Flake Eater*, are contained in **Case 7-2 Frosted Flakes.xls**. A random sample of 300 children aged 2–5 were sorted into four groups based on whether each watches at least 3 h of television on Saturday morning at least twice a month and whether each consumes Frosted Flakes at least twice a week. The number of *Children* indicates popularity of each *Saturday Morning Cartoons*, *Frosted Flake Eater* combination.[4]

1. Is there an association between watching *Saturday Morning Cartoons* and consumption of Frosted Flakes?

2. What is the probability that a *cartoon watcher* consumes Frosted Flakes?

3. How likely is each segment to consume Frosted Flakes?

4. Illustrate your results with a properly labeled PivotChart. Include a bottom line title.

5. What are the implications of results for Tony Grate?

---

[4]These data are fictitious, although designed to reflect a realistic scenario.

Chapter 8
Building Multiple Regression Models

Models are used to accomplish two complementary goals: *identification of key drivers of performance* and *prediction of performance under alternative scenarios*. The variables selected affect the explanatory accuracy and power of models, as well as forecasting precision. In this chapter, the focus is on variable selection, the first step in the process used to build powerful and accurate multiple regression models.

Multiple regression offers a major advantage over simple regression. Multiple regression enables us to account for the joint impact of multiple drivers. Accounting for the influence of multiple drivers provides a truer estimate of the impact of each one individually. In real world situations, multiple drivers together influence performance. Looking at just one driver, as we do with simple regression, we are very likely to conclude that its impact is much greater than it actually is. A single driver takes the credit for the joint influence of multiple drivers working together. For this reason, multiple regression provides a clearer picture of influence.

We use logic to choose variables initially. Some of the variables that logically belong in a model may be insignificant, either because they truly have no impact or because their influence is part of the joint influence of a correlated set of predictors that together drive performance. *Multicollinear* predictors create the illusion that important variables are insignificant. *Partial F test(s)* are used to decide whether seemingly insignificant variables contribute to variance explained. Using *partial F* tests does not cure multicollinearity, but acknowledges its presence and helps us assess the incremental worth of variables that may be redundant or insignificant.

If an insignificant predictor adds no explanatory power, it is removed from the model. It is either not a performance driver or it is a redundant driver, because other variables reflect the same driving dimension. Correlations help to distinguish whether it is multicollinearity that is producing insignificance for a variable.

8.1 Multiple Regression Models Identify Drivers and Forecast

Multiple regression models are used to achieve two complementary goals: identification of key *drivers* of performance and prediction of performance under alternative scenarios. This prediction can be either what would have happened had an alternate course of action been taken or what can be expected to happen under alternative scenarios in the future.

Decision makers want to know, given uncontrollable external influences, which controllable variables make a difference in performance. We also want to know the nature and extent of each of the influences when considered together with the full set of important influences. A multiple regression model will provide this information.

C. Fraser, *Business Statistics for Competitive Advantage with Excel 2010: Basics, Model Building, and Cases*, DOI 10.1007/978-1-4419-9857-6_8, © Springer Science+Business Media, LLC 2012

Once key drivers of performance have been identified, a model can be used to compare performance predictions under alternative scenarios. This *sensitivity analysis* allows managers to compare expected performance levels and to make better decisions.

8.2 Use Your Logic to Choose Model Components

The first step in model building happens before looking at data or using software. Using logic, personal experience, and others' experiences, we first decide which of the potential influences ought to be included in a model. From the set of variables with available data, which could reasonably be expected to influence performance? In most cases, a reason is needed for including each independent variable in a model. Independent variables tend to be related to each other in our correlated world, and models are unnecessarily complicated if variables that don't logically affect the dependent performance variable are included. This complication from correlated predictors, *multicollinearity*, is explored later in this chapter.

> *Example 8.1 Sakura Motors Quest for Cleaner Cars*
> The new product development group at Sakura Motors is in the midst of designing a new line of cars that will offer reduced greenhouse gas emissions for sale to drivers in global markets where air pollution is a major concern. They expect to develop a car that will emit only 5 tons of greenhouse gases per year.
> What car characteristics drive emissions? The management team believes that smaller, lighter cars with smaller, more fuel efficient engines will be cleaner. The US government publishes data on the fuel economy of car models sold in the United States (fueleconomy.gov), which includes *manufacturer*, *model*, engine size (*cylinders*), and gas mileage (*MPG*) for each category of car. This data source also includes *emissions* of tons of greenhouse gases per year. A second database, consumerreports.org, provides data on acceleration in *seconds* to go from 0 to 60 miles per hour, which reflects car model sluggishness, and two measures of size, *passengers* and curb *weight*. Management believes that responsiveness and size may have to be sacrificed to build a cleaner car.

The multiple linear regression model of *emissions* will include the car characteristics, *miles per gallon (MPG)*, *seconds* to accelerate from 0 to 60, *horsepower*, *liters*, *cylinders*, *passenger* capacity, and weight in *pounds (K)*, each thought to drive *emissions*:

$$\hat{emissions}_i = b_0 + b_1 MPG_i + b_2 seconds_i + b_3 horsepower_i + b_4 liters_i + b_5 passengers_i$$

$$+ b_6 pounds_i + b_6 cylinders_i + b_7 liters_i \,,$$

where $emi\hat{s}sions_i$ is the expected tons of annual emissions of the ith car model; b_0 is the intercept indicating expected emissions if *MPG, seconds, pounds (K), passengers, horsepower, cylinders,* and *liters* were zero; $b_1, b_2, b_3, b_4, b_5, b_6, b_7$ are the regression coefficient estimates indicating the expected marginal impact on emissions of a unit change in each car characteristic when other characteristics are at average levels; and $MPG_i, seconds_i, horsepower_i, cylinders_i, liters_i, passengers_i, pounds (K)_i$ are characteristics of the ith car model.

When more than one independent variable is included in a linear regression, the coefficient estimates, or parameter estimates, are *marginal*. They estimate the marginal impact of each predictor on performance, given average levels of each of the other predictors.

The new product development team asked the model builder to choose a sample of car models that represents extremes of emissions, worst and best. Thirty-five car models were included in the sample. These included imported and domestic cars, subcompacts, compacts, intermediates, full size sedans, wagons, SUVs, and pickups. Within this set there are considerable differences in all of the car characteristics, as shown in Fig. 8.1.

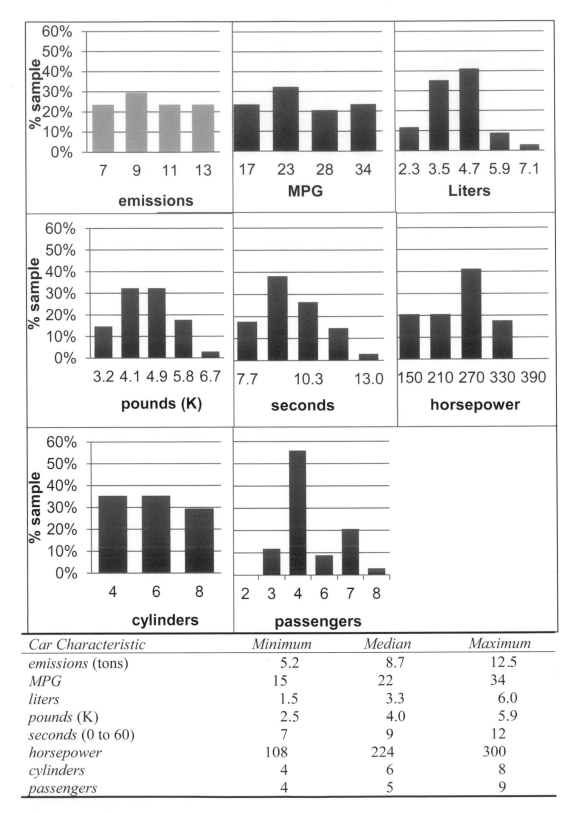

Car Characteristic	Minimum	Median	Maximum
emissions (tons)	5.2	8.7	12.5
MPG	15	22	34
liters	1.5	3.3	6.0
pounds (K)	2.5	4.0	5.9
seconds (0 to 60)	7	9	12
horsepower	108	224	300
cylinders	4	6	8
passengers	4	5	9

Fig. 8.1 Car characteristics in the Sakura Motors sample

8.3 Multicollinear Variables Are Likely When Few Variable Combinations Are Popular in a Sample

Inasmuch as these data come directly from the set of cars actually available on the market, many characteristic combinations do not exist. For example, there is no car with a 1.5 liter engine that weighs 4,000 pounds. The seven car characteristics are related to each other and come in particular combinations in existing cars. We are knowingly introducing correlated independent variables, also called *multicollinear* independent variables, into our model, because the characteristic combinations that are not represented do not exist.

Results from Excel are shown in Table 8.1.

Table 8.1 Multiple linear regression of emissions with seven car characteristics

SUMMARY OUTPUT					
Regression statistics					
R Square		.928			
Standard error		.644			
Observations		34			
ANOVA	*df*	*SS*	*MS*	*F*	*Significance F*
Regression	7	138	19.8	47.7	.0001
Residual	26	11	.4		
Total	33	149			

R Square is .928, or 93%, indicating that, *together*, variation in the seven car characteristics accounts for 93% of the variation in emissions. The *standard error* is .64, which indicates that forecasts of emissions would be within 1.3 tons of average actual emissions for a particular car configuration.

8.4 *F* Tests the Joint Significance of the Set of Independent Variables

F tests the null hypothesis that *R Square* is 0%, or, equivalently, that all of the coefficients are zero:

$$H_0: R\ Square = 0$$

versus

$$H_1: R\ Square > 0$$

or

H_0: All of the coefficients are equal to zero, $\beta_i = 0$.

versus H_1: At least one of the coefficients is not equal to zero.

The F test compares explained variation, SSR, per predictor, MSR (=SSR/regression df), with unexplained variation, SSE, for a given sample and model size, MSE (=SSE/residual df). The F statistic is compared to the F distribution with the same degrees of freedom. Figure 8.2 illustrates F distributions for 1, 2, 4, and 7 predictors with a sample of 30.

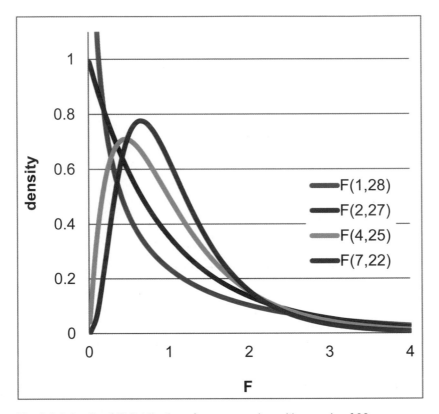

Fig. 8.2 A family of F distributions for a regression with sample of 30

The model F is 47.7, which lies to the extreme right of the $F_{7,26}$ distribution. The p $Value$, labeled $Significance$ F in Excel, is .0001, indicating that it is unlikely that we would observe these data patterns, were none of the seven car characteristics driving emissions. It may be that just one of the seven characteristics drives emissions or it may be that all seven are significant influences. With this set of seven predictors, some of the variations in $emissions$ have been explained.

8.5 Insignificant Parameter Estimates Signal Multicollinearity

To determine which of the seven car characteristics are significant drivers of emissions, we initially look at the significance of *t tests* of the individual regression parameter estimates. A *t* statistic in multiple regression is used to test the hypothesis that a marginal coefficient is zero.

When we have no information about the direction of influence, a two tail test of each marginal slope is used:

$$H_0: \beta_t = 0$$

versus $$\qquad\qquad\qquad H_1: \beta_t \neq 0.$$

In the more likely case that, when, from theory or experience, we know the likely direction of influence, a one tail test is used. When the suspected direction of influence is positive, the null and alternate hypotheses are

$$H_0: \beta_t \leq 0$$

versus $$\qquad\qquad\qquad H_1: \beta_t > 0.$$

Conversely, when the expected direction of influence is negative the hypotheses are

$$H_0: \beta_t \geq 0$$

versus $$\qquad\qquad\qquad H_1: \beta_t < 0.$$

Excel provides a two tail *t* statistic for each marginal slope by calculating the number of standard errors of each marginal slope from zero:

$$t_{residual\ df,i} = \beta_t / s_{\beta_i}.$$

Notice that a *t* statistic of a marginal slope in multiple regression is compared with the *t* distribution for the residual degrees of freedom. For each predictor in the model, we lose 1 degree of freedom. Excel provides the corresponding *p Value* for the *two tail t test* of each marginal slope. In the case that we want to use a one tail test, the *p Value* is divided by 2. The *t* distribution used in the *emissions* model, with 26 degrees of freedom, is shown in Fig. 8.3.

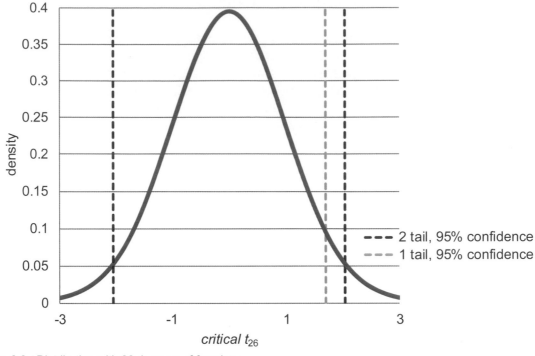

Fig. 8.3 *t* Distribution with 26 degrees of freedom

In the emissions model, Sakura analysts were confident that the impact of *MPG* on emissions ought to be negative, and that each of the influences of *horsepower*, *cylinders*, *liters*, *weight*, and *passengers* on *emissions* ought to be positive. For these six potential drivers, *one tail t tests* could be used. Sakura managers were not sure of the direction of influence of acceleration on *emissions*, and so a *two tail test* would be used for the *seconds* slope.

Table 8.2 Marginal slopes and their *t* tests

	Coefficients	Standard error	t Stat	p Value	one tail p Value[a]
Intercept	9.2	1.90	4.8	<.0001	
Seconds	.23	.099	2.3	.03	
MPG	−.23	.037	−6.2	.0001	<.0001
Liters	.41	.29	1.4	.17	.08
Cylinders	−.035	.19	−.2	.85	.43
Horsepower	−.00052	.0037	−.1	.89	.44
Pounds (K)	.54	.30	1.9	.08	.04
Passengers	−.086	.12	−.7	.48	.24

[a] *p* Values corresponding to one tail tests are not provided by Excel and have been added here

Excel t tests of the marginal slopes, shown in Table 8.2, suggest that only *seconds* to accelerate 0–60, *MPG*, and *pounds* drive differences in emissions. Neither engine size characteristics, *horsepower*, *cylinders*, and *liters*, nor car size characteristic, *passengers*, appears to influence *emissions*. Coefficient estimates for *cylinders*, *horsepower*, and *passengers* have the "wrong signs." Cars with more cylinders, larger, more powerful engines, and more passenger capacity are expected to emit more pollutants. These are surprising and nonintuitive results.

When predictors that ought to be significant drivers appear to be insignificant or when parameter estimates are of the wrong sign, we suspect *multicollinearity*. Multicollinearity, the correlation between predictors, thwarts driver identification. When the independent variables are themselves related, they jointly influence performance. It is difficult to tell which individual variables are more important drivers, since they vary together. Because of their correlation, the standard errors s_{b_i} of the marginal slope coefficient estimates, b_i, are inflated. Their influence is joint, and we are not very certain of each true influence in the population. The confidence intervals of the true partial slopes are large, because these are multiples of the standard errors of the partial slope estimates. Individual predictors seem to be insignificant although they may be truly significant. In some cases, coefficient signs may be "wrong."

8.6 Combine or Eliminate Collinear Predictors

We have two remedies for multicollinearity cloudiness:

- We can combine correlated variables.
- We can eliminate variables that are contributing redundant information.

Correlations between the predictors reveal that *horsepower*, *cylinders*, and *liters* are highly correlated with each other ($r_{horsepower,liters} = .76; r_{cylinders,liters} = .92; r_{cylinders,horsepower} = .77$) and with *seconds*, *MPG*, *pounds (K)*, and *passengers*, as shown in Table 8.3.

Table 8.3 Pairwise correlations between predictors

	MPG	Seconds	Liters	Horsepower	Cylinders	Pounds (K)
MPG	1					
Seconds	−.05	1				
Liters	−.81	−.17	1			
Horsepower	−.53	−.36	.76	1		
Cylinders	−.74	−.19	.92	.77	1	
Pounds (K)	−.77	−.01	.84	.72	.81	1
Passengers	−.53	−.05	.59	.55	.60	.70

Some of the correlated predictors can be eliminated, assuming that several reflect a common dimension. If *liters*, *horsepower*, and *cylinders* each reflect engine size, two are possibly redundant and may be represented by the third. The alternative is to combine correlated predictors, either by constructing an index from a weighted average of the correlated predictors or by forming ratios of pairs of correlated predictors.

An index of engine size could be made from a weighted average of *liters*, *cylinders*, and *horsepower*. *Factor analysis* is a statistical procedure that would choose the weights to form such an index. The challenge associated with use of an index is in its interpretation. Sakura managers need to know how much difference particular car characteristics make, and they may not be satisfied knowing that an *engine size index* influences *emissions*. Factor analysis is beyond the scope of this text, but does enable construction of indices from correlated predictors.

Ratios of correlated predictors are used when they make intuitive sense. For example, economic models sometimes use the ratio of *GDP* and *population* to make *GDP per capita*, an intuitively appealing measure of personal wealth.

We eliminate the seemingly redundant predictors to build a model for Sakura, although combining correlated predictors would be an acceptable alternative. This does not eliminate multicollinearity but reduces multicollinearity by removing correlated predictors.

Cars with larger engines have more power. *Horsepower* and *cylinders* are removed from the model, expecting that they are redundant measures of engine size. If explanatory power is not substantially reduced, we can designate *liters* as the measure of engine size that reflects *cylinders* and *horsepower*. *Liters* is the preferred predictor to retain, because its coefficient sign is as expected, whereas coefficient signs for both *cylinders* and *horsepower* are "wrong."

Passenger capacity is highly correlated with weight (*pounds (K)*): $r_{passengers, pounds} = .70$. Larger, more spacious cars weigh more. *Passengers* will be removed from the model, expecting that it is a redundant measure of car size. If explanatory power is not sacrificed, *pounds (K)* will reflect car size. *Pounds (K)* is chosen to represent car size, since its coefficient sign is as expected and significant, while the sign for the *passengers* coefficient is "wrong."

The revised *partial* model becomes

$$em\hat{i}ssions_i = b_0 + b_1 MPG_i + b_2 seconds_i + b_3 liters_i + b_4 pounds(K)_i.$$

Regression results using this *partial* model are shown in Table 8.4.

Table 8.4 Regression of emissions with four car characteristics

SUMMARY OUTPUT

Regression statistics

R Square		.926			
Standard error		.617			
Observations		34			
ANOVA	*df*	*SS*	*MS*	*F*	*Significance F*
Regression	4	138	34.5	90.8	.0000
Residual	29	11	.4		
Total	33	149			

	Coefficients	*Standard error*	*t Stat*	*p Value*	one tail *p Value*
Intercept	9.0	1.8	5.0	<.0001	
Seconds	.24	.087	2.8	.01	
MPG	−.23	.034	−6.7	<.0001	<.0001
Liters	.36	.20	1.8	.08	.04
Pounds (K)	.43	.24	1.8	.08	.04

The *partial* model *R Square*, .926, is less than one percentage point lower than the *full* model *R Square*, .929. With just four of the seven car characteristics, we can account for 93% of the variation in emissions. Little explanatory power has been lost, and the standard error has dropped from .644 to .617, reducing the margin of error in forecasts by 5% (=(1.32−1.26)/1.32). Model *F* is significant, suggesting that one or more of the four predictors influence emissions. All four of the predictors are significant drivers. All coefficient estimates have correct signs. As was the case in the full model, *emissions* are lower for smaller responsive cars with higher fuel economy. By reducing multicollinearity, it can now also be concluded that *emissions* are lower for cars with smaller engines.

8.7 *Partial F* Tests the Significance of Changes in Model Power

Can *horsepower*, *cylinders*, and *passengers* be eliminated without loss of explanatory and predictive power? Multicollinearity is reduced when we remove variables, increasing the certainty of parameter estimates for variables left in the model. With this small change in *R Square* (less than 1%), we do not need to test the significance of the change in *R Square*. When *R Square* does change by more than 1%, we use a *Partial F* test to assess the significance of the decline. The *Partial F* tests the hypothesis that the explanatory power of the *partial* model equals that of the *full* model:

$$H_0: R\ Square_{partial} = R\ Square_{full}$$

versus

$$H_1: R\ Square_{partial} < R\ Square_{full}$$

or

H_0: The variables removed add no explanatory power.

versus H_1: The variables removed add explanatory power.

If the null hypothesis is rejected, we must return the variables removed, because they contribute explanatory power to the model. If the null hypothesis cannot be rejected, the variables can be removed, because they add no additional explanatory power.

The *Partial F* statistic compares the change in *R Square per variable removed* to unexplained variation in the *full* model *per residual df* in the *full* model:

$$F_{k-g,N-1-k} = \frac{(R\,\text{square}_{full} - R\,\text{square}_{partial})/g}{(1 - R\,\text{square}_{full})/(N-1-k)},$$

where $R\,\text{square}_{full}$ is *R Square* from the larger model before variables are removed, $R\,Square_{partial}$ is *R Square* from the smaller model after variables are removed, g is the number of predictors removed from the larger model, N is the sample size, k is the number of predictors in the larger model, and $(N - 1 - k)$ is the *residual* degrees of freedom (*df*) from the larger model.

A larger change in *R Square* is expected if a larger number of variables are removed, so the change comparison is per predictor removed, g.

In the **Sakura Motors** model, *Partial F* to test the significance of incremental explanatory power of the three variables removed, *horsepower*, *cylinders*, and *passengers*, is

$$F_{3,26} = \frac{(.928 - .926)/3}{(1 - .928)/(34 - 1 - 7)} = \frac{.0017/3}{.072/26} = \frac{.00058}{.0028} = .21,$$

Partial F Significance $= .89$

For these degrees of freedom, 3 and 26, an *F* value of .21 includes only 11% ($=(1 - .89)\%$) of the *F* distribution area and is smaller than the 95% required for significance of .05. *R Square* did not change significantly when the three presumably redundant variables were eliminated. *Horsepower*, *cylinders*, and *passengers* do not add sufficient explanatory power to the model and can be removed. The partial model now becomes the full model.

The final multiple linear regression model of emissions is

$$\widehat{emissions}_i = 9.0^a + .24^a\,second - .23^a\,MPG_i + .36^a\,Liters_i + .43^a\,pounds \text{ (K)}$$

$R\,\text{square}^a = .93$

[a]*Significant at a .05 level or better.*

To determine whether our model satisfies the assumptions of linear regression, the distribution of residuals is examined, just as with a simple regression model. In Fig. 8.4, the residuals are approximately *Normal*.

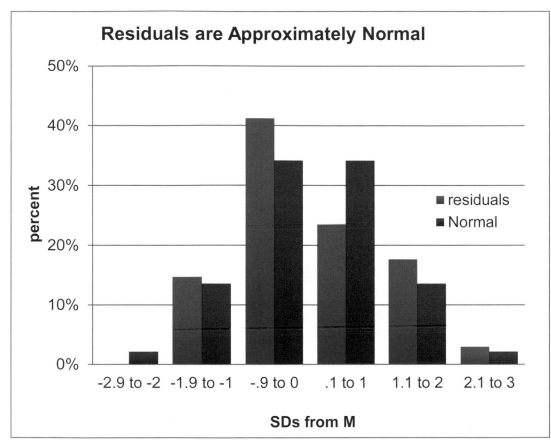

Fig. 8.4 Distribution of residuals

8.8 Sensitivity Analysis Quantifies the Marginal Impact of Drivers

We want to compare influences of the significant drivers to identify those making the greatest difference. We forecast emissions at average levels of each of the car characteristics. Then, we compare forecasts at minimum and maximum levels of each, holding the other three constant at mean levels. The sensitivity analysis is summarized in Table 8.5.

Table 8.5 Emissions response to car characteristics

MPG	Seconds to accelerate 0–60	Pounds (K)	Liters	Expected emissions	Improvement (reduction) in expected emissions
15.0	9.0	3.5	4.1	10.7	
33.5	9.0	3.5	4.1	6.5	4.2
22.6	**11.9**	3.5	4.1	9.7	
22.6	**6.7**	3.5	4.1	8.4	1.2
22.6	9.0	**6.0**	4.1	9.9	
22.6	9.0	**1.5**	4.1	8.3	1.6
22.6	9.0	3.5	**5.9**	9.8	
22.6	9.0	3.5	**2.5**	8.3	1.5

MPG: Within a representative range of values for each of the car characteristics, fuel economy makes the largest difference in emissions, as shown in Fig. 8.5.

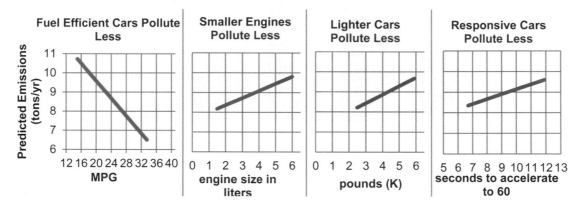

Fig.8.5 Predicted emissions by car characteristic

Improving fuel economy by 19 MPG, the sample range, is associated with an expected reduction in emissions of 4.2 tons per year. This is a large improvement, although not enough alone to meet the 5.0 tons per year goal. Fuel economy improvements will need to be made in conjunction with improvements in one or more of the other car characteristics.

The linear model suggests that improving average fuel economy by 4 MPG, from 25 to 29, would produce an expected average improvement in emissions of about 1 ton (.60–1.20 tons) per year, assuming other car characteristics were at mean levels, which is shown in Fig. 8.6:

$$\Delta MPG[b_{MPG} - 2s_{b_{MPG}}] \le \Delta MPG\beta_{MPG} \le \Delta MPG[b_{MPG} - 2s_{b_{MPG}}]$$

$$(29 - 25)[-.23 - 2(.034)] \le (29 - 25)\beta_{MPG} \le (29 - 25)[.23 + 2(.034)]$$

$$(4)(.30) \le (4)\beta_{MPG} \le (4)(.16)$$

$$-1.20 \le (4)\beta_{MPG} \le -.60.$$

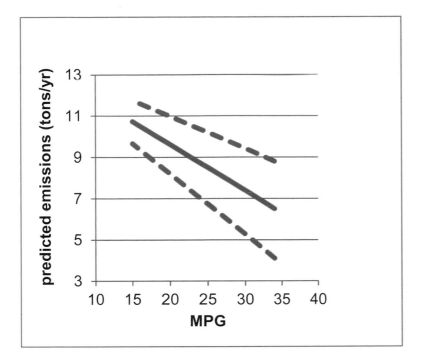

Fig. 8.6 95% prediction intervals by MPG

Pounds (K) and Liters: Reducing car weight by 2,500 pounds or reducing engine size by 3 liters improves expected emissions by about 1 ton per year. Even the combination of a lighter car with a smaller engine is probably not enough to reach the emissions goal of 5 tons per year. In combination with fuel economy improvements, either car weight or engine size improvements could make the goal attainable.

Seconds: Improving car responsiveness by reducing the time to accelerate from 0 to 60 by 4 seconds could improve expected emissions about 1 ton. Combined with any of the other car characteristics, responsiveness could help Sakura achieve their emissions goal, although acceleration alone makes the least difference in emissions.

The model provides clear indications for the new product development team. To improve emissions, they will need to design more responsive, lighter weight cars with smaller engines and superior fuel economy. Changing just one car characteristic will not be enough to meet the goal of 5 tons per year.

The Quantitative Analysis Director summarized model results in the following memo to Sakura Management.

MEMO

Re: Light, responsive, fuel efficient cars with smaller engines are cleanest
To: Sakura Product Development Director
From: Benjamin Nowak, Quantitative Analysis Director
Date: June 2010

Improvements in gas mileage and responsiveness, with reductions in weight, passenger capacity or engine size, will allow Sakura to achieve the emissions target of 5 tons per year.

A regression model of emissions was built from a representative sample of 34 diverse car models, considering fuel economy, acceleration, engine size, and car size.

Model results: Differences in fuel economy, weight, engine size, and acceleration account for 93% of the variation in emissions. Forecasts from car characteristics are expected to be no further than 1.2 tons from actual average emissions for a particular car.

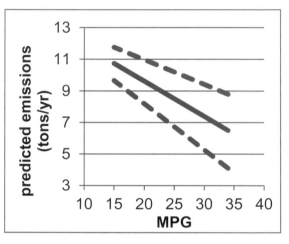

Fuel economy is the most powerful driver of emissions. Increasing gas mileage by four MPG is expected to reduce annual emissions by .6–1.2 tons per year.

Car and engine size and responsiveness matter, but make less of a difference. A 2.5 thousand pound reduction in weight is expected to reduce emissions by about 1 ton per year. Reducing engine size by 3 liters reduces expected emissions by about 1 ton per year.

$$emi\hat{s}sions_i = 9.0^a + .24^a seconds_i - .23^a MPG_i$$
$$+ .36^a liters_i + .43^a pounds(K)_i$$

$RSquare^a = .93$

[a]*Significant at a .05 level or better.*

Passenger capacity, horsepower, and cylinders matter, but were not included in the model, because these characteristics are similar to weight and engine size in liters. Reduction in passenger capacity, horsepower, or cylinders is expected to reduce emissions.

Responsiveness makes the smallest difference in emissions. Reducing acceleration from 0 to 60 by 4 seconds would improve emissions about 1 ton per year.

Conclusions: Fortunately, cleaner cars are also more fuel efficient and more responsive. This will allow Sakura to design cleaner models without sacrificing responsiveness. Improvements in fuel economy and responsiveness, with reductions in weight or engine size, will enable Sakura to meet the emissions target of 5 tons per year. To achieve emissions of 5 tons per year within existing characteristic ranges, more than one car characteristic must be changed. For example, improvements in MPG to 34, engine size to 1.5 liters, and weight to 2.5K pounds produce expected emissions of 5 tons per year.

Other considerations: Model results assume existing engine technology. With the development of cleaner, more fuel efficient, responsive technologies, even lower emissions could possibly be achieved.

8.9 Model Building Begins with Logic and Considers Multicollinearity

Novice model builders sometimes mistakenly think that the computer can choose those variables that belong in a model. Computers have no experience making decisions and can never replace decision makers' logic. (Have you ever tried holding a conversation with a computer?) The first step in superior model building is to use your head. Use logic and experience to identify independent variables that should influence the performance variable you are interested in explaining and forecasting. Both your height and GDP increased over the past 10 years. Given data on your annual height and annual GDP, the computer could churn out a significant parameter estimate relating variation in your height to variation in GDP (or variation in GDP to variation in your height). Decision makers must use their logic and experience to select model variables. Software will quantify and calibrate the influences that we know, from theory or experience, ought to exist.

It is a multicollinear world. Sets of variables together jointly influence performance. Using ratios of collinear predictors reduces multicollinearity. *Partial F tests* are used to confirm that eliminating redundancies does not reduce model power. Removing redundant predictors allows us to more accurately explain performance and forecast. Correlations are used to determine whether insignificant variables matter, but simply look as though they don't because of multicollinearity, or whether they simply do not matter.

From the logically sound set of variables, pruned to eliminate redundancies and reduce multicollinearity, we have a solid base for superior model building. To this we consider adding variables to account for seasonality or cyclicality in time series in Chap. 9 and the use of indicators to build in influences of segment differences, structural shifts, and shocks in Chap. 10. In Chap. 11, alternative nonlinear models are considered for situations where response is not constant.

Excel 8.1 Build and Fit a Multiple Linear Regression Model

Sakura Motors Quest for a Clean Car

Assist Sakura Motors in their quest for a less polluting car model, using data from bea.gov and consumerreports.org, which together provide information on individual car models. The dataset, **Excel 8.1 Sakura Motors.xls** contains data on 35 car models, representing US, European, and Asian manufacturers and a variety of sizes and styles.

Management is unsure which car characteristics influence *emissions*, but they suspect that fuel economy, *MPG*, acceleration capability (measured as *seconds* to accelerate from 0 to 60 mph), engine size, (*cylinders, liters, and horsepower*), car *passenger* capacity, and weight in *pounds (K)* may be significant influences. Smaller, lighter models with smaller, less powerful engines are expected to be cleanest.

Open the dataset and run multiple regression with the dependent variable *emissions* in **Input Y Range** and the independent variables, *MPG, seconds, cylinders, liters, horsepower, passengers*, and *pounds*, in the **Input X Range**.

Add a column for one tail *t tests* of *MPG, liters, horsepower, cylinders, pound (K)*, and *passengers*, and the fill in *p Values* by dividing Excel's *two tail p Values* by 2.

ANOVA							
	df	*SS*	*MS*	*F*	*Significance F*		
Regression	7	138.44	19.8	47.7	3E-13		
Residual	26	10.77	0.41				
Total	33	149.22					
	Coefficients	*Standard Error*	*t Stat*	*P-value*	*Lower 95%*	*Upper 95%*	*1 tail p value*
Intercept	9.16	1.90	4.8	0.0001	5.25	13.08	
MPG	-0.23	0.04	-6.2	0.0000	-0.30	-0.15	0.0000
seconds	0.23	0.10	2.3	0.0283	0.03	0.43	
liters	0.41	0.29	1.4	0.1698	-0.19	1.02	0.0849
pounds (K)	0.54	0.29	1.8	0.0762	-0.06	1.15	0.0381
horsepower	-0.0005	0.0037	-0.1	0.8889	-0.01	0.01	0.4444
cylinders	-0.035	0.188	-0.2	0.8535	-0.42	0.35	0.4267
passengers	-0.086	0.120	-0.7	0.4815	-0.33	0.16	0.2407

Multicollinearity symptoms: Although the model is significant (*Significance F* < .0001), only three of the car characteristics are significant (*p Value* <.05). We are not certain that liters, cylinders, passengers, and horsepower are influential, because their *p Values* >.05. Horsepower, cylinders, and passengers have "incorrect" negative signs. Cars with greater horsepower, more cylinders, and more passenger space ought to be bigger polluters. Together, the lack of significance of seemingly important predictors and the three sign reversals signal multicollinearity.

Look at the correlations to confirm suspicions that *liters, horsepower,* and *cylinders* are correlated (and together reflect car power) and that *pounds (K)* and *passengers* are correlated (and together reflect car size). This may allow elimination of two of the power variables and one of the size variables to reduce multicollinearity.

Run correlations between the car characteristics.

	MPG	seconds	liters	pounds (K)	horsepower	cylinders
MPG	1					
seconds	-0.049	1				
liters	-0.810	-0.171	1			
pounds (K)	-0.769	-0.013	0.835	1		
horsepower	-0.529	-0.363	**0.763**	0.718	1	
cylinders	-0.744	-0.189	**0.924**	0.808	**0.771**	1
passengers	-0.526	-0.049	0.593	**0.703**	**0.545**	0.602

Eliminating two of the three measures of power and one of the two measures of size will reduce multicollinearity.

Use Partial F to test significance of contribution to R Square.
Eliminate potentially redundant characteristics that appear to add little explanatory power. This does not mean that they are not important. More likely, they are closely related to other important characteristics and contribute redundant information. Characteristics with "wrong" signs in the full regression are removed first.

Run the partial model regression, changing the **Input X Range**, and add one tail *p Values*.

Regression Statistics							
Multiple R	0.962						
R Square	0.926						
Adjusted R	0.916						
Standard E	0.617						
Observatior	34						
ANOVA							
	df	*SS*	*MS*	*F*	*Significance F*		
Regression	4	138.2	34.5	90.8	6E-16		
Residual	29	11.0	0.38				
Total	33	149.2					
	Coefficients	Standard Error	t Stat	P-value	Lower 95%	Upper 95%	1 tail p value
Intercept	8.99	1.80	5.0	0.0000	5.30	12.68	
MPG	-0.23	0.034	-6.7	0.0000	-0.30	-0.16	0.0000
seconds	0.24	0.087	2.8	0.0100	0.062	0.42	
liters	0.36	0.20	1.8	0.0781	-0.043	0.76	0.039
pounds (K)	0.43	0.24	1.8	0.0806	-0.055	0.91	0.040

This partial model, with three fewer predictors, is significant, and all predictors are now significant with "correct" signs. The standard error is smaller than that in the full model, although *R Square* is lower.

Partial F compares reduction in *R Square*, per variable removed, to unexplained variation, divided by the *residual degrees of freedom* in the larger model.

	A	B	C
1	SUMMARY OUTPUT		
2			
3	Regression Statistics		
4	Multiple R	0.963233	Partial model RSquare
5	R Square	0.927817	0.92607731

Copy *R Square* in **B5** from the model (with only *MPG* and *seconds*) and paste it into the original full model output sheet.

Find the change in *R Square* due to removal of the three predictors.

	D5			f_x	=B5-C5
	A	B	C		D
1	SUMMARY OUTPUT				
2					
3	Regression Statistics				
4	Multiple R	0.963233	Partial model RSquare		change in RSquare
5	R Square	0.927817	0.92607731		0.001739917

Three variables were removed to build the partial regression model. This reduced the *residual dfs* by 3. Find the change in *R Square* per variable removed (3, the change in *residual dfs*), which will be the numerator of *Partial F*.

	E5			f_x	=D5/3	
	A	B	C	D		E
1	SUMMARY OUTPUT					
2						
3	Regression Statistics					
4	Multiple R	0.963233	Partial model RSquare	change in RSquare		per variable removed
5	R Square	0.927817	0.92607731	0.001739917		0.00058

For the denominator of the *Partial F* statistic, find the *Variation unexplained in the larger model*.

	F5			f_x	=1-B5		
	A	B	C	D		E	F
1	SUMMARY OUTPUT						
2							
3	Regression Statistics						
4	Multiple R	0.963233	Partial model RSquare	change in RSquare		per variable removed	variation not explained
5	R Square	0.927817	0.92607731	0.001739917		0.00058	0.072183

Find the proportion of *variation unexplained per residual df*, which is the denominator of the *Partial F* statistic.

	G5			f_x	=F5/B13		
	A	B	C	D	E	F	G
1	SUMMARY OUTPUT						
2							
3	*Regression Statistics*						
4	Multiple R	0.963233	Partial model RSquare	change in RSquare	per variable removed	variation unexplained	per residual df
5	R Square	0.927817	0.92607731	0.001739917	0.00058	0.072183	0.002776

Calculate the *partial F statistic* from the ratio of change in *R Square per variable removed* to *unexplained variation per residual df*.

	H5			f_x	=E5/G5			
	A	B	C	D	E	F	G	H
1	SUMMARY OUTPUT							
2								
3	*Regression Statistics*							
4	Multiple R	0.963233	Partial model RSquare	change in RSquare	per variable removed	variation unexplained	per residual df	partial F
5	R Square	0.927817	0.92607731	0.001739917	0.00058	0.072183	0.002776	0.208904

To find the level of significance of this *F* value, with 3 (variables omitted) and 26 (*residual df* in the larger model) degrees of freedom, use the Excel **F.DIST.RT(***F,numerator df, denominator df* **)** function.

	I5			f_x	=F.DIST.RT(H5,3,26)				
	A	B	C	D	E	F	G	H	I
1	SUMMARY OUTPUT								
2									
3	*Regression Statistics*								
4	Multiple R	0.963233	Partial model RSquare	change in RSquare	per variable removed	variation unexplained	per residual df	partial F	p value
5	R Square	0.927817	0.92607731	0.001739917	0.00058	0.072183	0.002776	0.208904	0.889306

The *Partial F p Value* is greater than the *critical p Value*, .05. The null hypothesis that the change in *R Square* from removal of three predictors is zero cannot be rejected. The three car characteristics removed were redundant and were not adding explanatory power to the model.

Look at residuals to check model assumptions. Excel gives us the residuals (predicted minus actual) in the regression output sheet.

To be sure that the model residuals are free of patterns and normally distributed, find skewness of the residuals and make a histogram of the residuals.

25	RESIDUAL OUTPUT					
26						
27	Observatior	cted emiss	Residuals	mean	standard deviation	sds from the mean (-3 to +3)
28	1	9.592942	0.107058	2.37718E-15	0.5781	-1.734
29	2	7.674988	0.225012			-1.156
30	3	7.921034	-0.22103			-0.578
31	4	11.59088	0.209116			2E-15
32	5	11.60328	-0.20328			0.5781
33	6	9.553118	-0.25312			1.1563
34	7	6.943534	1.356466			1.7344
35	8	12.54517	-0.54517			
60	33	5.454643	1.145357			
61	34	7.529488	-1.12949			
62			0.457325	skewness		

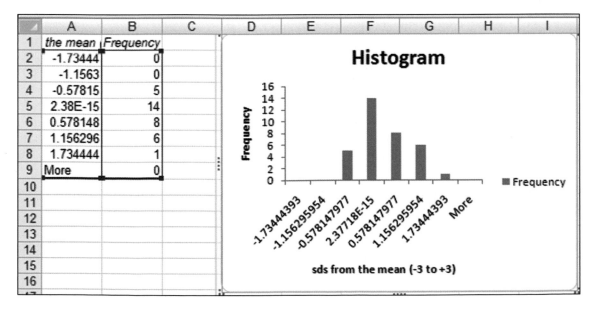

	A	B	C	D	E	F	G	H	I
1	the mean	Frequency							
2	-1.73444	0							
3	-1.1563	0							
4	-0.57815	5							
5	2.38E-15	14							
6	0.578148	8							
7	1.156296	6							
8	1.734444	1							
9	More	0							

The residuals are approximately Normal.

Excel 8.2 Use Sensitivity Analysis to Compare the Marginal Impacts of Drivers

For sensitivity analysis, identify a "low" and a "high" value for each of the four predictors, the minimum and maximum. For each, compare predictions given low and high values to find the range of response. To study marginal response to a predictor, vary only that predictor and set the remaining predictors at their mean values.

Find the sample mean for *MPG*, **Alt MUA**.

Find the sample maximum and minimum values using Excel functions **MAX(*array*)** and **MIN(*array*)**.

Select the three new cells, then fill in the sample mean, maximum, and minimum for *seconds*, *liters*, and *pounds (K)*, **Shift+>>>, Cntl+R**.

	C	D	E	F	G
	emissions	MPG	seconds	liters	pounds (K)
	6.4	29	8.8	3.5	4.2
mean		22.6	9.0	3.49	4.1
max		33.5	11.9	6	5.9
min		15	6.7	1.5	2.485

The benchmark or a "typical" car would achieve 22.6 *MPG*, accelerate from 0 to 60 in 9 *seconds* with a 3.5 liter engine, and would weigh 4.1 thousand pounds.

Comparing the difference in expected *emissions* when all but one driver are at mean levels allows us to isolate the impact of that driver. This reveals how relatively important each driver is, and which have the greater potential to reduce *emissions*. Within the existing range of car designs, a car could achieve the "best" gas mileage of 33.5 *MPG*, or it could have the worst gas mileage of 15 *MPG*. To find the impact of this range in gas mileage, make two new rows for hypothetical cars that are average in all other respects, but achieve either the minimum or the maximum gas mileage.

Add two new rows for hypothetical cars that are average in all respects except minimum or maximum *seconds*.

Add two new rows for hypothetical cars that are average in all respects except minimum or maximum *liters*.

Add two new rows for hypothetical cars that are average in all respects except minimum or maximum *pounds (K)*.

C	D	E	F	G
emissions	MPG	seconds	liters	pounds (K)
6.4	29	8.8	3.5	4.2
mean	22.6	9.0	3.49	4.1
max	33.5	11.9	6	5.9
min	15	6.7	1.5	2.485
best MPG	**33.5**	9	3.5	4.1
worst MPG	**15**	9	3.5	4.1
worst accel	23	**11.9**	3.5	4.1
best accel	23	**6.7**	3.5	4.1
largest engine	23	**9**	**6**	4.1
smallest engine	23	**9**	**1.5**	4.1
heaviest	23	9	3.5	**5.9**
lightest	23	9	3.5	**2.5**

J	K	L
passengers		Coefficients
7		8.9899904
5		-0.2283979
5		0.2395163
5		0.3605461
7		0.4269954

To find *emissions* predicted by the model for each hypothetical car, copy the coefficients from the regression output sheet **B16:B21** and paste into the Sakura sheet.

Use the regression equation formula to find *predicted emissions* using the car characteristic data and coefficient estimates.

Double click the lower right corner of the new cell to fill in the new column.

fx	=$I$2+$I$3*D2+$I$4*E2+$I$5*F2+$I$6*G2					
C	D	E	F	G	H	I
emissions	MPG	seconds	liters	pounds (K)	predicted emissions	Coefficients
9.7	20	8.2	3.5	4.555	9.6	8.99
7.9	24.5	6.7	3.2	3.565	7.7	-0.23
7.7	24	7.4	3	3.65	7.9	0.24
11.8	15.5	8.3	6	4.66	11.6	0.36
11.4	15.5	7.6	5.7	5.335	11.6	0.43

Select *predicted emissions* in the last data row and **Shift+down** through the new hypothetical car rows, then **Cntl+D** to fill in *predicted emissions* for the hypothetical cars.

C		D	E	F	G	H
emissions		MPG	seconds	liters	pounds (K)	predicted emissions
	6.4	29	8.8	3.5	4.2	7.5
mean		22.6	9.0	3.49	4.1	9.0
max		33.5	11.9	6	5.9	8.9
min		15	6.7	1.5	2.485	8.8
best MPG		**33.5**	9	3.5	4.1	6.5
worst MPG		**15**	9	3.5	4.1	10.7
worst accel		23	**11.9**	3.5	4.1	9.6
best accel		23	**6.7**	3.5	4.1	8.4
largest engine		23	9	**6**	4.1	9.8
smallest engine		23	9	**1.5**	4.1	8.2
heaviest		23	9	3.5	**5.9**	9.7
lightest		23	9	3.5	**2.5**	8.2

The difference between *predicted emissions* given maximum and minimum levels for a characteristic provide an estimate of the difference that a characteristic makes.

	6.6	33	6.9	2.4	3.475	5.5	difference
	6.4	29	8.8	3.5	4.2	7.5	between
mean		22.6	9.0	3.49	4.1	9.0	maximum
max		33.5	11.9	6	5.9	8.9	and
min		15	6.7	1.5	2.485	8.8	minimum
best MPG		33.5	9	3.5	4.1	6.5	
worst MPG		15	9	3.5	4.1	10.7	4.2
worst accel		23	11.9	3.5	4.1	9.6	
best accel		23	6.7	3.5	4.1	8.4	1.2
largest engine		23	9	6	4.1	9.8	
smallest engine		23	9	1.5	4.1	8.2	1.6
heaviest		23	9	3.5	5.9	9.7	
lightest		23	9	3.5	2.5	8.2	1.5

Scatterplots of marginal response: To see the impact of each driver, plot actual and predicted *emissions* of hypotheticals.

Focus on *MPG*.

Rearrange columns so that *MPG* and *Predicted* emissions are adjacent.
Cut *predicted emissions* column **K** and insert into column **E**.

Select *MPG* and *predicted emissions* of the two hypothetical rows and insert a scatterplot.
Choose a style and layout, and adjust axes.

emissions	MPG	predicted emissions	seconds	liters	pounds (K)	Coeffic
mean	22.6	9.0				
max	33.5	8.9				
min	15	8.8				
best MPG	33.5	6.5				
worst MPG	15	10.7				
worst accel	23	9.6				
best accel	23	8.4				
largest engine	23	9.8				
smallest engine	23	8.2				
heaviest	23	9.7				
lightest	23	8.2				

Fuel Efficient Cars Pollute Less

Repeat to see the impacts of acceleration, engine size, and car size. Choose the same minimum and maximum axes values for *predicted emissions* to make comparisons across car characteristics easiest:

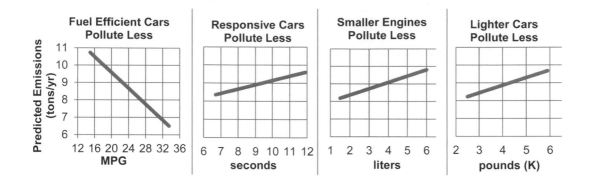

Lab Practice 8 Multiple Regression: Drivers of Preemie Diaper Fit Importance

Procter & Gamble managers were encouraged by concept test results of their Pampers Preemies. Test results revealed that

- Superior diaper fit, the benefit which differentiates Pampers Preemies, is an important attribute to preemie moms.
- The most promising target market is unique demographically.

Product manager Deb Henretta wants to know which key demographics are driving the importance of fit.

Use the data in **Lab Practice 8 Diaper Fit Drivers.xls** to build a multiple regression model that will provide this information.

1. Do the set of demographics, *age, income, family size,* and *number of other kids,* together drive *fit importance?* Y or N

    Your evidence will be the *significance* level (*p Value*) of your model *F test*: _____

2. Based on multiple regression evidence, which particular demographics drive *fit importance?*

	Age	Income	Family size	Other kids
Significant?	Y or N	Y or N	Y or N	Y or N
Evidence (*p Value*)				

3. Which coefficients have the "wrong" sign?

	Age	Income	Family size	Other kids
Unexpected sign	Y or N	Y or N	Y or N	Y or N

4. Is it possible that the demographics that seem insignificant really matter? Y or N

5. Find correlations between demographic variables and identify those that are highly correlated ($|r| > .7$).

	$|r| > .7$?		$|r| > .7$?
Age, income		*Income, family size*	
Age, family size		*Income, other kids*	
Age, other kids		*Family size, other kids*	

6.   Choose one of the pair of most strongly correlated demographics to represent the other and
     rerun regression.

     Is your model explanatory power just as good without the omitted demographic?

                                          Y or N

     Full model *R Square*: ___   Partial model *R Square*: ___   Change in *R Square*: ___

7.   In your partial model, which demographics drive *fit importance?*

	Age	Income	Family size	Other kids
Significant?	Y or N	Y or N	Y or N	Y or N
Evidence (*p Value*)				

8.   Which coefficients have the "wrong" sign in your partial model?

	Age	Income	Family size	Other kids
Unexpected sign	Y or N	Y or N	Y or N	Y or N

9.   Can Procter & Gamble managers safely assume that the demographic variable that was
     omitted is not a driver of *fit importance* and can be ignored? Y or N

10. Find the skewness of your residuals and make a histogram of your residuals. Are the
     residuals approximately *Normal*? Y or N

Use the final model coefficients and the regression equation to make *predicted fit importance*.

11. Find the sample *mean*, *maximum*, and *minimum* for each demographic variable in your
     model.
     Compare *predicted fit importance* between hypothetical moms who are (1) wealthiest and
     poorest, (2) with largest and smallest families, (3) oldest and youngest to find the expected
     marginal impact on differences, in each case holding two of the three drivers constant at
     the mean.

Demographic	Expected marginal impact on differences in fit importance
Age	
Income	
Family size	

12. Plot *predicted fit importances* of the two hypothetical preemie moms who differ along the
     most important demographic driver to illustrate your results.

Lab 8 Model Building with Multiple Regression: Pricing Dell's Navigreat

Dell has experience selling GPS systems built by other firms and plans to introduce a Dell system, the Navigreat. They would like information that will help them set a price.

The Navigreat has

- An innovative, *highly portable* design, weighing only 5 ounces, with a state of the art display
- A 3.5-in screen, neither large nor small, relative to competitors
- An innovative technology that guarantees precise routing time estimates

Dell executives believe that these features, *portability*, *weight*, *display* quality, *screen size*, and *routing time* precision, drive the price that customers are willing to pay for a GPS system.

Recent ratings by *Consumer Reports* provide data on the retail price of 18 competing brands, as well as

- *Portability* (1–5 scale), *weight* (ounces), and *display quality* (1–5 scale)
- *Screen size* (inches)
- *Routing time precision* (1–5 scale)

These data are in **Lab 8 Dell Navigreat.xls**. Also in the file, in row 21, are the attributes and expected ratings of the Navigreat.

Build a multiple regression model of GPS system *price*, including the characteristics thought by the management to be drivers of *price*.

Regression results: Is the model *R Square* significantly greater than 0? Y N

Evidence: *Significance F* = _____

Which of the potential drivers have slopes significantly different from 0?

	Portability	Weight	Display	Screen size	Routing time
Slope ≠ 0	Y or N	Y or N	Y or N	Y or N	Y or N
Evidence (*p Value*)					

Which of the drivers have slopes of unexpected sign?

	Portability	Weight	Display	Screen size	Routing time
Slope sign unexpected	Y or N	Y or N	Y or N	Y or N	Y or N

Confirm suspected multicollinearity: The GPS system physical design determines its *screen size*, *display quality*, *weight*, and *portability*. Run correlations to see if these characteristics are highly correlated.

	Highly correlated $(r_{x_1,x_2} > .5)$
Portability, weight	Y or N
Portability, display	Y or N
Portability, screen size	Y or N
Weight, display	Y or N
Weight, screen size	Y or N
Display, screen size	Y or N

Choose one of the set of correlated characteristics to represent the set, eliminating the other potentially redundant characteristics, and re-run the regression.

Is this partial model *R Square* significantly greater than 0?   Y   N

Evidence: *Significance F* = _____

Which of the potential drivers in this partial model have slopes significantly different from 0? (Cross out characteristics that you excluded in this reduced model.)

	Portability	Weight	Display	Screen size	Routing time
Slope ≠ 0	Y or N	Y or N	Y or N	Y or N	Y or N
Evidence (*p Value*)					

Which of the drivers have slopes of unexpected sign? (Cross out characteristics that you excluded in this partial model.)

	Portability	Weight	Display	Screen size	Routing time
Slope sign unexpected	Y or N	Y or N	Y or N	Y or N	Y or N

Find *Partial F* to decide whether the partial model's explanatory power is significantly lower than in the full model.

Full model *R Square* (1)	Partial model *R Square* (2)	Change in *R Square* (3) =(1) – (2)	Change per *g* predictors excluded (4) =(3)/*g*	%Variation unexplained by full model (5) =1 – (1)	%Variation unexplained per *Residual df*s (6) =(5)/(*N* – 1 – *k*)	*Partial F* (7) =(4)/(6)	*p Value* with *g* and (*N* – 1 – *k*) *df*s

Conclusion:

_____partial model *R Square* is significantly lower than full model *R Square*, and potentially redundant variables are jointly significant and cannot be excluded.

or   _____partial model *R Square* is equivalent to the full model *R Square*, and excluded variables are redundant or unimportant and can remain excluded.

Determine the improvement in predictive accuracy.

	Full model (1)	Reduced model (2)	Improvement in *margin of error* (3)=(2) – (1)
Standard error	$	$	
Approximate margin of error in 95% predictions	$	$	$

Assess residuals: Produce a residual histogram.
Are residuals approximately *Normal*? Y or N

Predict prices: Copy the *coefficients* and paste into the Navigreat sheet, and then use the regression equation to find *expected prices* for each of the GPS systems, including the Navigreat.
Copy the *standard error* and paste into the Navigreat sheet.
Find the *critical t* for 95% prediction intervals with your model *residual degrees of freedom*.
Find the *lower* and *upper 95% prediction intervals* for each model, including the Navigreat.

Will Dell be able to charge a retail price of $650 for the Navigreat? Y or N

Sensitivity analysis: Identify the most important driver of prices by comparing the differences in *expected prices* between four hypothetical GPS systems.
Add these four hypotheticals at the bottom of the file, and then extend *expected price, lower* and *upper 95% prediction* bounds to include these.

Screen size	Route time rating	95% prediction interval	Difference due to
Largest (5″)	Average (4 = "Good")	$	
Smallest (3.4″)	Average (4 = "Good")	$	Screen size: $_____
Average (3.8″)	Best (5 = "Excellent")	$	
Average (3.8″)	Worst (2 = "Poor")	$	Route time rating: $_____

If Dell wants to charge a retail price of $650 for the Navigreat, what product design modification ought to be made?_____

Assignment 8-1 Sakura Motor's Quest for Fuel Efficiency

The new product development team at Sakura Motors has decided that the new car they are designing will have superior gas mileage on the highway.

Use the data in **Assignment 8-1 Sakura Motors.xls** to build a model to help the team. Variables in the dataset include the following:

MPGHwy
Manufacturer's suggested retail base *price*
Engine size (liters)
Engine cylinders
Engine horsepower
Curb weight
Acceleration in seconds to go from 0 to 60
Percentage of owners satisfied who would buy the model again

Use your logic to choose car characteristics that should influence highway gas mileage.

Determine which car characteristics influence *highway gas mileage*. Use *partial F test(s)* to decide whether to remove apparently insignificant variables.

With sensitivity analysis, find the relative importance of significant influences on *highway fuel economy*.

Find the car characteristic levels that could be expected to achieve *40 miles per gallon* in highway driving. (Sakura is not limited to existing designs.)

Write a one page, single spaced memo presenting your model, sensitivity analysis, and design recommendations.

Present your final model in standard format.
What is the margin of error of model forecasts of MPG?
Discuss the relative importance of significant influences, including the expected difference in *fuel economy*, that differences in each could be expected to make if other characteristics were held at mean values.

Conduct a sensitivity analysis comparing expected fuel economy with best and worst levels of each predictor in your final model when other characteristics are at average levels. Discuss the relative importance of significant influences, referring to the following:

1. A table of Fuel Response to Car Characteristics that you have added to the second page of Attachments.

2. A scatterplot of the impact of each significant driver on fuel economy. This plot shows predicted fuel economy on the vertical axis by values of a driver.

There is a two page limit:

1. One single spaced page for your memo text with a single embedded scatterplot (of *predicted MPGHwy* by the most important car characteristic)
2. A second page of attachments showing your sensitivity analysis table and plots referred to in your memo.

Please use Times New Roman 12 pt font and round your statistics to two or three significant digits.

You do not need to include description of your *partial F test*.

Assignment 8-2 Starting Room Prices at Marriott

Marriott executives are evaluating the prices in their hotels and believe that *starting room price* differences ought to reflect differences in quality, reflected in *Star ratings* and *Guest ratings*. The file **Assignment 8-2 DCHotel Prices.xls** contains two quality ratings, *Stars* and *Guest Rating* and *Starting Room Price*, for the 51 branded hotels in Washington, DC. The hotel industry in Washington, DC, is representative of the hotel industry in cities throughout the United States:

1. Identify the hotel(s) that is (are) outliers based on *starting room price*.

2. Quantify how powerful the quality rating differences are in explaining differences in hotel prices.

3. Executives are considering a promotional campaign designed to increase the *Star* ratings of Marriott hotels. If *Star* ratings could be (significantly) increased, executives believe this would justify raising *starting room prices*. Does it make sense to implement the promotional campaign?

4. Provide an equation, including coefficient significance levels, that a manager could use to find *expected starting room prices*.

5. Among branded hotels in US cities, how much difference could the management expect between average *starting room prices* of hotels with *Guest ratings* of 4.2 and hotels with *Guest ratings* of 3.2?

6. If the management were to use your equation, the *predicted starting room price* of hotels ought to be no further than _____ from the actual *starting room price* of a hotel in a US city with 95% confidence.

Attach the regression sheet that shows your final model.

Assignment 8-3 Identifying Promising Global Markets

Harley–Davidson would like to identify the most promising global markets for motorcycle sales.
Some managers believe that motorcycle sales potential is greater in developed countries with higher GDP. Others believe that per capita GDP may a better indicator. Management believes that motorcycle sales potential will necessarily be greater in more populated countries. Some believe that population density may matter more, inasmuch as motorcycles may be preferred to cars for parking and commuting in larger cities.

Build a model to identify the drivers of motorcycle market potential. **Global Moto.xls** contains measures of *Motorcycle sales, GDP, per capita GDP, population*, and *population density* in 2009 for 20 countries with the highest motorcycle sales.

1. Identify outliers: Which countries have unusually high motorcycle sales? _____

2. Which economic and population variables drive motorcycle sales?
   ___ GDP   ___ per capita GDP   ___ population   ___ population density
   Explain how you reached your conclusions, including statistics that you used to decide.

3. Present your regression equation, including slope significance levels.

4. Illustrate the impact of the two most influential drivers, using the same scale for the y axis on both plots.

5. Compare predicted sales with actual sales in all countries in the sample to identify two markets with the greatest unrealized potential that Harley–Davidson should target.

6. H–D is considering expansion in the BRICK countries (Brazil, Russia, India, China, and South Korea). The column *BRICK* distinguishes these four countries from other countries in the sample.

   What economic or demographic characteristic(s) distinguish the BRICK countries from other countries in the sample? (Include the statistics that you used to test hypothesized differences between BRICK and other countries.)

7. In light of your model results and the distinguishing characteristic(s) of the BRICKs, should Harley–Davidson make expansion in the BRICKs a high priority?

   Explain the logic of your answer:

   Attach Excel output showing your final model (Summary table, ANOVA table, and Coefficient table), being sure that variable names in column A are showing.

Chapter 9
Model Building and Forecasting with Multicollinear Time Series

A regression model from time series data allows us to identify performance drivers and forecast performance given specific driver values, just as regression models from cross sectional data do. When decision makers want to forecast *future* performance, a time series of past performance is used to identify drivers and fit a model. A time series model can be used to identify drivers whose variation over time is associated with later variation in performance over time.

Three differences in the model building process distinguish cross sectional and time series models:

- The use of lagged predictors
- Addition of trend, seasonality, and cyclical variables
- The model validation process

In time series models, the links between drivers and performance are stronger if changes in the drivers precede change in performance. Therefore, lagged predictor variables are often used. Patterns of change in drivers that also occur in the dependent variable in later time periods are identified to choose driver lags. Time series models are built using predictor values from past periods to explain and forecast later performance. Figure 9.1 illustrates the differences in model building processes between cross sectional and time series models.

Most business performance variables and economic indicators are cyclical. Economies cycle through expansion and recession, and performance in most businesses fluctuates following economic fluctuation. Business and economic variables are also often seasonal. Cyclicality and seasonality are accounted for by adding cyclical and seasonal predictors.

Before a time series model is used to forecast future performance, it is validated:

- The two most recent observations are hidden while the model is built.
- The model equation is used to forecast performance in those two most recent periods.
- Model prediction intervals are compared with actual performance values in those two most recent periods, and if the prediction intervals contain actual performance values, this is evidence that the model has *predictive validity* and can be reliably used to forecast unknown performance in future periods.

C. Fraser, *Business Statistics for Competitive Advantage with Excel 2010: Basics, Model Building, and Cases*, DOI 10.1007/978-1-4419-9857-6_9, © Springer Science+Business Media, LLC 2012

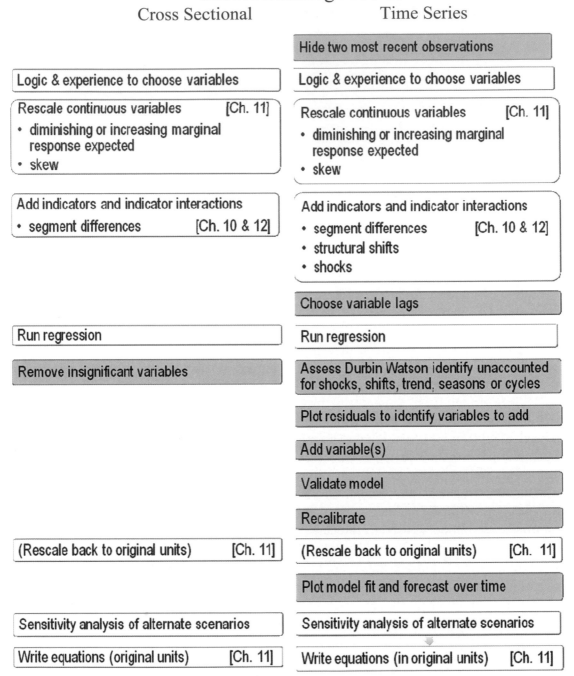

Fig. 9.1 Model building processes with cross sectional and time series data

9.1 Time Series Models Include Decision Variables, External Forces, Leading Indicators, and Inertia

Most successful forecasting models logically assume that performance in a period, Y_t, depends upon

- Decision variables under management control
- External forces, including
  - Shocks such as 9/11, Hurricane Katrina, change in presidential party
  - Market variables
  - Competitive variables
- Inertia from past performance
- Leading indicators of the economy, industry, or the market
- Seasonality
- Cyclicality

Ultimately, the multiple regression forecasting models contain several of these components, which together account for variation in performance. This chapter introduces trend, inertia, and leading indicator components of regression models built from time series.

Performance across time depends on decision variables and the economy. Decision variables, such as spending on advertising, sales effort, and research and development tend to move together. In periods of prosperity, spending in all three areas may increase; in periods where performance is sluggish, spending in all three areas may be cut. Firm strategy guides resource allocation to the various firm functions. As a result, it is common for spending and investment variables to be correlated in time series data.

Many economic indicators also move together across time. In times of economic prosperity, GDP is growing faster, consumer expectations increase, and investments increase. Increasing wealth filters down from the economy to consumers and stockholders, where some proportion of gains is channeled back into consumption of investments.

It is common for decision variables, past performance, and leading indicators to be correlated in time series data. This inherent correlation of performance drivers in time series data makes logical choice of drivers a critical component of good model building.

It is also often more promising to build models by adding variables, one at a time, looking at residuals for indications of the most promising variables to add next. Multicollinearity, including its consequences, diagnosis, and alternate remedies, is further considered in this chapter.

Example 9.1 Home Depot Revenues[5]

Home Depot executives were concerned in early 2010 that revenues were not yet recovering from the economic recession of 2008–2009. Quarterly revenues had been down from the same quarterly revenues the year before, since the second quarter of 2008. A slowdown in the US economy, combined with the financial crisis, had reduced lending, and new home sales had slowed down. Traditionally, Home Depot revenues have grown following growth in *new home sales*, because builders and homeowners buy construction materials, flooring, and appliances at Home Depot.

---

[5]This example is a hypothetical scenario based on actual data.

9.2 Indicators of Economic Prosperity Lead Business Performance

A *leading indicator* model links changes in a leading indicator and later performance:

$$revenues(B\$)_t = b_0 + b_1 GDP(T\$)_{q-l} + b_2 NewHomeSales(K)_{q-l},$$

where l denotes the length of lag or delay from change in GDP or new home sales to change in revenues.

Amanda, a recent business school graduate with modeling expertise, was asked to build a model of Home Depot revenues, which would both explain revenue fluctuations and forecast revenues in the next four quarters.

Home Depot executives wanted to know how strongly

- Growth in past GDP
- Growth in past *new home sales*

influenced revenues. After being briefed by the executives, Amanda created a model reflecting their logic. She considered as possible drivers in her model:

- *New home sales*$(K)_{q-l}$
- *GDP*$(B\$)_{q-l}$

Amanda also considered Lowe's past revenues. Lowes' business was similar to Home Depot's business. Both firms' revenues were seasonal and linked to the housing market, although Lowes offered installation services and Home Depot had not. Amanda was not sure whether Lowes revenues had a positive or a negative impact on Home Depot revenues. Whenever either firm advertised or promoted home improvement items, later sales at both tended to be higher. Nonetheless, the two firms were competing for the business of many of the same customers.

9.3 Hide the Two Most Recent Datapoints to Validate a Time Series Model

Before Amanda proceeded further, she excluded the two most recent observations from fourth quarter 2009 and first quarter 2010. These *holdout* observations would allow her to compare forecasts for the two most recent periods with actual revenues to *validate* her model. If the 95% prediction intervals from the model contained the actual revenues for both quarters, she would be able to conclude that her model was valid. She could then use the model to forecast with confidence.

Amanda used datapoints for revenue from fourth quarter 1999 through first quarter 2010. After excluding the two most recent datapoints, her regression would use 10 years of quarterly data.

9.4 Compare Scatterplots to Choose Driver Lags: Visual Inspection

The potential drivers each reflects economic conditions and moves together over time. Consequently, they are highly correlated, as Fig. 9.2 illustrates. Including all of the drivers in a multiple regression model at once would introduce a high degree of multicollinearity and make it difficult to identify each of their marginal impacts. To most effectively build a time series model, start with one driver, and then add additional drivers, one at a time.

Amanda plotted *Home Depot revenues*, *GDP*, *new home sales*, and *Lowes revenues* by quarter. She noted quarters in which *Home Depot revenues* were growing faster than average. These are shown in green. She colored quarters when growth was below average in blue. She added trend lines for reference. The trend is the average linear growth over the series. Her scatterplots are shown in Fig. 9.2.

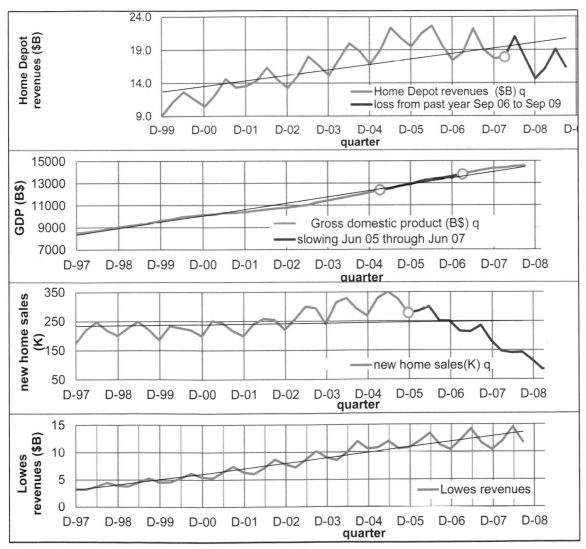

Fig. 9.2 Home Depot revenues and past GDP, new home sales, and Lowes revenues by quarter

Home Depot revenue shows an upward trend, with growth slowing in the more recent quarters. Past *GDP* shows a similar trend. *Home Depot revenue* is seasonal, as are both *new home sales* and *Lowes revenue.*

It seemed likely that recent *Home Depot revenue* losses were following the slowdown in *GDP.* The delay seemed to be about 12 quarters. *GDP* slowed in mid 2005. *Home Depot revenue* losses began in mid 2008. To see this more clearly, Amanda lined up the two plots with a 12 quarter difference, shown in Fig. 9.3.

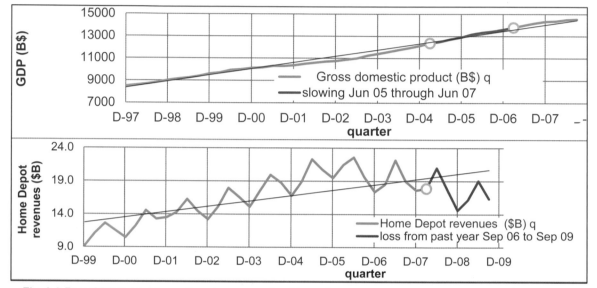

Fig. 9.3 Past GDP leads Home Depot revenues by 12 quarters

To confirm her conclusions from visual inspection, Amanda ran correlations between *Home Depot revenues* and 4-, 8-, and 12-quarter *GDP* lags, shown in Table 9.1. The 4 quarter lag was most highly correlated with revenues, although the correlation with the 12 quarter lag was just 1% smaller. The longer lag would allow for longer, 12 quarter forecasts. Amanda ran simple regression using the 12 quarter lag. Results are shown in Table 9.2.

Table 9.1 Correlations between Home Depot revenues and past GDP

	$GDP\,(B\$)_{q-4}$	$GDP\,(B\$)_{q-8}$	$GDP\,(B\$)_{q-12}$
Home Depot revenues $(\$B)_q$	.641	.623	.631

Table 9.2 Regression with past GDP

SUMMARY OUTPUT

Regression statistics

R Square	.40				
Standard error	2.70				
Observations	40				

ANOVA

	df	SS	MS	F	Significance F
Regression	1	183	183	25.1	.0000
Residual	38	277	7.3		
Total	39	460			

	Coefficients	Standard error	t Stat	p Value	Lower 95%	Upper 95%
Intercept	2.2	2.9	.7	.47	−3.8	8.1
GDP(B$)$_{q-12}$	.0014	.00028	5.0	.0000	.00083	.0019

Regression results suggest that past *GDP* drives *Home Depot revenues* 12 quarters later, accounting for 40% of quarterly variation over the past 10 years.

9.5 The Durbin Watson Statistic Identifies Positive Autocorrelation

The Durbin Watson (*DW*) statistic incorporates correlation between residuals across adjacent time periods which allows assessment of the presence of unaccounted for trend, cycles, or seasonality in the data. If there is an unaccounted cycle, seasonality, or trend, higher residuals are likely to be followed by similar higher residuals, and lower residuals are likely to be followed by similar lower residuals. In Amanda's model, past *GDP* accounts for trend, but not for the obvious seasonality in *Home Depot revenues*. The unaccounted for seasonality will be evident in residuals, and there will be positive autocorrelation.

DW indicates such *positive autocorrelation*, the correlation of residuals over time, which signals that a trend, seasons, or cycle has been ignored. The Durbin Watson statistic compares the sum of squared differences between pairs of adjacent residuals with the sum of squared residuals:

$$DW = \frac{\sum_{2}^{N}(e_q - e_{q-1})^2}{\sum_{1}^{N}e_q^2}.$$

If all of the trend, seasons, and cycles in the data have been accounted for, *DW* will be "high." Exactly how high depends on the length of time series, which is the number of observations used in the regression, and the number of independent variables, including the intercept. *DW* critical values are available online at stanford.edu/~clint/bench/dwcrit.htm, found by Googling "Durbin Watson critical values." (In this online table, sample size is indexed by *T*, and the number of independent variables, plus intercept, is indexed by *K*.)

There are two relevant critical values, a lower value and an upper value, dL and dU:

DW below the lower critical value, dL, indicates presence of positive autocorrelation from unaccounted for trend, cycle, or seasons that we would then attempt to identify and incorporate into the model.

DW above the upper critical value, dU, indicates lack of autocorrelation and freedom from unaccounted for trend, cycle, or seasons, which is the goal.

DW between dL and dU is the gray area, indicative of possible autocorrelation and presence of unaccounted for trend, cycle, or seasons. When *DW* is in the gray area, we look for a pattern in the residuals from unaccounted for trend, cycle, or seasons, knowing that there is a reasonable chance that a pattern may not be identified.

Figure 9.4 illustrates critical values for several sample and model sizes.

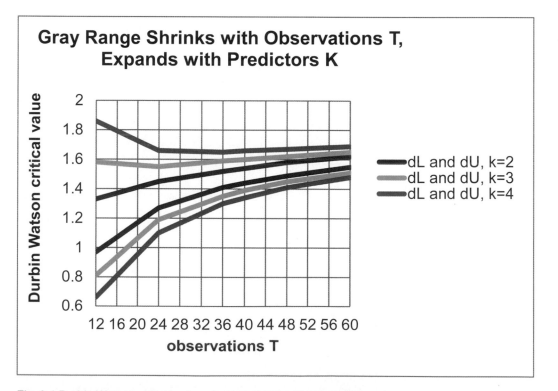

Fig. 9.4 Durbin Watson critical values by sample size *T* and predictors *k*

Notice that the gray area shrinks as sample size increases, but expands as the number of predictors increases. Amanda's initial model, with one driver, plus intercept, and a sample size of 40, has *DW* critical values of dL = 1.44 and dU = 1.54. The model *DW* statistic is .38, which is below dL, leading to the conclusion that the residuals are positively autocorrelated. The data contain trend, seasonality, or cycles not accounted for by the model. Figure 9.5 illustrates the positive autocorrelation between residuals and residuals lagged one quarter.

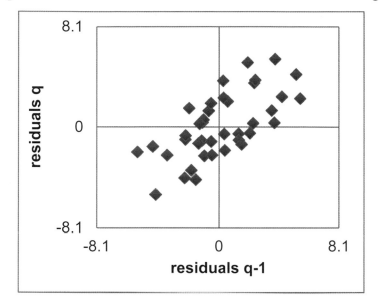

Fig. 9.5 Positive autocorrelation between residuals and residuals lagged one quarter

Examining the residuals is likely to provide clues to identify which variables can be added to account for the unaccounted for trend, season, or cycles.

9.6 Assess Residuals to Identify Unaccounted for Trend or Cycles

Model residuals should not show trend, seasonality, nor cyclicality. If a driver has been omitted, the residuals will not be pattern free. The residuals will provide clues to help identify which variable to add to the model.

Amanda plotted the residuals across quarters in Fig. 9.6 and observed a noticeable decline beginning late in 2006 and continuing through 2009, as well as seasonality. These patterns were similar to *new home sales* patterns. *New home sales* began to decline about four quarters earlier, in late 2005, and also showed seasonality. Lining up the two plots, a four quarter lag in *new home sales* promised to improve the model.

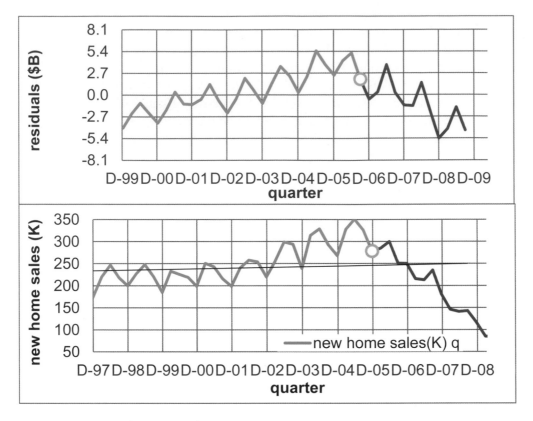

Fig. 9.6 Residuals (B$) and *GDP* ($B)

To back up her conclusions from visual inspection, Amanda compared correlations between the residuals and four- and eight-quarter *new home sales* lags, shown in Table 9.3.

Table 9.3 Correlations between residuals and past *new home sales*

	New home sales$(K)_{q-4}$	*New home sales*$(K)_{q-8}$
Residuals	.86	.63

The residual correlation with the four quarter lag was higher. Amanda added *new home sales* with a four quarter lag to the model and ran multiple regression. Results are shown in Table 9.4.

Table 9.4 Home Depot revenue regression with past new home sales and GDP

SUMMARY OUTPUT

Regression statistics

R Square	.85				
Standard error	1.36				
Observations	40				

ANOVA

	df	SS	MS	F	Significance F
Regression	2	392	196	105.2	.0000
Residual	37	37	1.9		
Total	39	460			

	Coefficients	Standard error	t Stat	p Value	Lower 95%	Upper 95%
Intercept	−10.5	1.9	−5.5	.0000	−14.3	−6.6
New home sales $(K)_{q-4}$	.046	.0043	10.6	.0000	.037	.054
Gross domestic product $(B\$)_{q-12}$	.0015	.00014	10.9	.0000	.0012	.0018

DW	T	k	dL	dU
1.57	40	3	1.39	1.60

Together, past *GDP* and *new home sales* drive *Home Depot revenues*. With both leading indicators, *R Square* is now 85%, an increase of 45%. The model *standard error* has been reduced from $2.7B to $1.4B, reducing the forecast margin of error from $5.5B to $2.7B.

Both drivers have positive slopes, as expected. The Durbin Watson statistic, 1.57, has improved, but does not clear the upper critical value, 1.60. Trend, seasons, or cycles probably remain to be accounted for.

Amanda examined a plot of the residuals, this time with a plot of Lowes quarterly revenues, shown in Fig. 9.7. *Lowes revenue* exhibits seasonality, resembling the residual pattern. In 2001–2003, and later, in 2007–2009, residuals move opposite *Lowes revenues* lagged six quarters. This inverse pattern suggested competition between Lowes and Home Depot. Residual correlations with four-, six-, and eight-quarter lags of *Lowes revenues* are shown in Table 9.5.

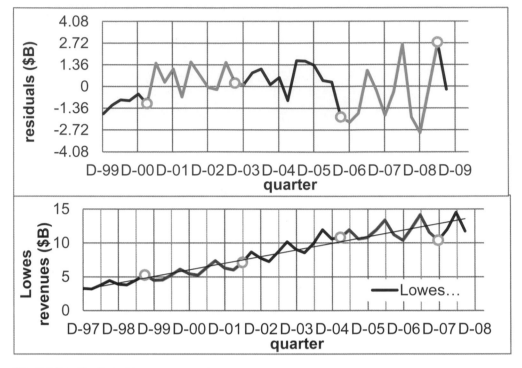

Fig. 9.7 Residuals and Lowes revenues

Table 9.5 Correlations between residuals and past Lowes revenues

	Lowes revenues (B$)$_{q-4}$	Lowes revenues (B$)$_{q-6}$	Lowes revenues (B$)$_{q-8}$
Residuals	.186	−.182	.076

The four quarter lag is strongest and positively correlated with residuals. However, the six quarter lag is smaller by less than 1%, and negative, reflecting the competition that was apparent in the visual inspection. Amanda added the six quarter lag of *Lowes revenues* to the model and ran a multiple regression, shown in Table 9.6.

Table 9.6 Regression with leading indicators and competition

	Coefficients	Standard error	t Stat	p Value	Lower 95%	Upper 95%
SUMMARY OUTPUT						
Regression statistics						
R Square	.90					
Standard error	1.11					
Observations	40					
ANOVA						

	df	SS	MS	F	Significance F	
Regression	3	416	139	113	.0000	
Residual	36	44	1.2			
Total	39	460				

	Coefficients	Standard error	t Stat	p Value	Lower 95%	Upper 95%
Intercept	−22.8	3.2	−7.2	.0000	−29.3	−16.4
Gross domestic product (B$)$_{q-12}$	.0031	.00037	8.4	.0000	.0023	.0038
New home sales (K)$_{q-4}$	.0580	.0045	13.0	.0000	.049	.067
Lowes revenues (B$)$_{q-6}$	−.856	.191	−4.5	.0000	−1.24	−.47
DW	dL	dU				
1.73	1.34	1.66				

The addition of past *Lowes revenues* improved the model. *R Square* is now 90%. *Lowes revenues* increased explanatory power by 5%. The model *standard error* $1.11B is now smaller by $.3B, reducing the margin of error in predictions by $.5B, to $2.2B.

All three marginal slopes differ from zero. The marginal slope for *Lowes revenues* is negative, indicating that the competitor's revenues are at the expense of *Home Depot revenues*.

The Durbin Watson statistic, $DW = 1.73$, now exceeds the upper critical value, dU = 1.66, providing evidence that trend, cycles, and seasonality have been accounted for.

9.7 Forecast the Recent Hidden Points to Assess Predictive Validity

With a significant model, logically correct coefficient signs, and residuals free of autocorrelation, Amanda could proceed to assess the predictive validity of her model by comparing actual *Home Depot revenues ($B)* in the two most recent quarters with the model's 95% prediction intervals.

(Recall that those two most recent quarters were hidden and not used in the regression to fit the model.) Validation evidence is shown in Table 9.7.

Table 9.7 Model predictions include actual values

Quarter	Lower 95% prediction	Home Depot revenues ($B)	Upper 95% prediction
D-09	11.3	14.6	15.8
M-10	12.5	16.9	17.0

9.8 Add the Most Recent Datapoints to Recalibrate

With evidence of predictive validity, Amanda used the model to forecast revenues in the next four quarters. Before making the forecast, she added the two most recent observations that were hidden to validate. The recalibrated model became

$$reve\hat{n}ues(\$B)_q = -22.6^a + .055^a \ new \ home \ sales \ (K)_{q-4} + .0031^a GDP \ (\$B)_{q-12}$$

$$-.84^a \ Lowes \ revenue \ (\$B)_{q-6}$$

$R \ Square: .90^a$
[a]Significant at .01.

Variation in past *new home sales*, US GDP, and *Lowes revenue*, together, account for 90% of the quarterly variation in *Home Depot revenues*. Using this multiple linear regression model, forecast quarterly revenues are expected to fall within $2.7B of predictions.

Changes in the housing market influence revenues, although less than economic fluctuation or competitor sales. Revenues follow the changes in the housing market 1 year later. In recent quarters, a decline of 20,000 (K) new homes sold would be typical. Following such a typical decline in a quarter, Home Depot revenues are expected to decline by about $1 billion (=.055($B/K) × 20 (K)) 1 year later.

The economy strongly influences revenues. Revenues follow longer economic cycles. For a typical increase in US GDP of $700 billion in a quarter, revenues are expected to increase by about $2 billion (=.0031($B/$B) × $700B) 4 years later.

Revenues are lower when revenues of Lowes, a major competitor, grow. Lowes revenues have lost an average of about $4 billion in recent quarters. Following similar losses by competitor Lowes, Home Depot revenues are expected to increase by about $3 billion (= −.84($B/$B) × $4B) six quarters later. Competitor sales are the largest single influence on Home Depot revenues.

Model forecasts are shown in Fig. 9.8 and Table 9.8.

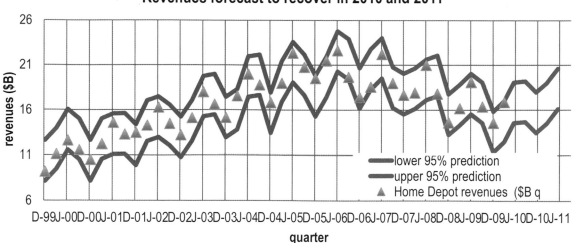

Fig. 9.8 Optimistic forecast for 2010–2011

Table 9.8 Quarterly revenue forecast

Quarter	95% Lower prediction (B$)	95% Upper prediction ($B)	Revenues ($B)$_{q-4}$	Forecast annual growth for quarter (%)
Jun-10	14.6	19.1	19.1	−12
Sep-10	14.7	19.2	16.4	3
Dec-10	13.5	18.0	14.6	8
Mar-11	14.5	19.0	16.9	−1
Jun-11	16.1	20.6	16.8 est^{a}	9 est^{a}

[a]Forecast growth for June 2011 is relative to revenues forecast for June 2010

Annual quarterly growth (from same quarter in the past year) averaged −6% over the past 4 years. Home Depot revenues are expected to recover in late 2010 and 2011, growing 2% over same quarter in the past year, on average.

Amanda summarized her model results for management.

MEMO

Re: Revenue Recovery Forecast for late 2010 and 2011
To: Home Depot Management
From: Amanda Chanel
Date: June 2010

Following past growth in the US economy and slowed competitor sales, quarterly revenues are expected to increase an average of 2% annually in the next five quarters.

A regression model of quarterly revenues was built from past US GDP, new home sales, and Lowes revenues. The model accounts for 90% of the variation in revenues and produces valid forecasts within $2.7 billion of actual revenues.

Model results. Revenues are driven by growth in the US economy, housing market, and competitor revenues. Shrinking new home sales are expected to reduce revenues by $1 billion in the each of the next five quarters; however, past annual growth in quarterly GDP of 5% is expected to increase revenues by $2 billion in each of the next five quarters. Slowed growth in Lowes revenues, the strongest driver, is expected to increase revenues by $3 billion in each of the next five quarters.

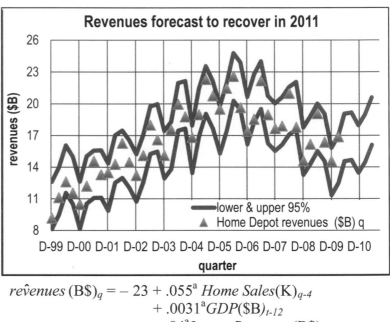

$$rev\hat{e}nues\,(B\$)_q = -23 + .055^a\,Home\,Sales(K)_{q\text{-}4}$$
$$+ .0031^a GDP(\$B)_{t\text{-}12}$$
$$- .84^a Lowes\,Revenues(B\$)_{q\text{-}6}$$

$RSquare:$ $.90^a$
[a]Significant at .01.

Forecast. Considering the positive impacts of economic growth and reduced competitor sales, with the negative impact of slowed new home sales, annual revenue growth of about 2% is expected in each of the next five quarters.

Conclusions. Home Depot Revenues are driven primarily by past competitor revenues and economic productivity, though also affected by the housing market. Recovering growth is forecast in late 2010 and 2011.

Quarter	Forecast ($B)		Forecast growth
10-II	14.6	to 19.1	−12%
10-III	14.7	to 19.2	3%
10-IV	13.5	to 18.0	8%
11-I	14.5	to 19.0	−1%
11-II	16.1	to 20.6	9%

Other Considerations. This linear model assumes constant response to changes in the economy, housing market, and competitor sales, which may not reflect actual response as well as a nonlinear model might.

9.9 Inertia and Leading Indicator Components Are Powerful Drivers and Often Multicollinear

Like cross sectional models, time series models allow identification of performance drivers and forecasts of performance. However, time series models differ from cross sectional models, and the model building process with time series contains additional steps.

- Often lagged predictors are used to make driver identification more certain and to enable forecasts.
- Lagged predictors tend to move together across time and are often highly correlated. Consequently, to minimize multicollinearity issues, model building begins with one predictor, and then others are added, considering their joint influence and incremental model improvement.
- Forecasting accuracy of time series models is tested, or validated, before they are used for prediction of future performance.

Predictors in time series models tend to be highly correlated, since most move with economic variables and most exhibit predictable growth (*trend*). Model building with time series begins with the strongest among logical predictors and additional predictors are added which improve the model.

Time series typically contain trend, business cycles, and seasonality that are captured with these components. Unaccounted for trend, cycles, or seasonality is detected through inspection of the residual plot and the Durbin Watson statistic. Leading indicators are often stable and predictable performance drivers. Competitive variables may account for trend, seasonality, or cycles common to a market.

Useful forecasting models must be valid. Holding out the two most recent performance observations allows a test of the model's forecasting capability. With successful prediction of the most recent performance, the model is validated, and the recalibrated model can be used with confidence to forecast performance in future periods.

Excel 9.1 Build and Fit a Multiple Regression Model with Multicollinear Time Series

Home Depot Revenues
Build a model of Home Depot quarterly revenues that potentially includes past economic growth, growth in the housing market, and variation in a competitor's revenues. The data are in **Excel 9.1 Home Depot Revenue.xls.**

Plot *Home Depot revenues* from December 1999 to September 2009 to see the pattern of movement over time. (Hide or ignore the two most recent datapoints, December 2009 and March 2010.)

In Excel scatterplots, time is measured in days. To set the quarter axis beginning and end points, format the axis, **Alt JAA**, setting the minimum to 36,500, the maximum to 40,200, major units to 366 (1 year), and minor units to 183.

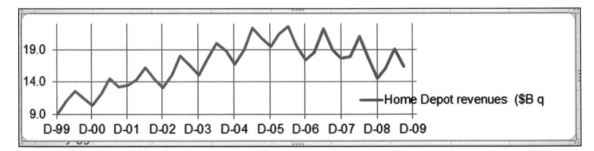

Consider a 2 year or a 3 year lag for *GDP*. Plot *GDP*, beginning 3 years earlier, in December 1996, and stopping 2 years earlier, in September 2007.

Use a minimum of 35,400 and a maximum of 39,500, for the quarter axis, with major units at 366 and minor units at 183.

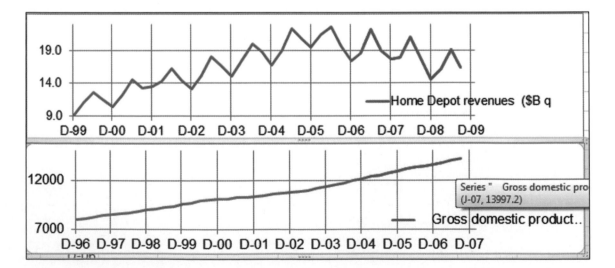

Consider 1- or 2-year lags for *new home sales* and *Lowes revenue*. Plot both, beginning 2 years earlier, in December 1997, and stopping 1 year earlier, in September 2008.

Use a minimum of 35,780 and a maximum of 39,900, for the quarter axis, with major units at 366 and minor units at 183.

Add trendlines, **Alt JAN**, and major grid lines, **Alt JAG**.

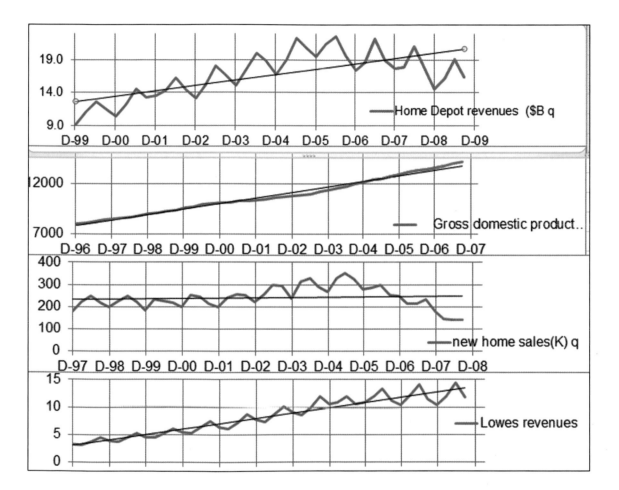

Home Depot revenue shows an upward trend, with growth slowing in the more recent quarters. A similar trend exists in past *GDP*. *Home Depot revenue* is seasonal, as is *new home sales revenue*. Begin model building with the economic indicator, *GDP*, expecting that the *Home Depot revenue* trend follows *GDP* trend.

To compare their patterns, find the quarters in which GDP growth slows. Add two columns, *growth in GDP from past year* and *change in GDP growth*. Find the year to year change each quarter in *growth in GDP from past year*. Compare year to year change in *GDP growth from past year*.

Highlighting cells in the *change in GDP growth from past year* column reveals slowing growth in GDP, beginning in June 2005.

	E41		f_x	=D41-D37	
	A	B	C	D	E
1	Quarter	Home Depot revenues ($B) q	Gross domestic product (B$) q	GDP growth from past year	change in GDP growth from past year
32	J-04	20.0	11778	770	364
33	S-04	18.8	11951	695	141
34	D-04	16.8	12145	728	79
35	M-05	19.0	12380	782	73
36	J-05	22.3	12517	738	-32
37	S-05	20.7	12742	791	96
38	D-05	19.5	12916	771	42
39	M-06	21.5	13184	804	22
40	J-06	22.6	13348	831	93
41	S-06	19.6	13453	711	-80
42	D-06	17.4	13612	696	-75
43	M-07	18.5	13796	612	-192
44	J-07	22.2	13997	649	-182
45	S-07	19.0	14180	727	16
46	D-07	17.7	14338	726	31

Add a second series of *GDP* points to your plot, from March 2005 to June 2007, **Alt JCE**.

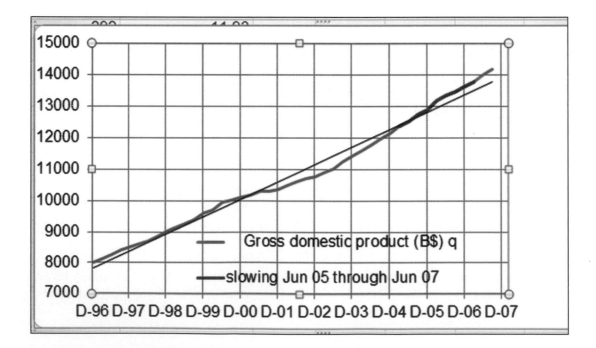

Find quarters of slower annual growth in *Home Depot revenues*.

Add a column, *HD Revenue growth from past year*, and highlight quarters with revenue losses, relative to past year.

	C47		f_x =B47-B43	
	A	B	C
1	Quarter	Home Depot revenues ($B) q	HD revenue growth from past year
39	M-06	21.5	2.5
40	J-06	22.6	0.3
41	S-06	19.6	-1.1
42	D-06	17.4	-2.1
43	M-07	18.5	-2.9
44	J-07	22.2	-0.4
45	S-07	19.0	-0.7
46	D-07	17.7	0.3
47	M-08	17.9	-0.6
48	J-08	21.0	-1.2
49	S-08	17.8	-1.2
50	D-08	14.6	-3.1
51	M-09	16.2	-1.7
52	J-09	19.1	-1.9
53	S-09	16.4	-1.4

Add a second series of *Home Depot revenue* points to your plot, from June 2006 to June 2009, **Alt JCE**.

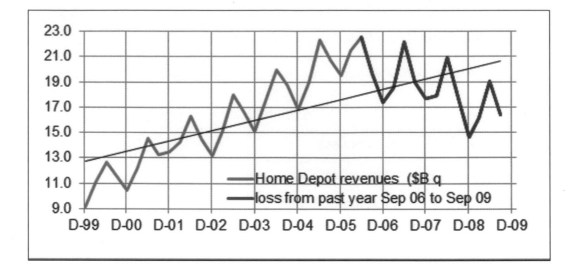

Compare 2- and 3-year lags of GDP by lining up the plots of GDP and revenues. To see a 2 year lag, match GDP in 1997 with *Home Depot revenues* in 1999.

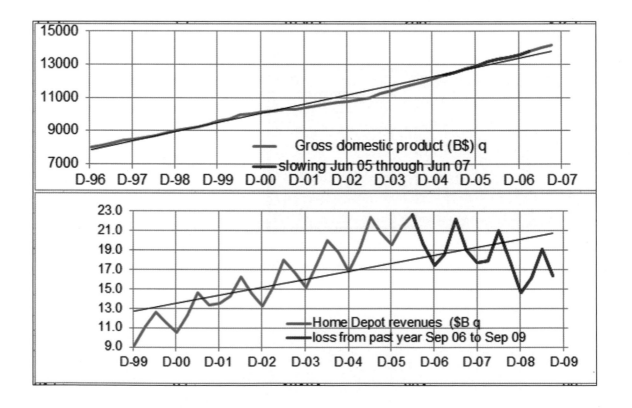

Home Depot revenues are declining during each of the slowing quarters of the 2 year lag. However, GDP improves in 2006, and we do not see an improvement 2 years later in revenues in 2008. Given this lack of match in the most recent quarters, the 2 year lag may not provide a valid model.

Line up GDP in 1996 with revenues in 1999 to see the 3 year lag.

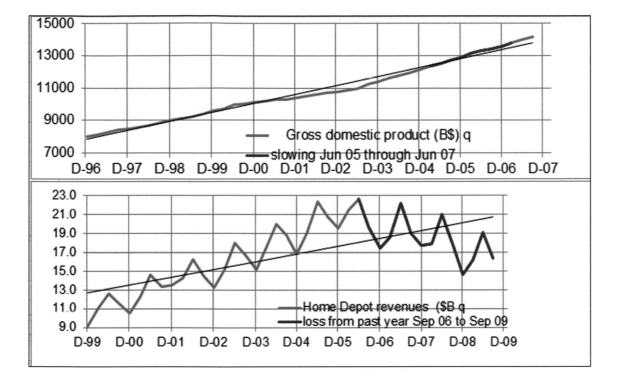

Home Depot revenues are declining in each of the slowed quarters of the 3 year GDP lag.

To back up visual inspection, create lagged columns that can be used to see correlations between the alternate lags and revenue.

Paste copies of *quarter*, *Home Depot revenues* ($B)$_q$, and *GDP*($B)$_q$. Paste a second copy of *GDP*($B)$_q$.

To make the 2 year lag, $GDP(\$B)_{q-8}$, add eight cells at the top of the column in the first pasted copy. Select the first eight data cells, **Alt HIID**.

J	K	L
Quarter	Home Depot revenues ($B) q	Gross domestic product (B$) q-8
D-95		
M-96		
J-96		
S-96		
D-96		
M-97		
J-97		
S-97		
D-97		7543
M-98		7638
J-98		7800
S-98		7893
D-98		8023

Create a 3 year lag by inserting 12 cells at the beginning of the second copy of GDP.

J	K	L	M
Quarter	Home Depot revenues ($B) q	Gross domestic product (B$) q-8	Gross domestic product (B$) q-12
D-95			
M-96			
J-96			
S-96			
D-96			
M-97			
J-97			
S-97			
D-97		7543	
M-98		7638	
J-98		7800	
S-98		7893	
D-98		8023	7543
M-99		8137	7638
J-99		8277	7800
S-99		8410	7893

Delete cells in the new columns before December 1999, so that all four new columns have data in each quarter. Select cells in rows of the four new columns from quarters before December 1999 and select **Alt HDDU**.

J	K	L	M
Quarter	Home Depot revenues ($B) q	Gross domestic product (B$) q-8	Gross domestic product (B$) q-12
D-99	9.2	8506	8023
M-00	11.1	8601	8137
J-00	12.6	8699	8277
S-00	11.5	8847	8410

Find the correlations between *Home Depot revenues* and the 8- and 12-quarter lags of *GDP*, **Alt AY11 C**. (Exclude the two most recent quarters, December 2009 and March 2010.)

	Home Depot revenues ($B) q
Home Depot revenues ($B) q	1
Gross domestic product (B$) q-8	0.62
Gross domestic product (B$) q-12	0.63

The 12 quarter lag has a slightly higher correlation with revenues and will allow longer, 12 quarter forecasts. Evidence from both the visual inspection and correlations suggests that the longer, 12 quarter lag of GDP will produce a better model.

Run the simple regression of *Home Depot revenues* with the 12 quarter *GDP* lag.

	A	B	C	D	E	F	G
2							
3	*Regression Statistics*						
4	Multiple R	0.630791					
5	R Square	0.397897					
6	Adjusted R Square	0.382052					
7	Standard Error	2.700987					
8	Observations	40					
9							
10	ANOVA						
11		*df*	*SS*	*MS*	*F*	*gnificance F*	
12	Regression	1	183.2014	183.2014	25.11214	1.28E-05	
13	Residual	38	277.2226	7.29533			
14	Total	39	460.424				
15							
16		*Coefficient*	*andard Err*	*t Stat*	*P-value*	*Lower 95%*	*Upper 95%*
17	Intercept	2.159728	2.930257	0.737044	0.465622	-3.77227	8.091724
18	Gross domestic product (B$) q-12	0.001383	0.000276	5.011202	1.28E-05	0.000825	0.001942

Excel 9.2 Assess Autocorrelation of the Residuals

If past *GDP* has been growing at the same rate as revenues, we will have accounted for trend in the data. However, *revenues* are also highly seasonal, unlike the more consistent movement in *GDP*. Unaccounted for trend, cycles, or seasonality produces positive autocorrelation in the residuals. The Durbin Watson statistic will allow us to assess positive autocorrelation in the residuals.

Next to the residuals in the regression page, find the Durbin Watson statistic using the two Excel functions, **SUMXMY2(***array1,array2***)** and **SUMSQ(***array***)**. **SUMXMY2** sums the squared differences between adjacent residuals. For *array1*, enter all but the last residual, and for *array2*, enter all but the first residual. **SUMSQ** sums the squared residuals. Enter all of the residuals in this array.

$$=\textbf{SUMXMY2}(array1,array2)/\textbf{SUMSQ}(array).$$

Consult the online DW critical value table. Google "Durbin Watson critical values" to find the Stanford University site: stanford.edu/~clint/bench/dw05a.htm.

Copy and paste the critical values for sample size, $T = 40$, and two independent variables (including the intercept), $K = 2$, next to the Durbin Watson statistic.

D25		f_x	=SUMXMY2(C25:C63,C26:C64)/SUMSQ(C25:C64)				
	A	B	C	D	E	F	G
22	RESIDUAL OUTPUT						
23							
24	Observation	ne Depot re	Residuals	DW			dL and dU
25		1 13.25899	-4.08499	0.629228		40. 2.	1.44214 1.54436
26		2 13.41671	-2.30471				

DW for the model is less than the lower critical value. Evidence suggests that the residuals contain unaccounted for trend, cycles, or seasonality.

Excel 9.3 Plot Residuals to Identify Unaccounted for Trend, Cycles, or Seasonality

The next step is to make a scatterplot of the residuals to identify trend, cycles, or seasonality that we can account for by adding one or more variables to the model.

Paste a copy of the residuals next to quarters in the Home Depot page and make a scatterplot over quarters.

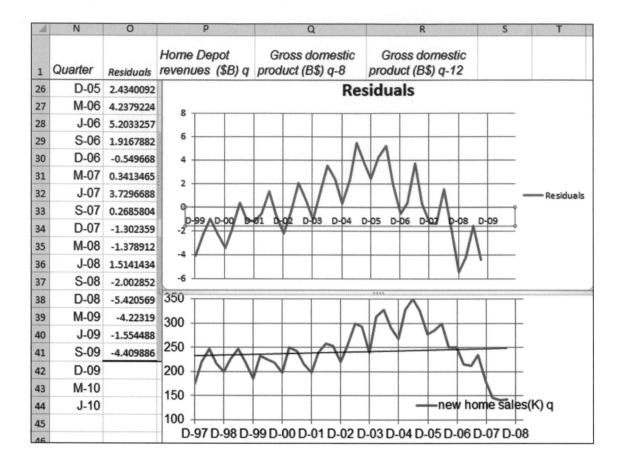

The residuals contain a cycle and seasonality, very similar to the *new home sales* plot. In this case, the plots are lined up to illustrate an eight period lag in *new home sales*.

Compare plots lining up the residuals with a four quarter lag in *new home sales*.

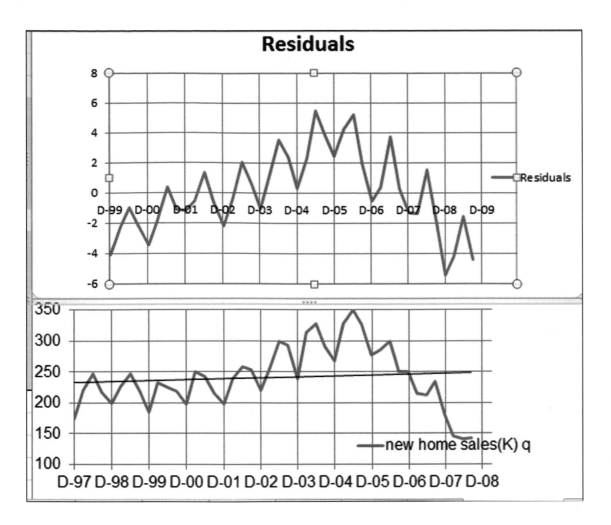

In order to compare correlations with the four- and eight-period *new home sales* lags and residuals, paste two copies of *new home sales*, with cells beginning in December 1998 for a four quarter lag, and with cells beginning in December 1997 for an eight quarter lag.

N	O	P	Q	R	S	T
Quarter	Residuals	Home Depot revenues ($B) q	Gross domestic product (B$) q-8	Gross domestic product (B$) q-12	new home sales(K) q-4	new home sales(K) q-8
D-99	-4.084995	9.2	8506	8023	200	174
M-00	-2.304706	11.1	8601	8137	227	220
J-00	-0.992109	12.6	8699	8277	248	247
S-00	-2.249244	11.5	8847	8410	221	218

Move the residual column next to the *new home sales* lags and find their correlations, **Alt AY11 C**.

	A	B	C
1		*new home sales(K) q-4*	*new home sales(K) q-8*
2	new home sales(K) q-4	1	
3	new home sales(K) q-8	0.659	1
4	Residuals	0.863	0.625

The correlation is greater between residuals and the four quarter lag, as visual inspection also suggested. Add the 4 quarter lag of *new home sales* to the regression with the 12 quarter lag of *GDP*, and find *DW* for the regression with two drivers.

	A	B	C	D	E	F	G	H
1	SUMMARY OUTPUT							
2								
3	*Regression Statistics*							
4	Multiple F	0.922201						
5	R Square	0.850455						
6	Adjusted I	0.842372						
7	Standard I	1.364155						
8	Observati	40						
9								
10	ANOVA							
11		*df*	*SS*	*MS*	*F*	*gnificance F*		
12	Regressio	2	391.57	195.785	105.2088	5.41E-16		
13	Residual	37	68.85397	1.860918				
14	Total	39	460.424					
15								
16		*Coefficients*	*andard Err*	*t Stat*	*P-value*	*Lower 95%*	*Upper 95%*	*ower 9*
17	Intercept	-10.4512	1.900154	-5.5002	2.98E-06	-14.3013	-6.60114	-14.3
18	Gross d(	0.001524	0.00014	10.87952	4.39E-13	0.00124	0.001808	0.0(
19	new hom(	0.045577	0.004307	10.58163	9.61E-13	0.03685	0.054304	0.03
20								
21								
22								
23	RESIDUAL OUTPUT							
24								
25	*)bservatione Depot re*	*Residuals*	*DW*			*dL and dU*		
26	1	10.88951	-1.71551	1.569308		40. 3.	1.39083	1.59999
27	2	12.29381	-1.18181					

Durbin Watson is now much higher, 1.57, but does not yet exceed the upper critical value, 1.60.

Management believes that *Lowes revenues* also drive *Home Depot revenues*. It is likely that *Lowes revenue* may contain the unaccounted for pattern in residuals. Compare a plot of the new residuals with the plot of *Lowes revenues*.

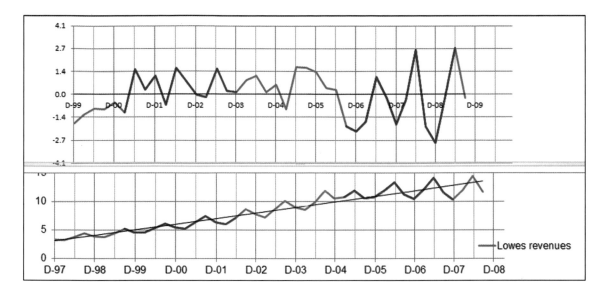

Lowes revenues exhibit seasonality, resembling residuals. In 2001–2003, and later, in 2007–2009, residuals move opposite *Lowes revenues* lagged six quarters. (We have added series in red, above, to illustrate.)

To compare lags with correlations, paste in three copies of *Lowes revenues*:

- The first beginning in December 1998, for a four quarter lag
- The second beginning in June 1997, for a six quarter lag
- The third beginning in December 1997, for an eight quarter lag

Find correlations between the three alternate lags and the residuals.

	A	B	C	D
1		Lowes Revenues (B$) q-4	Lowes Revenues (B$) q-6	Lowes Revenues (B$) q-8
2	Lowes Revenues (B$) q-4	1		
3	Lowes Revenues (B$) q-6	0.848	1	
4	Lowes Revenues (B$) q-8	0.977	0.872	1
5	Residuals	0.186	-0.182	0.076

The four- and six-quarter lags have equivalent correlations with residuals. Given the evidence of competition using the six period lag in the visual inspection, add this longer lag to the multiple regression.

Run multiple regression with all three predictors, calculate *DW*, and copy and paste the Durbin Watson critical values.

	A	B	C	D	E	F	G
2							
3	*Regression Statistics*						
4	Multiple R	0.950737					
5	R Square	0.903901					
6	Adjusted R Square	0.895892					
7	Standard Error	1.108633					
8	Observations	40					
9							
10	ANOVA						
11		*df*	*SS*	*MS*	*F*	*gnificance F*	
12	Regression	3	416.1775	138.7258	112.8708	2.28E-18	
13	Residual	36	44.24645	1.229068			
14	Total	39	460.424				
15							
16		*Coefficient*	*andard Err*	*t Stat*	*P-value*	*Lower 95%*	*Upper 95%*	*owe*
17	Intercept	-22.8392	3.170111	-7.20454	1.77E-08	-29.2685	-16.4099	-29
18	Gross domestic product (B$) q-12	0.003095	0.000369	8.384753	5.5E-10	0.002346	0.003843	0.0
19	new home sales(K) q-4	0.058015	0.00447	12.97908	3.83E-15	0.04895	0.067081	0.
20	Lowes Revenues (B$) q-6	-0.85619	0.191349	-4.47452	7.39E-05	-1.24426	-0.46812	-1.
21							
22							
23							
24	RESIDUAL OUTPUT						
25							
26	*Observation*	*e Depot re*	*Residuals*	*DW*			*dL and dU*
27	1	10.36607	-1.19207	1.725735		40. 4. 1.33835 1.65889	

Adding past *Lowes revenue* increased explanatory power. *R Square* is now .90, up from .85. The standard error is smaller by $.3B, making the margin of error in predictions $2.2B, an improvement in accuracy of $.5B.

All of the coefficient estimates are significant, and all are of the expected sign.
The *DW* statistic is higher, 1.73, clearing the upper critical value, dU = 1.66.

With a significant model, correct signs for the three drivers identified by management as relevant, an acceptable *R Square*, standard error, and residuals free of significant positive autocorrelation, we are ready to validate the model to see whether it produces accurate forecasts.

Excel 9.4 Test the Model's Forecasting Validity

To test model validity, copy the regression *coefficients* and paste into the original Home Depot page.

Calculate *predicted Home Depot revenues* using the regression equation.

S2		f_x =$T$2+$T$3*P2+$T$4*Q2+$T$5*R2		
P	**Q**	**R**	**S**	**T**
Gross domestic product (B$) q-12	new home sales(K) q-4	Lowes Revenues (B$) q-6	predicted Home Depot revenues ($B) q	Coefficients
8023	200	3.77	10.4	-22.8392001
8137	227	4.44	11.7	0.003094856
8277	248	3.91	13.8	0.05801541
8410	221	3.79	12.8	-0.85619174

f_x =T.INV.2T(0.05,36)	
V	**W**
Coefficient: critical t	
-22.839	2.03

To make the 95% *lower* and *upper prediction intervals*, find the *critical t* for 36 *residual df* using the Excel function **T.INV.2T** (*probability,df*), entering .05 for *probability* and a 95% level of confidence.

f_x =W2*X2		
W	**X**	**Y**
critical t	s	me
2.03	1.11	2.25

Copy the regression *standard error* from **B7** and paste next to *critical t*, and then calculate the *margin of error* by multiplying the *standard error* by the *critical t*.

Make the 95% *lower* and 95% *upper* prediction interval bounds from the *predictions* by subtracting and adding the margin of error. (Lock the reference to the margin of error cell with Fn 4.)

	R2				f_x	=Q2-$W$2	
	R	S	T	U	V	W	
	lower 95%	upper 95%	Coefficients	critical t	s	me	
1							
2	8.1	12.6	-22.8392001	2.03	1.11	2.25	
3	9.5	14.0	0.003094856				
4	11.6	16.1	0.05801541				
5	10.5	15.0	-0.85619174				
6	8.1	12.6					
7	10.5	15.0					

	L	M	N	O	P	Q	R	S
1	Quarter	Home Depot revenues ($B) q	Gross domestic product (B$) q-12	new home sales(K) q-4	Lowes Revenues (B$) q-6	predicted Home Depot revenues ($B) q	lower 95%	upper 95%
42	D-09	14.6	13612	116	14.51	13.6	11.3	15.8
43	M-10	16.9	13796	85	11.73	14.7	12.5	17.0

The model prediction intervals contain both of the two most recent, holdout revenues, providing evidence that the model is valid for forecasting.

Excel 9.5 Recalibrate to Forecast

Recalibrate the model by rerunning the regression, this time including the two most recent quarters.

SUMMARY OUTPUT					
Regression Statistics					
Multiple R	0.947256				
R Square	0.897294				
Adjusted R Square	0.889185				
Standard Error	1.120893				
Observations	42				

ANOVA					
	df	*SS*	*MS*	*F*	*gnificance F*
Regression	3	417.108	139.036	110.6622	7.91E-19
Residual	38	47.74321	1.2564		
Total	41	464.8512			

	Coefficients	*andard Err*	*t Stat*	*P-value*	*Lower 95%*	*Upper 95%*
Intercept	-22.5504	3.049327	-7.3952	7.23E-09	-28.7234	-16.3774
Gross domestic product (B$) q-12	0.003128	0.000355	8.801737	1.05E-10	0.002409	0.003848
new home sales(K) q-4	0.055186	0.003928	14.04883	1.25E-16	0.047234	0.063139
Lowes Revenues (B$) q-6	-0.84418	0.180728	-4.67098	3.69E-05	-1.21004	-0.47831

Copy and paste the recalibrated *coefficients* over the validation coefficients, which will update predictions.

V	W	X
Coefficient.	*critical t*	*s*
-22.55	2.02	1.12
0.00313		
0.05519		
-0.8442		

Copy and paste the recalibrated *standard error* over the validation standard error, and update *t* to reflect 38 *residual df*s, which will update the 95% prediction interval columns.

Excel 9.6 Illustrate the Fit and Forecast

To see the model fit and forecast, plot *Home Depot revenues ($B)* and 95% *predicted lower* and *upper* values by *quarter*.

Rearrange columns.

Select and cut the prediction interval columns, then insert into columns following *quarter*. Make a scatterplot from the *quarter*, 95% *lower*, and 95% *upper* columns, using rows through June 2011. (The forecast reaches through June 2011, and goes not further, since the last observation on *new home sales*$_{q-4}$ is in June 2011.)

Add the series of the actual revenues using **Alt JCE**.

	L	M	N	O	P	Q	R	S	
1	Quarter	lower 95%	upper 95%	Home Depot revenues ($B) q	Gross domestic product (B$) q-12	new home sales(K) q-4	Lowes Revenues (B$) q-6	predicted Home Depot revenues ($B) q	Coe
37	S-08	17.7	22.3	17.8					
38	D-08	13.6	18.2	14.6					
39	M-09	14.7	19.3	16.2					
40	J-09	16.0	20.5	19.1					
41	S-09	15.0	19.6	16.4					
42	D-09	11.9	16.4	14.6					
43	M-10	13.1	17.7	16.9					
44	J-10	15.2	19.7						
45	S-10	15.3	19.8						
46	D-10	14.1	18.6						
47	M-11	15.1	19.7						
48	J-11	15.4	19.9						

Recolor one of the 95% prediction intervals so that both are of the same color.

Change the *Home Depot revenues ($B)* series from a line to markers.

Format axes to reduce white space and add gridlines.

Select a style and design and add a chart title that summarizes your conclusion.

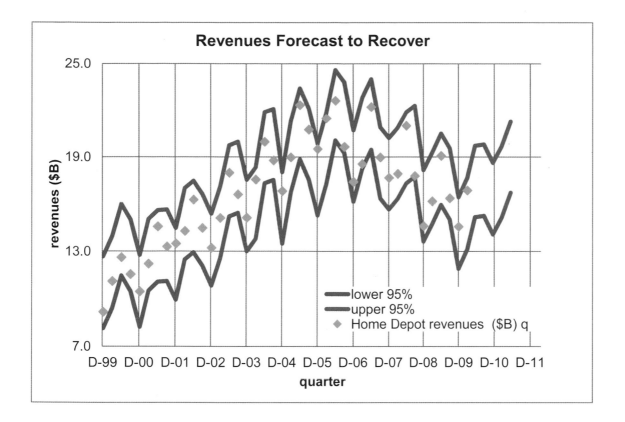

Excel 9.7 Assess the Impact of Drivers

Use the regression equation to look at the impact of each of the drivers on model forecasts.
Impact of drivers:

To see the impact of each of the drivers, first find the maximum and minimum levels that have occurred in the recent past. Use the past values used as input to the five quarter forecast as the "recent past," which will mean the past 3 years for GDP, the past six quarters for Lowes revenue, and the past year for new home sales.

Make six hypothetical quarters, setting two of the three variables to their values in the most recent period, March 2010, and the third variable to the max and min of values from recent quarters:

The six hypothetical quarters will resemble March 2010, except that

- *New home sales* will instead be either the max or the min from the past four quarters for the first two.
- *GDP* will instead be either the max or the min from the past 12 quarters for the next two.
- *Lowes revenues* will instead be either the max or the min from the past six quarters for the last two.

Select the last *predicted revenues* cell (June 2011), **Shift+dn** through the six new rows and **Cntl+D** to fill in the hypothetical predictions.

Find the absolute value of difference between each pair of predictions to learn the impact of each driver.

	R	S	T	U	V	W	X
	Gross domestic product (B$) q-12	new home sales(K) q-4	Lowes Revenues (B$) q-6	predicted Home Depot revenues ($B) q	Coefficients	critical t	s
	13796	85	11.73	15.4			
	13997	84	9.98	17.4			
	14180	104	11.83	17.5			
	14338	104	13.84	16.3			
	14374	83	11.38	17.4			
	14498	87	11.73	17.7	difference (B$)	due to	
	14601	85	11.73	17.92			
	13997	85	11.73	16.03	1.9	GDP ($B)	
	13796	104	11.73	16.44			
	13796	83	11.73	15.30	1.1	new home sales (K)	
	13796	85	14	13.61			
	13796	85	10	16.87	3.3	Lowes revenues (B$)	

Percent Growth Forecast: It is useful to state forecast quarterly growth as the annual percent from the same quarter last year. Find forecast quarterly growth, comparing forecasts for June 2010–2011 with June 2009–2010. In a quarter, find the difference expected from the past year as a percentage of past year revenues, using actual revenues as the baseline for the four quarters June 2010–March 2010 and using predicted revenues as the baseline for the last quarter June 2011:

$$=(predicted\ revenue_q - revenue_{q-4})/revenue_{q-4}\ \text{for June 2010–March 2010}$$

and

$$=(predicted\ revenue_q - predicted\ revenue_{q-4})/predicted\ revenue_{q-4}\ \text{for June 2011.}$$

Convert the four proportions to percents using **Alt Home Percent**.

T48			f_x	=(S48-S44)/S44					
	L	M	N	O	P	Q	R	S	T
1	Quarter	lower 95%	upper 95%	Home Depot revenues ($B) q	Gross domestic product (B$) q-12	new home sales(K) q-4	Lowes Revenues (B$) q-6	predicted Home Depot revenues ($B) q	Coefficients
40	J-09	16.0	20.5	19.1	13348	141	10.38	18.2	
41	S-09	15.0	19.6	16.4	13453	143	12.01	17.3	
42	D-09	11.9	16.4	14.6	13612	116	14.51	14.2	
43	M-10	13.1	17.7	16.9	13796	85	11.73	15.4	annual growth fo
44	J-10	15.2	19.7		13997	84	9.98	17.4	-9%
45	S-10	15.3	19.8		14180	104	11.83	17.5	7%
46	D-10	14.1	18.6		14338	104	13.84	16.3	12%
47	M-11	15.1	19.7		14374	83	11.38	17.4	3%
48	J-11	15.4	19.9		14498	87	11.73	17.7	1%

Lab Practice 9 Starbucks in China

Read the *first page* of the case description on page 305 and then build a *valid* leading indicator model of Starbucks revenues to forecast revenues in 2007–2009 from data in **Case 9-3 Starbucks Revenue.xls**. Instead of writing a memo, answer the following questions.

I.   Assess your model and its implications.

1. What percent of variation in Starbucks *revenues* can be explained with variation in past *Chinese per capita GDP* and past *revenues?*_____

2. What is the margin of error for your forecasts?_____

3. Is your model free of unaccounted for trend and cycles? Y Maybe N
Evidence: _____

4. Is your model valid? Y or N
Evidence:_____

II.   Recalibrate if your model is valid.

5. Following each increase of **$.3K (three hundred dollars)** in *Chinese per capita GDP*, the expected change in *revenues* is _____ years later.

6. Is there evidence of Starbucks customer loyalty? N or Y

If yes, what is the extent of this loyalty?. . .What range of increase in *revenues* is expected following cach *revenue* increase of **$1B?** _____ years later.

7. Is Starbucks likely to make its claim of *revenues* of **\$13 billion by 09?** Y or N

   Evidence:_____

III.   Embed a scatterplot of your fit and forecast by year with your regression equation, *R Square*, and *significance* levels.

IV.   By what percent has annual *revenue* grown in the 5 years 2002–2006, and what are the expected annual growth percents in the years 2007–2009?

	% Revenue growth		*Expected % revenue growth*
2001–2002	_____	2006–2007	_____
2002–2003	_____	2007–2008	_____
2003–2004	_____	2008–2009	_____
2004–2005	_____		
2005–2006			

Lab 9: HP Revenue Forecast

Meg Whitman, Hewlett Packard's new CEO, would like to promise shareholders that worldwide revenues will reach $150 billion by 2012. A competing consulting firm produced the above forecast, which indicates that HP revenues will reach this target.

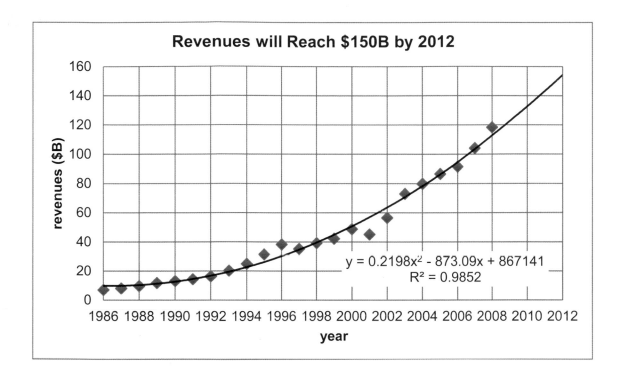

The consultant used Excel to fit a trend. Although pleased with this optimistic forecast, Meg Whitman would like a second opinion. You have been hired to confirm that this seems likely. Whitman is concerned by Chinese competitors who are gaining ground as China industrializes.

Data are in **Lab 9 HP forecast**.xls and contain annual *HP revenues* in billion dollars, *GDP* in trillion dollars, *Dell revenues* in billion dollars, and *Chinese per capita GDP* in thousand dollars for nearly 25 years, 1985–2008 or 2009.

Make scatterplots to see *GDP* leading *HP Revenues*. To see how *GDP* leads *HP revenues*, make scatterplots of each year, and add trendlines to both. Add an X to cells for years in which you see slowed growth.

Year	93	94	95	96	97	98	99	00	01	02	03	04	05	06
GDP slowed														
HP slowed														

Following slowing of *GPD*, *HP* sometimes slows ____ 2 ____3 ____ 4 years later.

Copy *Year* and *HP revenues* into new columns and then delete four cells of each for years 1986–1989.

Add lagged indicators. Add in years 1990–2012 seven new columns:

GDP_{t-2} and GDP_{t-4}, $Dell_{t-3}$ and $Dell_{t-4}$, *Chinese per capita* GDP_{t-2}, *Chinese per capita* GDP_{t-3}, and *Chinese per capita* GDP_{t-4}.

Decide whether you will use one or two tail tests of the coefficients by specifying expected signs:

	Expected coefficient sign			*One or two tail test?*
	Negative	*Positive*	*Negative or positive*	
Past GDP				1 2
Past Dell revenues				1 2
Past Chinese per capita GDP				1 2

Find the correlations between *HP revenue* and each of the seven lagged variables, using years 1990–2006, and then choose the lag with the highest correlation with expected sign to run a simple leading indicator regression.

Assess autocorrelation. Look up the Durbin Watson critical values in http://www.stanford.edu/~clint/bench/dw05a.htm dL: ____ dU: ____

Find the model Durbin Watson value using the residuals: _____

Conclude: The model ___has ___ may have ___ is free of unaccounted for trend or cycles.

Copy the residuals into the HP sheet and find correlations with the two Dell lags and three Chinese per capita GDP lags.

Choose the lagged variable with the highest correlation with residuals with expected sign to add to your regression.

Compare *R Squares* and *standard errors*:

	R Square	Standard error
Model with *GDP*		($B)
Model with *GDP* and additional variable		($B)

Look up the Durbin Watson critical values: dL:____ dU: ____

Find the model Durbin Watson value: ____

   Conclude: The model ___has ___ may have ___ is free of unaccounted for trend or cycles.

Copy the residuals into the HP sheet and find correlations with the three lags for the variable not yet in the model.

Choose the lagged variable with the highest correlation with residuals (and expected sign) to add to your regression.

Compare *R Squares* and *standard errors*:

	R Square	Standard error
Model with *GDP* and additional variable		(B$)
Model with *GDP*, *Dell*, and *Chinese per capita GDP*		(B$)

What does the coefficient sign for the lagged *Dell revenues* variable tell us?

_____

What does the coefficient sign for the lagged *Chinese per capita GDP* variable tell us?

_____

Look up the Durbin Watson critical values: dL:____ dU: ____

Find the model Durbin Watson value: ____

   Conclude: The model ___has ___ may have ___ is free of unaccounted for trend or cycles.

Validate your model: Copy the *coefficients* and *standard error* into the HP sheet and use the regression equation to make *predicted HP revenues* and *lower* and *upper 95% prediction intervals*.

Do prediction intervals contain the hidden *HP revenues* for 2007 and 2008?   Y or N

Recalibrate by running the regression again with years through 2008.
Can Chairman Whitman claim that HP revenues will reach $150 billion by 2012?   Y or N

Explain why Chairman Whitman should rely on your results instead of the competing consultant's

results: _____

Case 9-1 Revitalizing Dell (Harvard Business School case 9-710-442)

Read the case, and then build a model of Dell revenues. Based on model results, identify the revenue drivers and their relative importance.

Data are in **Case 9-1 Dell Revenue Forecast.xls** and contain

> *PC Market Worldwide (MM), PC Market Worldwide (B$)*, and *Average PC Price ($)* from Exhibit 1
> *Dell Revenues (B$)*, from Exhibit 7
> *Average Dell PC Price ($)* and *Average HP PC Price ($)*, from Exhibit 9
> *Ratings of Dell PCs* and *Ratings of HP PCs*, from Exhibit 11
> plus *Hewlett Packard Revenues ($B)*, and *Internet Users in China (MM)*

for years 1989, 1990, or 1991–2008.

Proposed Steps

1. Hide the two most recent *Dell revenues* from 2007 and 2008. Build your model without these two cells so that you will be able to test your model's forecasting validity.

2. Create new columns to use in your model which begin in 1993, adding lags for the potential drivers. Consider 3- and 4-year lags for the *PC Market* drivers, 2 year lags for the *Price*, *Ratings*, and *HP Revenues* drivers, and 2- and 3-year lags for *Internet Users in China*.

3. Decide what coefficient signs you expect and whether you will use a one or two tail test for each potential driver.

4. Use visual inspection and correlations to identify the strongest lagged driver and run a simple regression of *Dell Revenue*.

5. Check model significance, the coefficient sign, and autocorrelation using the Durbin Watson statistic.

6. If your model is not significant, choose a different driver, rerun, and reassess. If your model is significant, choose a second lagged driver by looking at pattern in residuals and correlations. Add only those potential drivers that have correlations with correct signs.

7. Repeat this process to complete your model, considering all of the potential drivers thought important by the management. You may add additional variables to improve the model's explanatory power, even though you have accounted for trend, cycles, and seasonality.

8. If all of the slopes have correct signs, are significant, and your model is at least possibly free from unaccounted for trend and cycles (dL < *DW*), validate your model. Your model is valid if the actual *Dell revenues* in 2007 and 2008 fall within your 95% *prediction intervals* for both years.

9. If your model is not valid, try one or more different lags, rerun, and reassess. If your model is valid, recalibrate by adding data from 2007 and 2008 and then rerunning.

10. Make a scatterplot showing your fit and forecast: 95% *prediction intervals* through 2010 or 2011 plus actual *Dell Revenues* through 2008, with years on the *x* axis.

11. Conduct sensitivity analysis to find the expected change in revenues following a realistic change in each driver. (A realistic change would be the average of changes in the past 5 years.)

Deliverables

Present your final model in a one page, single spaced memo to Dell executives. Explain that you built a forecasting model from time series, using a 20 year series of data from case data, annual reports, and the China Internet Network Information Center. Embed your scatterplot and include your regression equation with *R Square* and significance levels.

- Note the margin of prediction error in your presentation of results.
- Note that you validated your model, and briefly explain how you did this, and what validity means (in plain English).
- Explain what your regression equation means, including examples of the impact of *realistic* changes in each of the drivers and noting the influence(s) of important, but redundant driver(s) which matter, but which are not in the equation, if any.
- Present your forecast for 2010.

Be sure to specify units throughout, and be sure to round to two or three significant digits. Do not include discussion of your model building process. This is not of interest to Dell executives. They care only about the impact of drivers and your forecast. You should not use terms such as *multicollinearity*, 95% *certain*, *significant* in your text.

For quantitatively savvy readers, please attach the two Excel sheets showing your model (1) before recalibration with your Durbin Watson analysis and (2) after recalibration. Please attach only these two sheets. Please make sure that your columns are wide enough to show complete variable labels.

Case 9-2 Mattel Revenues Following the Recalls

> Despite recent press reports that recalls of toys manufactured in China will curb revenues, Mattel management is claiming that revenue growth will double in 2007 and 2008, reaching $6 billion by 2008.
>
> Mattel management is counting on the growing number of preschool and elementary children to fuel revenues. More children ought to translate to more toy sales.
>
> Management is aware that toys are luxuries and sales are likely to be linked to past growth in GDP. Mattel managers are also aware that when children choose Hasbro toys, products of their strongest competitor, Mattel has traditionally lost sales.

Build a *valid* leading indicator model of Mattel revenues to forecast revenues in 2007 and 2008 from data in **Case 9-2 Mattel.xls**. The dataset contains *Mattel Revenues* (B$) in billion dollars, *US GDP* (*$T*) in trillion dollars, *4 year old population* (*MM*) in millions, *7 year old population* (*MM*) in millions, and *Hasbro revenues* (*$B*) in billion dollars for years 1985–2006 with population estimates through 2008. Use years 1989–2004 to build your model.

First, choose *GDP* from 2 or 3 years prior and include this in a regression with *4-* and *7-year olds*.

Next, choose *Hasbro revenues* from 2 or 3 years prior.

Write a one page memo to present your results to management. Include in your memo

1. Percent of variation in *Mattel revenues* explained with variation in past *GDP*, *4-* and *7-year old* populations, and past *Hasbro revenues*.

2. Margin of error for your forecasts.

3. The range in *revenue* increase which Mattel can expect following each increase of **$1T (one trillion dollars)** in *GDP*.

   (Be sure to specify units and when the increase can be expected.)

4. The change in *revenue* which Mattel could expect if an additional **1M (one million)** babies were born 4 years ago.

5. The change in *revenue* expected if an additional **1M (one million)** babies were born 7 years ago.

6. The expected **revenue** change if **Hasbro revenues** increase by **$1B (one billion)**, on average. (Be sure to specify **units and time** of the expected **change.**)

7. Whether or not your model free of unaccounted for trend and cycles? (Use a footnote to refer to the statistic that you used to draw your conclusion.)

8. The range in *revenues* forecast in 2007 and 2008, with 95% confidence.

9. Likelihood that Mattel will meet its claim to achieve $6 billion by 2008.

10. Annual *revenue* growth percent average in the past 5 years, 2002–2006, and expected annual annual growth percent in the next 2 years.

11. Model validity.

Embed a scatterplot of your fit and forecast, including your regression equation, *R Square*, and significance levels.

Case 9-3 Starbucks in China

Despite recent press that their revenue growth is stagnating, Starbucks management is claiming that revenues will grow by 20% annually, reaching $13 billion by 2009. Starbucks management is counting on the growing coffee consumption in China to fuel revenues. In China, Starbucks coffee is considered a luxury. More and more Chinese will be able to afford the treat, as per capita GDP continues to grow. Two recent articles explain this:

A Tall Espresso Con Panna costs $1.63, while a small coffee of the day is $1.50. And a Mocha Frappuccino Grande sets you back a substantial 3.63 at the crowded Starbucks stores of Beijing, Shanghai, and Tianjin. Wait a second – isn't the mainland better known for leaves steeped in water, as demonstrated by the phrase "all the tea in China?" There's no shortage of tea in the country that invented it, but the fact is that java beans are a new sensation for the relatively well-off urban Chinese, who now earn on average $1,312 per year, up 9.6% this year. [Rural Chinese won't likely be drinking Seattle's finest anytime soon, however; rural incomes, still less than a third of their urban counterparts, this year grew 6.2% to $407.]

In the seven years since H&Q Asia – the former controlling shareholder of Beijing Mei Da Coffee – opened the first Starbucks shop in Beijing in 1999, the Seattle phenomenon has grown to 190 stores in 19 cities in mainland China. "It's not just a drink in China. It's a destination. It's a place to be seen and a place to show how modern one is," adds Technomic Asia's Kedl. And with China's economy growing in double digits, there are likely to be lots more young urban and modern Chinese ready to sip java in a sleek new Starbucks. (Business Week Online, October 26, 2006)

Starbucks Corp. executives have forecast that about 20 percent of its international growth will occur in China this year, which has the potential for more than 200 million customers. There already are more than 500 Starbucks Coffee outlets in China, about 300 of which have opened in the past two years, and Martin Coles, president of Starbucks 'international division, told a telephone conference of financial analysts that the chain would add 200 more there by 2008. Chairman Howard Schultz, emphasizing Starbucks' current presence in Beijing and 17 provinces, said he anticipates the brand will continue to do well in Hong Kong and gain strength in Taiwan. "We are dreaming very big in China ," he said. (Nation's Restaurant News, May 21, 2007)

Starbucks managers also believe that their loyal customers will continue to return to purchase their favorite coffees, in spite of growing competition.

Build a *valid* leading indicator model of Starbucks revenues to forecast revenues in 2007–2009 from data in **Case 9-3 Starbucks Revenue.xls**. The dataset contains *Starbucks Revenues* (B$) in billion dollars, and *China GDP per capita* (*$T*) in trillion dollars for years 1988–2006, with estimates of *China GDP per capita* through 2008. Use years 1991–2004 to build your model.

First, choose Chinese per capita GDP from 2 or 3 years prior.

Next, choose Starbucks revenues from 2 or 3 years prior. (Prior revenues reflect inertia in consumer behavior or the tendency for Starbucks customers to remain loyal, rather than switch to other coffee sources.)

Write a one page memo presenting your results to management. Be sure to include in your memo the following:

1. Percent of variation in Starbucks *revenues* which can be explained with variation in past *Chinese per capita GDP* and past *Starbucks revenues*.

2. The margin of error for your forecasts.

3. Following each increase of **$1K (one thousand dollars)** in *Chinese per capita GDP*, the expected change in *revenues.* (Be sure to specify uni**ts and the expected time o**f th*e change.)*

4. *Evidence of* Starbucks customer loyalt*y and* the extent of this loyalty. The range of increase in *Starbucks revenues* expected, following each *revenue* increase of **$1B (one billion dollars)**.

5. Whether or not your model is free of unaccounted for trend and cycles. (Use a footnote to include the statistic that you used to draw your conclusion.)

6. The *range* in revenues forecast in 2007, 2008, and 2009 with 95% confidence.

7. Likelihood that Starbucks' will match its claim to achieve *revenues* of **$13 billion by 2009**.

8. Average annual *revenue* growth percent in the past 5 years, 2002–2006, and expected annual growth percent the next 3 years.

9. Model validity.

Embed a scatterplot of your fit and forecast with your regression equation *R Square* and *significance* levels.

Case 9-4 Harley–Davidson Revenue Forecast

H–D management would like to explain the recent downturn in revenues and forecast revenues in the next 2–3 years.

The managers believe that gas prices drive revenues. The direction of this influence is not clear. While higher gas prices probably encourage customers to switch from cars to motorcycles, higher gas prices probably also motivate customers to buy lighter, more fuel efficient motorcycles.

Asian competitors' revenues have also slowed. Harley–Davidson has historically been the market leader. Are Asian firms, such as Honda and Kawasaki, selling motorcycles that would have been Harley–Davidsons? Or is the recent slump attributable to the recent recession? Motorcycles are probably a luxury item, and potential buyers may have decided to postpone purchases until the economy improves.

The primary customer segment has traditionally been Baby Boomers, born between 1946 and 1964. After seeing *Easy Rider* in the 1960s, many Baby Boomers wanted their own Harley and the associated image and lifestyle. The oldest Baby Boomers are now approaching retirement and may be less interested in buying motorcycles.

Build a valid model to explain the recent decline in H–D revenues and to forecast revenues in 2010–2011 or 2012.

Harley–Davidson Revenues.xls contains annual data for years 1993–2009 on Harley–Davidson revenues (M$), US gas prices ($ per gallon), Honda revenues (B¥), US GDP, and US male population by age (K).

1. Using visual inspection, identify the driver (with a 2- or 3-year lag) which best explains variation in revenues.

2. Use that driver in your model, and then assess the residuals. With 95% certainty, are residuals free of unaccounted for trend and cycles?

3. Using visual inspection, find the six potential drivers (with 2- or 3-year lag) which best explain residual variation.

4. Add a second driver and assess your model:
   a. Will you use a one or two tail test for each of the two drivers?

   b. Does variation in the two drivers explain some of the variation in revenues? Site the statistic and its value that you used to decide.

   c. If managers use the model to forecast, forecasts should be no further than \$_____B from the forecast 95% of the time.

   d. With 95% certainty, are residuals free of unaccounted for trend and cycles?

5. Explain how you will test the model's validity.

6. Is your two driver model valid?

7. Present your regression equation, including slope significance levels.

8. Which of the two drivers is more important?

9. Explain how you evaluated driver importance, including the numbers that you compared.

10. Summarize your results, qualitatively (without numbers), explaining to management which potential influences are responsible for the recent decline in revenues.

11. Illustrate your fit and forecast graphically. (You do not need to include the regression equation which you already provided.)

12. Attach Excel output showing your final model (Summary table, ANOVA table, and Coefficient table), being sure that variable names in column A are showing.

Chapter 10
Indicator Variables

In this chapter, 0–1 *indicator* or *"dummy"* variables are used to incorporate segment differences, shocks, or structural shifts into models. In cross sectional data, indicators can be used to incorporate the unique responses of particular groups or segments. In time series data, indicators can be used to account for external shocks or structural shifts. Indicators also offer one option to account for seasonality or cyclicality in time series.

Analysis of variance sometimes is used as an alternative to regression when potential drivers are categorical or when data are collected to assess the results of an experiment. In this case, the categorical drivers could be represented with indicators in regression or analyzed directly with analysis of variance.

This chapter introduces the use of indicators to analyze data from conjoint analysis experiments. Conjoint analysis is used to quantify customer preferences for better design of new products and services.

Model variable selection begins with the choice of potential drivers from logic and experience. Indicators are added to account for segment differences, shocks, shifts or seasonality, and, in time series models, if autocorrelation remains, trend, inertia, and a leading indicator or an indicator variable may be added to remedy the autocorrelation. The addition of indicators in the variable selection process is considered in this chapter.

10.1 Indicators Modify the Intercept to Account for Segment Differences

To compare two segments, a 0–1 indicator can be added to a model. One segment becomes the baseline, and the indicator represents the amount of difference from the base segment to the second segment. Indicators are like switches that turn on or off adjustments in a model intercept.

Example 10.1 Hybrid Fuel Economy. In a model of the impact of car characteristics on fuel economy:

$$\hat{MPG} = b_0 + b_1 Hybrid + b_2 Emissions + b_3 Horsepower$$
$$= 48 + 8.8 Hybrid - 2.3 Emissions - .025 Horsepower .$$

The coefficient estimate of 8.8 MPG for the *hybrid* indicator modifies the intercept. For conventional cars, the *hybrid* indicator is 0, making the intercept for conventional cars 48 MPG:

$$\hat{MPG} = 48 + 8.8(0) - 2.3 Emissions - .025 Horsepower$$
$$= 48 \qquad - 2.3 Emissions - .025 Horsepower .$$

For hybrids in the sample, the *hybrid* indicator is 1, which adjusts the intercept for hybrids to 56.8 MPG by adding 8.8 MPG to the baseline 48 MPG:

$$\hat{MPG} = 48 + 8.8(1) - 2.3\,Emissions - .025\,Horsepower$$
$$= 56.8 \qquad\quad - 2.3\,Emissions - .025\,Horsepower\,.$$

The adjustment is switched on when *hybrid* = 1, but remains switched off if *hybrid* = 0.
The parameter estimate for the indicator tells us that on average, hybrid gas mileage is 8.8 MPG higher than conventional gas mileage.

Example 10.2 Yankees v Marlins Salaries[6]. The Yankees General Manager has discovered that the hot rookie whom the Yankees are hoping to sign is also considering an offer from the Marlins. The General Manager would like to know whether there is a difference in salaries between the two teams. He believes that, in addition to a possible difference between the two teams, *Runs* by players ought to affect salaries.

We will build a model of baseball salaries, including *Runs* and an indicator for Team. This variable, *Yankees*, will be equal to 1 if a player is in the Yankees Team, and equal to 0 if the player is a Marlin. The Marlins is the baseline team. Data are shown in Table 10.1, and regression results are shown in Table 10.2.

---

[6] This example is a hypothetical scenario based on actual data

Table 10.1 Baseball team salaries

Player	Team	Yankee	Runs	Salary (M$)
Castillo	Marlin	0	72	5.2
Delgado	Marlin	0	81	4.0
Pierre	Marlin	0	96	3.7
Gonzalez	Marlin	0	45	3.4
Easley	Marlin	0	37	.8
Cabrera	Marlin	0	106	.4
Aguila	Marlin	0	11	.3
Treanor	Marlin	0	10	.3
Rodriguez	Yankee	1	111	21.7
Jeter	Yankee	1	110	19.6
Sheffield	Yankee	1	94	13.0
Williams	Yankee	1	48	12.4
Posada	Yankee	1	60	11.0
Matsui	Yankee	1	97	8.0
Martinez	Yankee	1	41	2.8
Womack	Yankee	1	46	2.0
Sierra	Yankee	1	13	1.5
Giambi	Yankee	1	66	1.3
Flaherty	Yankee	1	8	.8
Crosby	Yankee	1	10	.3
Phillips	Yankee	1	7	.3

Table 10.2 Multiple regression of baseball salaries

SUMMARY OUTPUT

Regression statistics

R Square	.57					
Standard Error	4.2					
Observations	35					

ANOVA	df	SS	MS	F	Significance F	
Regression	2	754	377	21.3	.0000	
Residual	32	566	18			
Total	34	1,320				

	Coefficients	Standard error	t Stat	p Value	Lower 95%	Upper 95%
Intercept	−3.90	1.56	−2.5	.02	−7.06	−.73
Yankee	6.31	1.43	4.4	.0001	3.40	9.22
runs	.104	.020	5.1	.0000	.062	.15

From the regression output, the model is

$$\hat{Salary}(\text{M\$}) = -3.90^a + 6.31^b \, Yankee + .104^b \, Runs$$

R Square: .57[b]
[a]Significant at .05
[b]Significant at .01.

The coefficient estimate for the Yankee indicator is \$6.31 M. The intercept for Yankees is \$6.31 M greater than the intercept for Marlins. The rookie can expect to earn \$6.31 million more if he signs with the Yankees.

His expected salary, with 40 runs last season, is

- As a Marlin, setting the *Yankee* indicator to zero:

$$\hat{Salary}(\text{M\$}) = -3.90 + .104(40) = -3.90 + 4.16 = .26(\text{M\$}) = \$260,000.$$

- As a Yankee, setting the *Yankee* indicator to one:

$$\hat{Salary}(\text{M\$}) = -3.90 + 6.31 + .104(40) = 2.41 + 4.16 = 6.57(\text{M\$}) = \$6,570,000.$$

The *Yankee* indicator modifies the intercept of the regression line, increasing it by \$6.31 M.

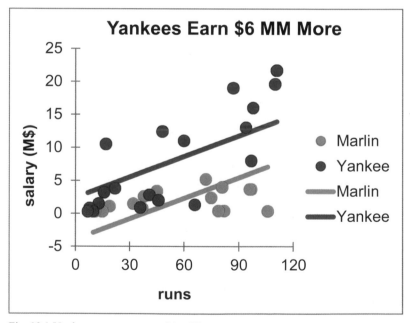

Fig. 10.1 Yankees expect to earn \$6 million more

In Fig. 10.1, the intercept represents the baseline Marlins segment; the indicator adjusts the intercept to reflect the difference between Yankees and Marlins.

It does not matter which team is the designated baseline. The model will provide identical results either way.

10.2 Indicators Estimate the Value of Product Attributes

New product development managers sometimes use *conjoint analysis* to identify potential customers' most preferred new product design and to estimate the relative importance of product attributes. The conjoint analysis concept assumes that customers' preferences for a product are the sum of the values of each of the product's attributes, and that customers *trade off* features. A customer will give up a desired feature if another, more desired, feature is offered. The offer of a more desired feature compensates for the lack of a second, less desired, feature.

> *Example 10.3 New PDA Design.* As an example, consider preferences for PDAs. Management believes that customers choose PDAs based on desired size, design, keypad, and price. For a new PDA design, they are considering
>
> - Three sizes: bigger than shirt pocket, shirt pocket, and ultrathin shirt pocket
> - Three designs: single unit, clamshell, and slider
> - Three keypads: standard, touch screen, and QWERTY
> - Three prices: $150, $250, and $350
>
> Management believes that price is a quality signal and that customers suspect the quality of less expensive PDAs.
>
> The least desirable, baseline configuration is expected to be bigger than shirt pocket, single unit, with standard keypad at the lowest price.

To find the *part worth utilities*, or the value of each cell phone feature, indicators are used to represent features that differ from the baseline. The conjoint analysis regression model is

$$PDA\ \hat{preference}_i = b_0 + b_1 shirt\ pocket\ size_i \quad + b_2 ultra\ thin\ shirt\ size_i$$
$$+ b_3 clamshell_i \qquad + b_4 slider_i$$
$$+ b_5 touch\ screen_i \qquad + b_6 QWERTY_i$$
$$+ b_7 \$250_i \qquad + b_8 \$350_i$$

for the *ith* PDA configuration.

b_0 is the intercept, which reflects preference for the baseline configuration, b_1, b_2, b_3, b_4, b_5, b_6, b_7, and b_8 are estimates of the *part worth utilities* of features.

The conjoint analysis process assumes that it is easier for customers to rank or rate products or brands, rather than estimating the value of each feature. For price preferences, this may be particularly true. It will be easier to customers to rate hypothetical PDA designs than it would be for customers to estimate the value of a $250 PDA, relative to a $150 PDA.

The four PDA attributes could be combined in 81 ($=3^4$) unique ways. Eighty-one hypothetical PDAs would be too many for customers to accurately evaluate. From the 81, a set of nine are carefully chosen so that the chance of each feature is equally likely (33%), and each feature is uncorrelated with other features. Slider designs, for example, are equally likely to be paired with each of the three sizes, each of the three keypads, and each of the three prices. This will eliminate multicollinearity among the indicators used in the regression of the conjoint model. Such a subset of hypothetical combinations is an *orthogonal array* and is shown in Table 10.3.

Table 10.3 Nine hypothetical PDA designs in an orthogonal array

Size	Shape	Keypad	Price
Bigger than shirt pocket	Single unit	Standard	$150
Bigger than shirt pocket	Clamshell	Touch screen	$250
Bigger than shirt pocket	Slider	QWERTY	$350
Shirt pocket	Single unit	Touch screen	$350
Shirt pocket	Clamshell	QWERTY	$150
Shirt pocket	Slider	Standard	$250
Ultrathin shirt pocket	Single unit	QWERTY	$250
Ultrathin shirt pocket	Clamshell	Standard	$350
Ultrathin shirt pocket	Slider	Touch screen	$150

Three customers rated the nine hypothetical PDAs after viewing concept descriptions with sketches. The configurations judged extremely attractive were rated 9 and those judged not at all attractive were rated 1. The regression with eight indicators is shown in Table 10.4.

Table 10.4 Regression of PDA preferences

Regression statistics

R Square	.747				
Standard error	1.644				
Observations	27				

ANOVA	Df	SS	MS	F	Significance F
Regression	8	143	17.9	6.6	.0004
Residual	18	49	2.7		
Total	26	192			

	Coefficients	Standard error	t Stat	p Value	Lower 95%	Upper 95%
Intercept	1.00	.95	1.1	.31	−.99	2.99
shirt pocket	.78	.78	1.0	.33	−.85	2.41
ultrathin shirt pocket	1.89	.78	2.4	.03	.26	3.52
clamshell	−1.56	.78	−2.0	.06	−3.18	.07
slider	−1.44	.78	−1.9	.08	−3.07	.18
touch screen	4.22	.78	5.4	.0000	2.59	5.85
QWERTY	3.78	.78	4.9	.0001	2.15	5.41
$250	1.67	.78	2.2	.05	.04	3.30
$350	1.67	.78	2.2	.05	.04	3.30

PDA size, keypad, and price features influence preferences, while design options do not. The preferred PDA is *ultrathin* and fits in a *shirt pocket*, features a *touch screen* or *QWERTY keypad*, and is priced at $250 or $350.

The *coefficients* estimate the part worth utilities of the PDA features. Expected preference for the ideal design is the sum of the part worth utilities for features included. Of all possible configurations of the four attributes, an ultrathin PDA that fits in a shirt pocket, with the simplest single unit design, with a touch screen, at the highest price is the ideal PDA. Design does not affect preferences, so the least expensive option would be used, and the two higher prices are equivalent to customers, so the higher, more profitable price would be charged:

$$\widehat{PDA\ preference}_j = 1.00 + 0.78\ shirtpocket_j + 1.89\ ultrathin\ shirtpocket_j$$
$$-1.56\ clamshell_j - 1.44\ slider_j$$
$$+4.22\ touch\ screen_j + 3.78\ QWERTY_j$$
$$+1.67\$250_j + 1.67\ \$350_j$$

$$
\begin{aligned}
= 1.00 &+0.78\,(0) &&+1.89\,(1)\\
&-1.56\,(0) &&-1.44\,(0)\\
&+4.22\,(1) &&+3.78\,(0)\\
&+1.67\,(0) &&+1.67\,(1)
\end{aligned}
$$

$$=8.78.$$

The part worth utilities from coefficient estimates are shown in Fig. 10.2 and Table 10.5.

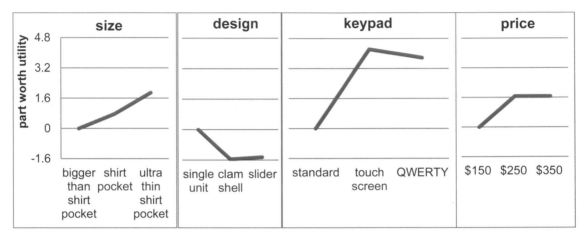

Fig. 10.2 PDA part worth utilities

Preferred *ultraslim shirt pocket size* adds 1.89 (= 1.89 − 0) to the preference rating, a *touch screen* adds 4.22 (= 4.22 − 0), and a price of *$250* adds 1.67 (= 1.67 − 0). The preferred design makes no significant difference, 1.56 (= 1.56 − 0).

The range in part worth utilities for each attribute is an indication of that attribute's importance. Preference depends most on the keypad configuration, which is more than twice as important as size or price.

Table 10.5 Relative importance of PDA attributes

Attribute	Part worth utility of least preferred	Part worth utility of most preferred	Part worth utility range	Attribute importance
Size	0	1.9	1.9	1.9/9.4 = .20
Shape	−1.6	0	1.6	1.6/9.4 = .17
Keypad	0	4.2	4.2	4.2/9.4 = .18
Price	0	1.7	1.7	1.7/9.4 = .45
	Sum of part worth utility ranges:		9.4	

Conjoint analysis has been used to improve the designs of a wide range of products and services, including:

- Seating, food service, scheduling, and prices of airline flights
- Offer of outpatient services and prices for a hospital
- Container design, fragrance, and design of an aerosol rug cleaner
- Digital camera pixels, features, and prices

Conjoint analysis is versatile and the attributes studied can include characteristics that are difficult to describe, such as fragrance, sound, feel, or taste. It is difficult for customers to tell us how important color, package design, or brand name is in shaping preferences, and conjoint analysis often provides believable, valid estimates.

10.3 ANOVA Identifies Segment Mean Differences

Analysis of variance tests hypotheses regarding equivalence of segment, group, or category means. With analysis of variance, managers can compare mean performance across categories. The following are questions that managers might use analysis of variance to address:

> Does job satisfaction differ across divisions?
> Does per capita demand differ across global regions?
> Do preferences differ across flavors?
> Do rates of return differ across portfolios?
> Does customer loyalty differ across brands?

In each of these scenarios, the question concerns performance differences across categories or groups: divisions, global regions, flavors, portfolios, or brands. Analysis of variance compares performance variation across groups with performance variation within groups, and more across group variation is evidence that the group performance levels differ.

> *Example 10.4 Background music to create brand interest.* A brand manager suspects that the background music featured in a brand's advertising may affect the level of interest in the advertised brand. Several background options are being considered, and those options differ along two categories, or *factors*.

Three vocals options are

1.	Backgrounds which feature vocals
2.	Backgrounds with brand related vocals substituted for original vocals
3.	Backgrounds with vocals removed

Three orchestration options are

1. Saxophone
2. Saxophone and percussion
3. Saxophone and piano

The hypotheses that the brand manager would like to test are

$H_{vocals0}$:   Mean interest ratings following exposure to ads with alternate vocals options are equivalent.

$$\mu_{original} = \mu_{brand\_specific} = \mu_{no\_vocals}$$

versus

$H_{vocals1}$:   At least one mean interest rating following exposure to ads with alternate vocals differs.

and

$H_{orchestration0}$:   Mean interest ratings following exposure to ads with alternate orchestrations are equivalent:

$$\mu_{saxophone} = \mu_{saxophone+percussion} = \mu_{saxophone+piano}.$$

versus

$H_{orchestration1}$:   At least one mean interest rating following exposure to ads with alternate orchestrations differs.

   To determine whether vocals and orchestration of backgrounds affect brand interest ratings, the ad agency creative team designed nine backgrounds for a brand ad. Since the ad message, visuals, and length of ad could also influence interest, the agency creatives were careful to make those ad features identical across the nine versions. By using ads that were identical, except for their musical backgrounds, any difference in resulting brand interest could be attributed to the difference in backgrounds.

   Nine consumers were randomly selected and then randomly assigned to one of the nine background *treatments*, or combination of *vocals* and *orchestration*. Each viewed the brand advertisement with one of the nine backgrounds and then rated their interest in the brand using a scale from 1 ("not at all interested") to 9 ("very interested"). The data are shown in Table 10.6 and Fig. 10.3.

Table 10.6 Brand interest ratings by vocals and orchestration levels

Vocals option	Orchestration			
	Sax	Sax & percussion	Sax & piano	Mean
None	9	6	7	7.3
Original	6	4	5	5.0
Brand specific	5	4	3	4.0
Mean	6.7	4.7	5.0	5.4

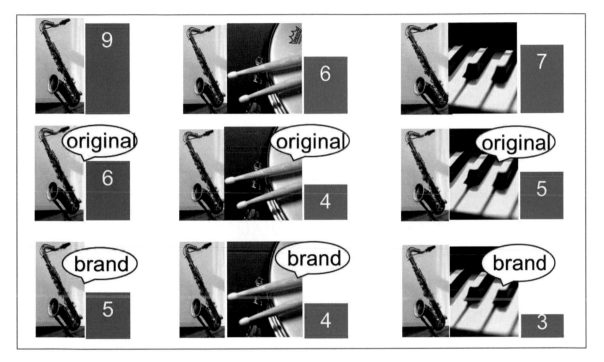

Fig 10.3 Brand interest by background instrumentation and vocals

In analysis of variance, an *F* statistic tests a null hypothesis of equivalent category means across *factor levels*, and those *F* statistics are ratios of variation explained by a *factor* to unexplained explained variation, each adjusted by its degrees of freedom.

In the background music example, variation across *vocals* levels is

$$SSB_{vocals} = [n_{none} \times (\overline{X}_{none} - \overline{\overline{X}})^2 + n_{original} \times (\overline{X}_{original} - \overline{\overline{X}})^2 + n_{brand} \times (\overline{X}_{brand} - \overline{\overline{X}})^2]$$
$$= [\quad 3 \times (7.3 - 5.4)^2 \quad + \quad 3 \times (5.0 - 5.4)^2 \quad + \quad 3 \times (4.0 - 5.4)^2]$$
$$= 17.6.$$

There are three *vocals* levels. The degrees of freedom for variation across *vocals* levels is two, comparing two of the levels to the third baseline. Mean variation is

$$MSB_{vocals} = SSB_{vocals}/df_{vocals}$$
$$= 17.6/2$$
$$= 8.8.$$

Variation across *orchestration* levels is

$$SSB_{orchestration} = [n_{sax} \times (\overline{X}_{sax} - \overline{\overline{X}})^2 + n_{sax+perc} \times (\overline{X}_{sax+perc} - \overline{\overline{X}})^2 + n_{sax+piano} \times (\overline{X}_{sax+piano} - \overline{\overline{X}})^2]$$
$$= [\quad 3 \times (6.7-5.4)^2 \quad + 3 \quad \times (4.7-5.4)^2 \quad + 3 \quad \times (5.0-5.4)^2]$$
$$= 6.9.$$

And mean variation between *orchestration* levels is

$$MSB_{orchestration} = SSB_{orchestration}/df_{orchestration}$$
$$= 6.9/2$$
$$= 3.4.$$

To compare mean variation across *vocals* levels and *orchestration* levels with mean variation within *vocals* and *orchestration* levels, the variation within levels is calculated by subtracting SSB_{vocals} and $SSB_{orchestration}$ from total variation, SST:

$$SST = [(9 - 5.4)^2 + (6 - 5.4)^2 + (7 - 5.4)^2$$
$$+ (6 - 5.4)^2 + (4 - 5.4)^2 + (5 - 5.4)^2$$
$$+ (5 - 5.4)^2 + (4 - 5.4)^2 + (3 - 5.4)^2]$$
$$= 26.2.$$

Of the total variation of 26.2, 17.6 has been explained by differences across *vocals* levels, and 6.9 has been explained by differences across *orchestration* levels, leaving 1.8 unexplained from variation within levels:

$$SSW = SST - SSB_{vocals} - SSB_{orchestration}$$
$$= 26.2 - 17.6 \quad - 6.9$$
$$= 1.8.$$

Mean unexplained variation is

$$MSW = SSW/(N - df_{vocals} - df_{orchestration} - 1)$$
$$= 1.8/4$$
$$= .4.$$

To test each of the two sets of hypotheses, the corresponding F statistic is calculated from the ratio of *mean squares between*, MSB_{vocals} or $MSB_{orchestration}$, and *mean square within*, MSW:

$$F_{vocals_{2,4}} = MSB_{vocals}/MSW$$
$$= 8.8/.4$$
$$= 19.8$$

$$F_{orchestration_{2,4}} = MSB_{orchestration} / MSW$$
$$= 3.4/.4$$
$$= 7.8.$$

With 2 and 4 degrees of freedom, the *critical F* for 95% confidence is 6.9. Both *F* statistics exceed the *critical F* and have *p Values* of .008 and .04. Based on the sample data, there is evidence that the *vocals* alternatives are not equally effective in backgrounds, and that the *orchestration* alternatives are also not equally effective. Both null hypotheses are rejected.

Excel provides the *F* statistics and their *p Values*, as well as factor level means.

Table 10.7 ANOVA results from Excel

ANOVA: Two factor without replication

SUMMARY	Count	Sum	Average	Variance		
None	3	22	7.3	2.3		
Original	3	15	5.0	1.0		
Brand specific	3	12	4.0	1.0		
Sax	3	20	6.7	4.3		
Sax & percussion	3	14	4.7	1.3		
Sax & piano	3	15	5.0	4.0		
ANOVA						
Source of variation	SS	df	MS	F	p Value	F crit
Rows	17.6	2	8.8	19.8	.008	6.9
Columns	6.9	2	3.4	7.8	.04	6.9
Error	1.8	4	.4			
Total	26.2	8				

In the sample, ads with no *vocals* produced highest average brand *interest ratings*, $\overline{X}_{none} = 7.3$, and ads with *brand specific vocals* produced lowest average *interest ratings*, $\overline{X}_{brand} = 4.0$. The F_{vocals} *test* allows the conclusion that at least one of the *vocals* factor levels differs. Therefore, it is possible that (1) no *vocals* (*none*) is more effective than either option with vocals, (2) *brand specific vocals* are less effective than either *original* or no *vocals*, or (3) all three levels may differ. To determine which of the three levels differ, *multiple comparisons*, which resemble *t tests*, would be used, though Excel does not offer this ability.

Ads with *sax* produced highest average brand *interest ratings*, $\overline{X}_{sax} = 6.7$, and ads with *sax+percussion orchestration* produced the lowest average *interest ratings*, $\overline{X}_{sax+perc} = 4.7$. The $F_{orchestration}$ test allows the conclusion that at least one of the *orchestration* factor levels differs; however, from analysis of variance results, it is not possible to determine which of the three levels are unique.

10.4 ANOVA and Regression with Indicators Are Complementary Substitutes

The F statistics used to test hypotheses with analysis of variance and with regression are similar. Both compare variation explained by model drivers or factors with unexplained variation. Analysis of variance enables us to determine whether each factor matters. For example, both *vocals* and *instrumentation* in ad backgrounds matter, and at least one *vocal* option and at least one *instrumental* option are more effective in generating *brand interest* following ad exposure. Regression enables us to determine whether at least one of the factors, either one or both, matters. Regression also identifies particular indicators that produce higher or lower expected performance relative to the baseline. To illustrate, a regression model of *vocals* and *orchestration* background influences on *brand interest* is shown below. For each of the two factors, three alternative levels are compared. One level becomes the baseline, and two indicators are used to test for differences from the baseline level:

$$\hat{Interest} = b_0 + b_{brand\_specific} brand\_vocals + b_{no\;vocals} no\_vocals + b_{sax+perc} sax + perc$$
$$+ b_{sax+piano} sax + piano,$$

where *original vocals* with orchestration for *saxophone* are the baseline levels, and *brand vocals, no vocals, sax+perc*, and *sax+piano* are 0–1 indicators.

Regression results are given in Table 10.8.

Table 10.8 ANOVA table from multiple regression

SUMMARY OUTPUT					
Regression statistics					
R Square	.932				
Standard error	.667				
Observations	9				
ANOVA	*df*	*SS*	*MS*	*F*	*Significance F*
Regression	4	24.4	6.1	13.8	.01
Residual	4	1.8	.4		
Total	8	26.2			

The model F statistic, 13.8, has a *p Value* (=.01) less than the *critical p Value* of .05. Sample evidence allows the conclusion that at least one of the *vocals* or *orchestration* options is driving the level of brand *interest*. From the model *R Square*, we learn that differences in *vocals* and *orchestration* together account for 93% of the variation in brand *interest* ratings. (While analysis of variance does not explicitly provide *R Square*, it is easily found from analysis of variance output as the ratio of explained variation, the sum of squares due to the factors, and total variation.) Variation explained by the two factors in analysis of variance is equivalent to variation explained by the model in regression:

$$SSB_{vocals} + SSB_{orchestration} = SSR$$
$$17.6 + 6.9 \quad = 24.4$$

and

$$R\ Square = (SSB_{vocals} + SSB_{orchestration})/SST$$
$$= (\ 17.6 \quad + 6.9\ )/26.2$$
$$= .932.$$

Regression enables identification of indicators that differ from the baseline:

Table 10.9 Indicator coefficient estimates, *t* statistics, and *p* Values

	Coefficients	Standard error	t Stat	p Value	Lower 95%	Upper 95%
Intercept	8.6	.50	17.2	.0001	7.2	9.9
Original vocals	−2.3	.54	−4.3	.01	−3.8	−0.8
Brand vocals	−3.3	.54	−6.1	.004	−4.8	−1.8
Sax & perc	−2.0	.54	−3.7	.02	−3.5	−.5
Sax & piano	−1.7	.54	−3.1	.04	−3.2	−.2

The coefficient estimates for *original vocals* and *brand vocals* are significant. *Vocals*, whether *original* or *brand* specific, produces expected brand *interest* ratings that are lower by two rating scale points than the options without, *none*.

Both the coefficient estimate for *sax+perc* and *sax+piano* are significant. Adding either percussion or piano to the background reduces expected brand *interest* ratings by about two rating scale points.

When a regression model is built using indicators, part worth graphs can be used to illustrate results. In the background music example, *part worth interest ratings* can be compared, as Fig. 10.4 illustrates.

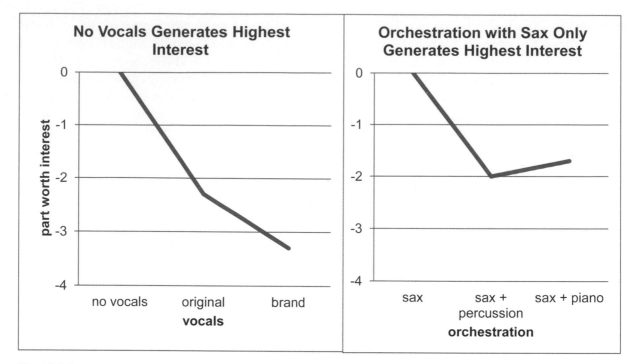

Fig. 10.4 Interest part worths

   From analysis of variance, we learn that both *vocals* and *orchestration* influence *interest* ratings. From regression, we learn that together, *vocals* and *orchestration* options account for 93% of the variation in *interest* ratings, and that backgrounds with *no vocals* instead of *original* or *brand vocals,* and *saxophone* alone, instead of a combination with either piano or percussion, are expected to generate the highest ratings, about four scale points higher than backgrounds with vocals and either piano or percussion.

   Multiple regression with indicators and analysis of variance are substitutes, though they each offer particular advantages. Multiple regression is designed to accommodate both categorical and continuous drivers, and interest is twofold: (1) identify performance drivers, including differences across groups, and (2) forecast performance under alternate scenarios. Regression accounts for the impact of continuous drivers by building them into a model. Analysis of variance is designed to identify performance differences across groups. Where possible, continuous drivers are controlled by choosing groups that have equivalent profiles, often in the context of an experiment.

10.5 ANOVA and Regression in Excel

Regression's dual goals of (1) identification of drivers and quantification of their influence and (2) forecasting performance under alternate scenarios provide more information than analysis of variance in Excel, where output is primarily geared toward hypothesis tests of the factors. However, other, more specialized software, such as *SAS, JMP,* and *SPSS,* offer more powerful and versatile analysis of variance features, including multiple comparisons. Marketing researchers and psychometricians sometimes use *analysis of covariance* to account for variation in experiments that has not been controlled and to compare factor levels to identify those that differ.

Analysis of variance is particularly well suited for use with experimental data, and since experiments tend not to be routinely conducted by managers experimental data collection and analysis are often outsourced to marketing research firms. Because Excel is targeted for use by managers, analysis of variance in Excel is basic. In Excel, there is the additional limitation that *replications,* the number of datapoints for each combination of factor levels, must be equivalent. In the background music experiment, for example, had 15 consumers been randomly selected to view 1 of the 9 ads, data from only 9 consumers could be used in analysis of variance with Excel. Six of the ads would have been viewed by two consumers each, and three of the ads would have been viewed by only one consumer. Data from six consumers would have to be ignored in order to use analysis of variance in Excel. Since all of the data could be used in regression with indicators, regression is a more useful choice in Excel, and it allows both hypothesis tests and forecasts under alternate scenarios.

10.6 Indicators Quantify Shocks in Time Series

Example 10.5 Tyson's Farm Worker Forecast[7]. Tyson's management would like to forecast quarterly self employed workers in agriculture. They believe that these self employed workers, family farmers, are leaving the farm to find more profitable work elsewhere, and that this hypothetical exodus may have been accelerated by the Stimulus program of 2009. Stimulus legislation enacted in 2009 offered benefits to wage and salary workers, but not to self employed workers, encouraging some self employed to take wage and salary jobs. This might result in a permanent shrinking of the self employed worker segment or the segment might switch back to self employment once Stimulus benefits expire.

Tyson's meet labor demand left unsatisfied by hiring agricultural workers. They have asked Mark, their master model builder, to build a model to forecast quarterly self employed agriculture workers. In months where the number of workers is expected to be down from the prior year, they will hire additional workers. If these gaps are large enough, they will implement a lobbying campaign to lesson restrictions on illegal immigrant workers who would work for lower wages.

Choice of the first predictor. Since Mark was working with a time series, he first chose a logically appealing leading indicator of *self employed workers*: *unpaid family workers* in agriculture. Self employed farmers often relied on unpaid family members. If *unpaid family workers* were leaving agriculture to work in paid jobs elsewhere, this might drive *self employed workers* to leave agriculture the following year. Mark began with this single predictor to minimize multicollinearity issues.

---

[7] This example is a hypothetical scenario based on actual data.

Choice of lag. In order to forecast *self employed agriculture workers* from *unpaid family workers*, Mark needed to lag the leading indicator. He hid the two most recent observations, April and May 2010, to later validate the model, since he wanted to be sure that his model could be relied upon to produce solid forecasts. Then, to confirm that 12 months was the appropriate lag for *unpaid family workers*, he plotted *self employed workers* and *unpaid family workers* using data from the Bureau of Labor, January 2004 through May 2010, shown in Fig. 10.5.

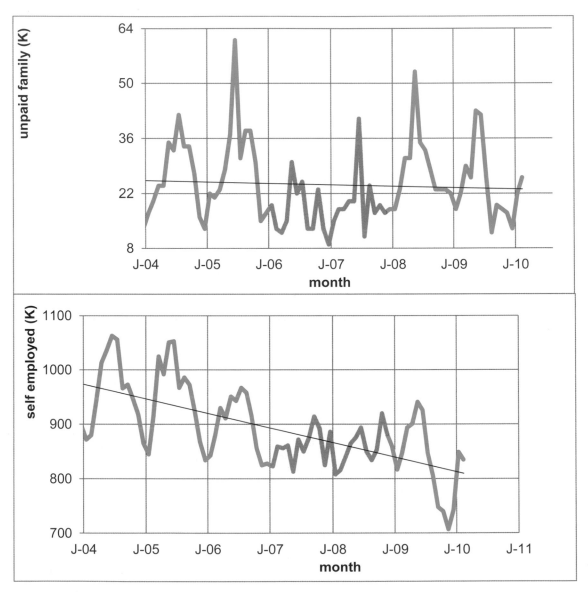

Fig. 10.5 Self employed and unpaid family workers in agriculture, January 04 through May 10

The scatterplots confirmed that agricultural labor follows an annual cycle that corresponds to planting and harvesting. In 2006, there were fewer *unpaid family workers,* and all but one datapoints lie below the trendline. One year later, in 2007, there were fewer *self employed workers,* and all but two datapoints lie below the trendline. Twelve months is the traditional growing cycle in agriculture, and the year with an unusually low number of *self employed workers* lags by 1 year the year with an unusually lower number of *unpaid family workers.* Mark chose a 12 month lag for *unpaid family workers* for the regression model using datapoints for *unpaid family workers* from April 2005 with datapoints for *self employed workers* from April 2006. His regression is shown in Table 10.10.

Table 10.10 Regression of self employed workers in agriculture

SUMMARY OUTPUT

Regression statistics

R Square	.188					
Standard error	52.2					
Observations	48					
ANOVA	df	SS	MS	F	Significance F	
Regression	1	28,913	28,913	10.6	.002	
Residual	46	125,285	2,724			
Total	47	154,197				
	Coefficients	Standard error	t Stat	p Value	Lower 95%	Upper 95%
Intercept	804	19	42.7	.0000	766	842
Unpaid family workers q-12	2.36	.72	3.3	.0021	.90	3.81
DW:	.78					

The model, shown in Table 10.10, is significant (*Significance F =.002),* though the *R Square,* .19, is low for time series data. The coefficient estimate is positive as expected: *self employed workers* leave agriculture following the exit of *unpaid family workers.*

Assessment of autocorrelation. Since time series often contain trend, cycles, and seasonality, those must be accounted for. If these systematic variations in the data are present, but unaccounted for, they will be present in the model residuals. The Durbin Watson statistic will identify the presence of unaccounted for trend, cycles, or seasonality in the residuals. Mark found that the residuals are autocorrelated ($DW = .78 < dL_{48,2} = 1.49$). Trend, cycles, or seasonality is present in the data and has not been accounted for. Mark plotted the residuals in Fig. 10.6 to identify potential trend, cycle, or seasonality variables.

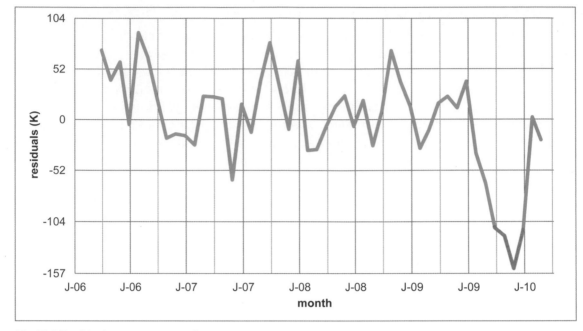

Fig 10.6 Residuals are not pattern free

There was evidence that the Stimulus legislation of 2009 had affected *self employed workers*, as management had hypothesized. Late in 2009, the number of workers had fallen noticeably. However, workers appeared to have returned by early 2010.

Mark added an indicator of the financial industry *Stimulus 09* to his model, setting the indicator equal to 1 in months September 2009 through January 2010, and setting the indicator to zero in all other months. The expanded regression model, with the *Stimulus 09* indicator, is shown in Table 10.11.

Table 10.11 Regression with Credit Crunch 09 Indicator

SUMMARY OUTPUT

Regression statistics

R Square	.640					
Standard error	35.1					
Observations	48					

ANOVA	df	SS	MS	F	Significance F	
Regression	2	98,737	49,368	40.1	.0000	
Residual	45	55,461	1,232			
Total	47	154,197				

	Coefficients	Standard error	t Stat	p Value	Lower 95%	Upper 95%
Intercept	818	13	63.9	.0000	792	843
Unpaid family $_{m-12}$	2.35	.49	4.8	.0000	1.37	3.33
Stimulus 09	−125	17	−7.5	.0000	−158	−91

DW : 1.66

R Square is now much higher, .64, and the standard error is now much smaller. Forecasts can be expected to fall within 71K (=2.0×35.1K) workers. The coefficient signs are as Mark expected. The number of *self employed workers* follows the number of *unpaid family workers* a year later. The *Stimulus 09* had a sizeable, though temporary, negative impact on the number of *self employed workers*. The residuals are now free of autocorrelation. *DW* is 1.66, which exceeds dU$_{48,3}$ = 1.62 for this sample of 48 months and a model with three variables, including intercept.

Model validity. To assess the model's validity, Mark compared the two most recent, hidden observations with the 95% mean prediction intervals, shown in Table 10.12.

Table 10.12 Model validation

Month	95% Lower prediction (K)	Self employed workers (K)	95% Upper prediction (K)
Apr-10	815	837	956
May-10	808	848	949

The model correctly predicts the number of *self employed workers* in the two most recent months.

With this evidence of model validity, Mark recalibrated the model by adding these two most recent months, which had been hidden to build the model and validate. The model became

Self emplôyed workers (K)$_t$ = 817(K)^a–123(K)^a *Stimulus 09$_t$*
$\qquad\qquad\qquad\qquad$ +2.29^a *unpaid family workers(K)$_{t-12}$*

R Square: .62
^{a}*Significant at .01.*

In months before September 2009 and after January 2010, setting the *Stimulus 09* indicator to 0, the expected number of *self employed workers* in agriculture is

Self emplôyed workers (K)$_t$ = (817(K)–123(K)(0))+2.29 *unpaid family workers (K)$_{t-12}$*
$\qquad\qquad\qquad$ = 817(K) $\qquad\qquad\qquad$ +2.29 *unpaid family workers(K)$_{t-12}$* .

In months September 2009 through January 2010, the *Stimulus 09* indicator is 1, and the expected number of *self employed workers* is

Self emplôyed workers (K)$_t$ = (817(K)–123(K)(1)) +2.29 *unpaid family workers(K)$_{t-12}$*
$\qquad\qquad\qquad$ = 694(K) $\qquad\qquad\qquad$ +2.29 *unpaid family workers(K)$_{t-12}$* .

The *Stimulus 09* indicator shifts the regression intercept and line down by 123(K) workers, as Fig. 10.7 illustrates.

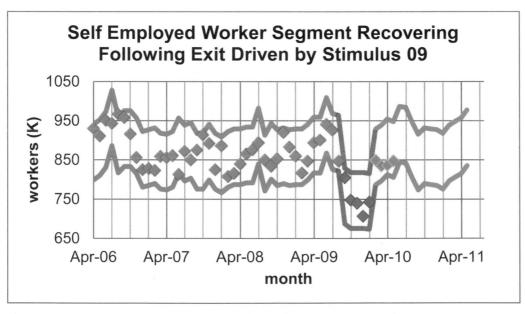

Figure 10.7 Self employed workers segment recovering

After accounting for the temporary depressing impact of the 2009 Stimulus program, the model forecasts mild growth in the self employed segment in 2010 and 2011. Figure 10.8 shows the percent increase each month over the same month in the past year.

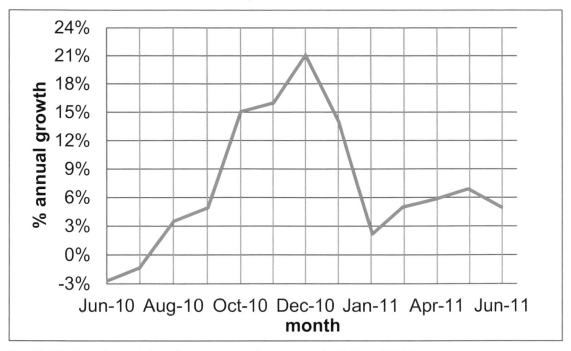

Fig. 10.8 Projected growth in *self employed worker* segment in 2010 and 2011

Mark would report to management the following:

MEMO

Re: Supply of Self Employed Workers Stable Following 09 Contraction
To: Tyson Directors of Planning and Legal Affairs
From: Mark Weisselburg, Director, Econometric Forecasting and Analysis
Date: June 2011

Following an unusually large exit of self employed workers in 2009, the segment recovered in 2010 and is expected to grow by eight percent over same month 2009.

Econometric Model. A model was built with data from the Bureau of Labor on self employed and unpaid family workers in agriculture. Using a 50 month series which excluded the two most recent months, the model correctly forecast the number of self employed workers in the two most recent months.

Model Results. Variation in past year unpaid family workers and Stimulus programs account for 62% of the variation in monthly self employed workers. The model forecast margin of error is 71,000 workers. Following a decline of 1,000 unpaid family workers, the number of self employed workers is expected to decline by 2,000 the following year. 2009 Stimulus programs led to the monthly exit of 120 (K) self employed workers, September 09 through January 10. Self employed workers returned, and numbers are expected to average 8% above 09 levels.

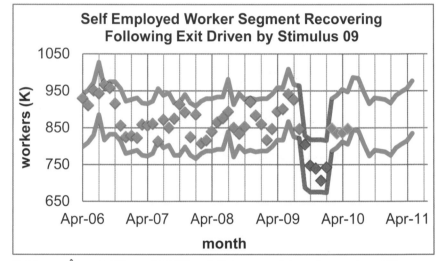

$$\textit{Self empl}\hat{\textit{o}}\textit{yed (K)}_t = 694(K) + 2.29 \textit{ unpaid family workers(K)}_{t-12}$$
following '09 Stimulus
$$= 817(K) + 2.29 \textit{ unpaid family workers(K)}_{t-12}$$
in future months

Forecasts for the next 12 months are:

Month	J-10	J-10	A-10	S-10	O-10	N-10	D-10	J-11	F-11	M-11	A-11	M-11
lower	850	850	810	780	790	790	790	780	800	810	820	840
upper	990	990	950	920	930	930	930	920	940	950	960	980

Conclusion. The number of self employed agriculture workers is expected to show modest growth in the next 12 months.

Other Considerations. This model accounts for less than two thirds of the monthly variation in self employed workers and does not account for changing prices, wages, or the pool of wage and salary workers, which may affect the self employed segment.

10.7 Indicators Allow Comparison of Segments and Scenarios, Quantify Shocks, and Offer an Alternative to Analysis of Variance

Indicators adjust the intercept in linear models to allow for differences in average levels of diverse segments or scenarios. Incorporating indicators in time series models allows us to gauge the impact of structural shifts and to estimate response levels that would have manifested had shocks not occurred. Similarly, if a shock is expected to recur, its indicator can be to one in future periods to forecast the expected change should the shock occur again.

Indicators are used to analyze conjoint analysis data, and estimate the part worth utilities, or the value of each product feature. The part worth utility estimates enable new product development managers to identify most preferred product designs and the most important attributes driving preferences.

Analysis of variance enables a manager to determine whether or not group or category means differ. Building a regression model with indicators offers an alternative to analysis of variance that also allows identification of the particular groups or categories that differ, as well as the extent of the difference. In Excel, regression with indicators may provide more information for decision making than the basic analysis of variance alternative.

Excel 10.1 Use Indicators to Find Part Worth Utilities and Attribute Importances from Conjoint Analysis Data

Three customers from the target market rated nine hypothetical PDA designs, shown in **Table 10.3,** using a scale from 1 (=least preferred) to 9 (=most preferred). The data are in **Excel 10.1 PDA conjoint.xls.**

Use indicators to estimate the part worth utilities of *size, shape, keypad,* and *price* attribute options for PDAs.

Baseline hypothetical. The baseline PDA is *bigger than shirt pocket*, with *single unit* design, *standard* keypad, at a retail price of *$150.* The first hypothetical PDA design in **Table 10.3**, and in rows **2, 11,** and **20** of the file, corresponds to the baseline.

Add indicators for differences from baseline. Add four indicators, two for each PDA attribute: *shirt pocket, ultraslim shirt pocket, clamshell, slider, QWERTY, touch screen, $250*, and *$350.*

Enter a zero or a one in each of these columns for each of the nine hypotheticals.
The baseline hypothetical, for example, will have zeros in all eight columns, since it is not *shirt pocket* or *ultraslim shirt pocket size*, it does not feature a *clamshell* or *slider design*, it does not have a *QWERTY* or *touch screen keypad* and it is not priced at *$250* or $350:

size	design	key pad	price	rating	shirt pocket	ultra thin shirt pocket	clamshell	slider	touch screen	QWERTY	$250	$350
bigger than shirt pocket	single unit	standard	$150	1	0	0	0	0	0	0	0	0
bigger than shirt pocket	clamshell	touch screen	$250	5	0	0	1	0	1	0	1	0
bigger than shirt pocket	slider	QWERTY	$350	5	0	0	0	1	0	1	0	1
shift pocket	single unit	touch screen	$350	7	1	0	0	0	1	0	0	1
shift pocket	clamshell	QWERTY	$150	3	1	0	1	0	0	1	0	0
shift pocket	slider	standard	$250	3	1	0	0	1	0	0	1	0
ultra thin shirt pocket	single unit	QWERTY	$250	8	0	1	0	0	0	1	1	0
ultra thin shirt pocket	clamshell	standard	$350	5	0	1	1	0	0	0	0	1
ultra thin shirt pocket	slider	touch screen	$150	5	0	1	0	1	1	0	0	0

Select and copy the indicator values for the nine hypotheticals in the first customer's rows and then paste into the other two customers' rows.

	H shirt pocket	I ultra thin shirt pocket	J clamshell	K slider	L touch screen	M QWERTY	N $250	O $350
2	0	0	0	0	0	0	0	0
3	0	0	1	0	1	0	1	0
4	0	0	0	1	0	1	0	1
5	1	0	0	0	1	0	0	1
6	1	0	1	0	0	1	0	0
7	1	0	0	1	0	0	1	0
8	0	1	0	0	0	1	1	0
9	0	1	1	0	0	0	0	1
10	0	1	0	1	1	0	0	0
11	0	0	0	0	0	0	0	0
12	0	0	1	0	1	0	1	0
13	0	0	0	1	0	1	0	1
14	1	0	0	0	1	0	0	1
15	1	0	1	0	0	1	0	0
16	1	0	0	1	0	0	1	0
17	0	1	0	0	0	1	1	0
18	0	1	1	0	0	0	0	1
19	0	1	0	1	1	0	0	0

Run a regression of *rating*, with the eight indicators.

▲	A	B	C	D	E	F	G
1	SUMMARY OUTPUT						
2							
3	*Regression Statistics*						
4	Multiple R	0.86					
5	R Square	0.75					
6	Adjusted R Square	0.63					
7	Standard Error	1.64					
8	Observations	27					
9							
10	ANOVA						
11		*df*	*SS*	*MS*	*F*	*ignificance F*	
12	Regression	8	143.33	17.92	6.63	0.0004	
13	Residual	18	48.67	2.70			
14	Total	26	192				
15							
16		*Coefficients*	*Standard Error*	*t Stat*	*P-value*	*Lower 95%*	*Upper 95%*
17	Intercept	1.00	0.95	1.05	0.3061	-0.99	2.99
18	shirt pocket	0.78	0.78	1.00	0.3290	-0.85	2.41
19	ultra thin shirt pocket	1.89	0.78	2.44	0.0254	0.26	3.52
20	clamshell	-1.56	0.78	-2.01	0.0600	-3.18	0.07
21	slider	-1.44	0.78	-1.86	0.0788	-3.07	0.18
22	touch screen	4.22	0.78	5.45	0.0000	2.59	5.85
23	QWERTY	3.78	0.78	4.87	0.0001	2.15	5.41
24	250	1.67	0.78	2.15	0.0454	0.04	3.30
25	350	1.67	0.78	2.15	0.0454	0.04	3.30

Coefficients for *shirt pocket*, *clamshell*, and *slider* are not significant. With conjoint analysis data, the indicators reflect an orthogonal design in which the product features' presence or absence is uncorrelated. Since multicollinearity will not affect results, there is no need to remove the insignificant indicators.

Part worth utilities. The *coefficients* are estimates of the part worth utilities, the value of each feature. Size, price, and keypad options drive preferences, while design options do not. The most preferred PDAs would be those combining the features with highest part worth utilities: ultrathin shirt pocket size, with a touch screen or QWERTY keypad, at a price of $250 or $350.

To find the *expected rating of the ideal design*, add the coefficients corresponding to these features. For an ultrathin shirt pocket size, single unit, with touch screen at $350

=SUM(B17,B19,B22,B25).

B26		f_x	=SUM(B17,B19,B22,B25)

	A	B	C
15			
16		Coefficients	Standard Err
17	Intercept	1.00	0.5
18	shirt pocket	0.78	0.1
19	ultra thin shirt pocket	1.89	0.1
20	clamshell	-1.56	0.1
21	slider	-1.44	0.1
22	touch screen	4.22	0.1
23	QWERTY	3.78	0.1
24	250	1.67	0.1
25	350	1.67	0.1
26	expected rating of ideal	8.78	

Attribute importances. To find the *attribute importances*, make a table of the part worth utilities, including the baselines. (Format cells in your *feature* column as text so that Excel will treat these cells as categories.)

		Coefficients	attribute	feature	part worth utility
16			attribute	feature	part worth utility
17	Intercept	1.00	size	bigger than shirt pocket	0.0
18	shirt pocket	0.78		shirt pocket	0.8
19	ultra thin shirt pocket	1.89		ultra thin shirt pocket	1.9
20	clamshell	-1.56	design	one piece	0.0
21	slider	-1.44		clam shell	-1.6
22	touch screen	4.22		slider	-1.4
23	QWERTY	3.78	keyboard	standard	0.0
24	250	1.67		touch	4.2
25	350	1.67		QWERTY	3.8
26	expected rating of ideal	8.78	price	$150	0.0
27				$250	1.7
28				$350	1.7

To see the difference that each feature makes, plot the part worth utilities for each attribute.

To see the preference difference due to alternate sizes, make a line plot, **Alt i**N**sert li**N**e.**

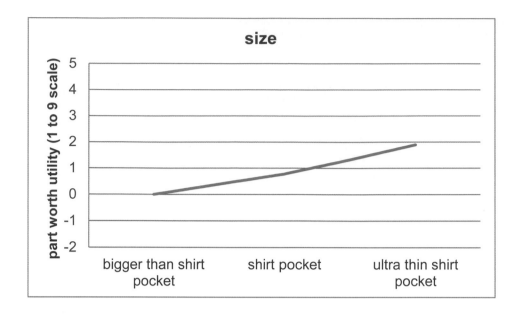

So that attributes can be compared, reformat the vertical axis range, **Alt JA** <u>A</u>xis <u>V</u>ertical, from the most negative to the most positive part worth utility, −2 to 5, choosing a value for major unit, such as 1, and specify that the horizontal axis crosses at the axis value −2:

Make plots of part worth utilities for the other three attributes and reformat the vertical axes similarly.

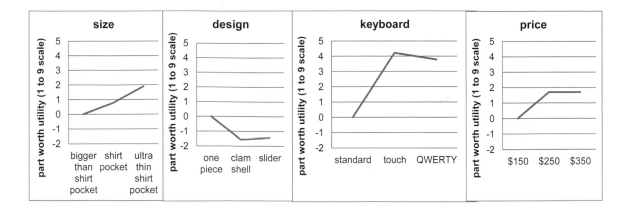

The importance of each attribute is the difference between the part worth utilities of the most and least preferred *attribute options:*

J	K	L	M
attribute	feature	part worth utility	attribute importance
size	bigger than shirt pocket	0.0	
	shirt pocket	0.8	
	ultra thin shirt pocket	1.9	1.9
design	one piece	0.0	
	clam shell	-1.6	
	slider	-1.4	1.6
keyboard	standard	0.0	
	touch	4.2	
	QWERTY	3.8	4.2
price	$150	0.0	
	$250	1.7	
	$350	1.7	1.7

To find the *standardized attribute importances*, first find the total of all attribute importances, **SUM(M19:M28)**, and then divide the attribute importances by the SUM. (Use Fn 4 to lock the SUM cell reference.)

N28	▾	f_x	=M28/$M$29
K	**L**	**M**	**N**
feature	part worth utility	attribute importance	standardized attribute importance
bigger than shirt pocket	0.0		
shirt pocket	0.8		
ultra thin shirt pocket	1.9	1.9	20%
one piece	0.0		0%
clam shell	-1.6		0%
slider	-1.4	1.6	17%
standard	0.0		0%
touch	4.2		0%
QWERTY	3.8	4.2	45%
$150	0.0		0%
$250	1.7		0%
$350	1.7	1.7	18%
		9.4	

Excel 10.2 Add Indicator Variables to Account for Segment Differences or Structural Shifts

> *Indian Imports of U.S. Products.* Build a model of India's annual imports of U.S. products using time series. A leading indicator of India's economic productivity and political leadership are thought to drive imports. Party leadership alters import policies and is likely to affect India's imports of U.S. products.

Data included in the time series, *year*, *Indian Imports(B$)$_t$*, and past year *Indian GDP per capita ($K)$_{t-1}$*, are in **Excel 10.2 Indian Imports.xls.**

Add a Party leadership indicator. To represent India's political leadership, the earliest period of leadership under the BJP Party will be our baseline.

To see how imports have differed under leadership of the alternate Congress Party, add a *Congress$_t$* indicator variable equal to 1 in years 1993–1996 and 2004–2007.

The indicator will modify the baseline intercept, quantifying differences in the level of *Indian imports* from the baseline leadership under BJP.

Since the *Congress* indicator will modify the intercept, it will simplify interpretation of results if the *Congress* indicator column is the first independent variable column. Add the new *Congress* in column **C.**

	A	B	C	D
1	year t	India Imports (B$) t	Congress t	past year GDP per capita ($K) t-1
2	1985	1.64	0	0.286
3	1986	1.54	0	0.291
4	1987	1.46	0	0.313
5	1988	2.50	0	0.338
6	1989	2.46	0	0.363
7	1990	2.49	0	0.354
8	1991	2.00	0	0.375
9	1992	1.92	0	0.326
10	1993	2.78	1	0.322
11	1994	2.29	1	0.307
12	1995	3.30	1	0.343
13	1996	3.33	0	0.382
14	1997	3.61	0	0.402
15	1998	3.56	0	0.432
16	1999	3.69	0	0.427
17	2000	3.67	0	0.446
18	2001	3.76	0	0.460
19	2002	4.10	0	0.465
20	2003	4.98	0	0.480
21	2004	6.11	1	0.548
22	2005	7.96	1	0.622
23	2006		1	0.687
24	2007		1	0.722

Run a regression of *Indian Imports*$_t$ with *Congress*$_t$ and past year *GDP per capita*$_{t-1}$ excluding the two most recent years, 2004 and 2005. (The two most recent years are excluded, since later we will want to test the model's validity for reliable forecasting.)

	A	B	C	D	E	F	G
1	SUMMARY OUTPUT						
2							
3	*Regression Statistics*						
4	Multiple R	0.94					
5	R Square	0.89					
6	Adjusted R Square	0.88					
7	Standard Error	0.34					
8	Observations	19					
9							
10	ANOVA						
11		*df*	*SS*	*MS*	*F*	*Significance F*	
12	Regression	2	15.29	7.65	64.7	2.15E-08	
13	Residual	16	1.89	0.12			
14	Total	18	17.19				
15							
16		*Coefficients*	*Standard Error*	*t Stat*	*P-value*	*Lower 95%*	*Upper 95%*
17	Intercept	-3.17	0.54	-5.84	3E-05	-4.32	-2.02
18	Congress t	0.82	0.23	3.54	0.0027	0.33	1.31
19	past year GDP per capita ($K) t-1	15.87	1.40	11.36	5E-09	12.91	18.83

Assess autocorrelation. Since data are time series, find *DW* to confirm that trend, cycles, and seasonality have been accounted for with the leading indicator and the political shift indicator.

D26		f_x	=SUMXMY2(C26:C43,C27:C44)/SUMSQ(C26:C44)			
	A	B	C	D	E	F
23	RESIDUAL OUTPUT					
24						
25	Observation	d India Impo	Residuals	DW		dL dU
26		1 1.3696766	0.272223443	1.71	19. 3. 1.07430 1.53553	
27		2 1.4385846	0.097615448			

The Durbin Watson statistic is 1.71, which exceeds dU$_{3,19}$ = 1.54. The residuals are free of unaccounted for trend or cycles.

Model validation. To test the model's validity, select and copy the coefficient estimates and paste them into the *Indian imports* worksheet, and then use the regression equation to find *predicted Indian imports.*

	E2			f_x	=$H$2+$H$3*C2+$H$4*D2			
◢	A	B	C	D	E	F	G	H
1	year t	India Imports (B$) t	Congress t	past year GDP per capita ($K) t-1	predicted imports (B$) t			Coefficients
2	1985	1.64	0	0.286	1.37			-3.17
3	1986	1.54	0	0.291	1.44			0.82
4	1987	1.46	0	0.313	1.80			15.87
5	1988	2.50	0	0.338	2.19			
6	1989	2.46	0	0.363	2.59			
7	1990	2.49	0	0.354	2.45			
8	1991	2.00	0	0.375	2.78			
9	1992	1.92	0	0.326	2.01			
10	1993	2.78	1	0.322	2.75			
11	1994	2.29	1	0.307	2.52			
12	1995	3.30	1	0.343	3.09			
13	1996	3.33	0	0.382	2.89			
14	1997	3.61	0	0.402	3.21			
15	1998	3.56	0	0.432	3.69			
16	1999	3.69	0	0.427	3.60			
17	2000	3.67	0	0.446	3.91			
18	2001	3.76	0	0.460	4.13			
19	2002	4.10	0	0.465	4.20			
20	2003	4.98	0	0.480	4.44			
21	2004	6.11	1	0.548	6.34			
22	2005	7.96	1	0.622	7.53			
23	2006		1	0.687	8.54			
24	2007		1	0.722	9.11			

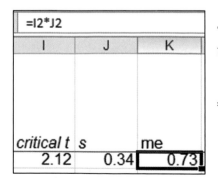

=I2*J2		
I	J	K
critical t s		me
2.12	0.34	0.73

To make the lower and upper prediction interval bounds, make the *margin of error* from the product of the *critical t* value for 16 residual degrees of freedom using

=**T.INV.2T**(*alpha, df*)

and the regression *standard error* (copied from **B7** in the regression sheet).

Find the 95% *lower* and *upper prediction interval* bounds by subtracting and adding the margin of error to *predicted* values. (Remember to lock the margin of error cell reference with **Fn 4.**)

	F2		f_x	=E2-$K$2							
	A	B	C	D	E	F	G	H	I	J	K
1	year t	India Imports (B$) t	Congress t	past year GDP per capita ($K) t-1	predicted imports (B$) t	lower 95% prediction	upper 95% prediction	oefficient	critical t	s	me
2	1985	1.64	0	0.286	1.37	0.64	2.10	-3.17	2.12	0.34	0.73
3	1986	1.54	0	0.291	1.44	0.71	2.17	0.82			
4	1987	1.46	0	0.313	1.80	1.07	2.53	15.87			
5	1988	2.50	0	0.338	2.19	1.47	2.92				
6	1989	2.46	0	0.363	2.59	1.86	3.32				
7	1990	2.49	0	0.354	2.45	1.72	3.18				
8	1991	2.00	0	0.375	2.78	2.05	3.51				
9	1992	1.92	0	0.326	2.01	1.28	2.73				
10	1993	2.78	1	0.322	2.75	2.02	3.48				
11	1994	2.29	1	0.307	2.52	1.80	3.25				
12	1995	3.30	1	0.343	3.09	2.36	3.82				
13	1996	3.33	0	0.382	2.89	2.16	3.62				
14	1997	3.61	0	0.402	3.21	2.48	3.94				
15	1998	3.56	0	0.432	3.69	2.96	4.41				
16	1999	3.69	0	0.427	3.60	2.87	4.33				
17	2000	3.67	0	0.446	3.91	3.18	4.64				
18	2001	3.76	0	0.460	4.13	3.40	4.86				
19	2002	4.10	0	0.465	4.20	3.47	4.93				
20	2003	4.98	0	0.480	4.44	3.71	5.17				
21	2004	6.11	1	0.548	6.34	5.62	7.07				
22	2005	7.96	1	0.622	7.53	6.80	8.25				
23	2006		1	0.687	8.54	7.82	9.27				
24	2007		1	0.722	9.11	8.38	9.84				

Confirm that the model is valid by comparing actual *Indian imports* in 2004 and 2005 with *lower* and *upper 95% prediction* intervals for 2004 and 2005.

Recalibrate by running the regression, this time including the two most recent years of data.

SUMMARY OUTPUT					
Regression Statistics					
Multiple R	0.98				
R Square	0.96				
Adjusted R Square	0.95				
Standard Error	0.34				
Observations	21				

ANOVA					
	_df_	_SS_	_MS_	_F_	_Significance F_
Regression	2	47.74	23.87	204.14	4.268E-13
Residual	18	2.10	0.12		
Total	20	49.84			

	Coefficients	_Standard Error_	_t Stat_	_P-value_	_Lower 95%_	_Upper 95%_
Intercept	-3.32	0.35	-9.35	2E-08	-4.06	-2.57
Congress t	0.84	0.18	4.69	0.0002	0.47	1.22
past year GDP per capita ($K) t-1	16.25	0.90	18.11	5E-13	14.37	18.14

Recalibrated forecasts. Copy and paste the recalibrated coefficient estimates into the _Indian imports_ sheet to update _predicted imports_.

Copy the recalibrated _standard error_ and paste into the _Indian imports_ sheet and change the error degrees of freedom in the _critical t_ formula to 18 to update the margin of error and the _lower_ and _upper 95% prediction_ intervals.

A	B	C	D	E	F	G	H	I	J	K
	India Imports (B$) t		*past year GDP per capita ($K) t-1*	*predicted imports (B$) t*	*lower 95% prediction*	*upper 95% prediction*	*oefficient*	*critical t*	*s*	*me*
year t		*Congress t*								
1985	1.64	0	0.286	1.33	0.61	2.05	-3.32	2.10	0.34	0.72
1986	1.54	0	0.291	1.40	0.68	2.12	0.84			
1987	1.46	0	0.313	1.77	1.05	2.49	16.25			
1988	2.50	0	0.338	2.18	1.46	2.90				
1989	2.46	0	0.363	2.58	1.87	3.30				
1990	2.49	0	0.354	2.44	1.72	3.16				
1991	2.00	0	0.375	2.78	2.06	3.50				
1992	1.92	0	0.326	1.98	1.27	2.70				
1993	2.78	1	0.322	2.75	2.03	3.47				
1994	2.29	1	0.307	2.52	1.80	3.24				
1995	3.30	1	0.343	3.10	2.38	3.82				
1996	3.33	0	0.382	2.89	2.17	3.61				
1997	3.61	0	0.402	3.22	2.50	3.94				
1998	3.56	0	0.432	3.70	2.99	4.42				
1999	3.69	0	0.427	3.62	2.90	4.33				
2000	3.67	0	0.446	3.94	3.22	4.66				
2001	3.76	0	0.460	4.16	3.44	4.88				
2002	4.10	0	0.465	4.23	3.51	4.95				
2003	4.98	0	0.480	4.47	3.76	5.19				
2004	6.11	1	0.548	6.43	5.71	7.15				
2005	7.96	1	0.622	7.64	6.92	8.36				
2006		1	0.687	8.68	7.96	9.40				
2007		1	0.722	9.26	8.54	9.98				

Sensitivity analysis. To plot and compare imports with the model forecasts under both leadership scenarios, insert two new columns, *predicted Indian imports under Congress* and *predicted Indian imports under BJP.*

Copy *predicted Indian imports* (which assume *Congress* leadership) and use shortcuts to paste with **values and formats** (but not formulas):

Alt Home **V S**pecial val**U**es

into a new column *predicted Indian imports under Congress*.

	C	D	E	F	G	H	I	J	K	L
	Congress t	past year GDP per capita ($K) t-1	predicted imports (B$) t	lower 95% prediction	upper 95% prediction	oefficient	critical t	s	me	predicted imports (B$) t under Congress
	0	0.286	1.33	0.61	2.05	-3.32	2.10	0.34	0.72	1.33
	0	0.291	1.40	0.68	2.12	0.84				1.40
	0	0.313	1.77	1.05	2.49	16.25				1.77
	0	0.338	2.18	1.46	2.90					2.18
	0	0.363	2.58	1.87	3.30					2.58
	0	0.354	2.44	1.72	3.16					2.44
	0	0.375	2.78	2.06	3.50					2.78
	0	0.326	1.98	1.27	2.70					1.98
	1	0.322	2.75	2.03	3.47					2.75
	1	0.307	2.52	1.80	3.24					2.52
	1	0.343	3.10	2.38	3.82					3.10
	0	0.382	2.89	2.17	3.61					2.89
	0	0.402	3.22	2.50	3.94					3.22
	0	0.432	3.70	2.99	4.42					3.70
	0	0.427	3.62	2.90	4.33					3.62
	0	0.446	3.94	3.22	4.66					3.94
	0	0.460	4.16	3.44	4.88					4.16
	0	0.465	4.23	3.51	4.95					4.23
	0	0.480	4.47	3.76	5.19					4.47
	1	0.548	6.43	5.71	7.15					6.43
	1	0.622	7.64	6.92	8.36					7.64
	1	0.687	8.68	7.96	9.40					8.68
	1	0.722	9.26	8 54	9 98					9.26

Make *predicted Indian imports under BJP* by changing the *Congress* indicator to zero in 2006 and 2007.

Use shortcuts to copy *predicted Indian imports* and paste (without formulas) into *predicted Indian imports under BJP.*

Congress t	past year GDP per capita ($K) t-1	predicted imports (B$) t	lower 95% prediction	upper 95% prediction	oefficient	critical t	s	me	predicted imports (B$) t under Congress	predicted imports (B$) t under BJP
0	0.286	1.33	0.61	2.05	-3.32	2.10	0.34	0.72	1.33	1.33
0	0.291	1.40	0.68	2.12	0.84				1.40	1.40
0	0.313	1.77	1.05	2.49	16.25				1.77	1.77
0	0.338	2.18	1.46	2.90					2.18	2.18
0	0.363	2.58	1.87	3.30					2.58	2.58
0	0.354	2.44	1.72	3.16					2.44	2.44
0	0.375	2.78	2.06	3.50					2.78	2.78
0	0.326	1.98	1.27	2.70					1.98	1.98
1	0.322	2.75	2.03	3.47					2.75	2.75
1	0.307	2.52	1.80	3.24					2.52	2.52
1	0.343	3.10	2.38	3.82					3.10	3.10
0	0.382	2.89	2.17	3.61					2.89	2.89
0	0.402	3.22	2.50	3.94					3.22	3.22
0	0.432	3.70	2.99	4.42					3.70	3.70
0	0.427	3.62	2.90	4.33					3.62	3.62
0	0.446	3.94	3.22	4.66					3.94	3.94
0	0.460	4.16	3.44	4.88					4.16	4.16
0	0.465	4.23	3.51	4.95					4.23	4.23
0	0.480	4.47	3.76	5.19					4.47	4.47
1	0.548	6.43	5.71	7.15					6.43	6.43
1	0.622	7.64	6.92	8.36					7.64	7.64
0	0.687	7.84	7.12	8.56					8.68	7.84
0	0.722	8.42	7.70	9.14					9.26	8.42

Rearrange columns so that predictions under the two alternative scenarios follow *year*, and then make a scatterplot to compare.

year t	predicted imports (B$) t under Congress	predicted imports (B$) t under BJP	India Imports (B$) t	Congress t	past year GDP per capita ($K) t-1	predicted imports (B$) t	lower 95% prediction	upper 95% prediction	oefficient
1985	1.33	1.33	1.64						
1986	1.40	1.40	1.54						
1987	1.77	1.77	1.46						
1988	2.18	2.18	2.50						
1989	2.58	2.58	2.46						
1990	2.44	2.44	2.49						
1991	2.78	2.78	2.00						
1992	1.98	1.98	1.92						
1993	2.75	2.75	2.78						
1994	2.52	2.52	2.29						
1995	3.10	3.10	3.30						
1996	2.89	2.89	3.33						
1997	3.22	3.22	3.61						
1998	3.70	3.70	3.56						
1999	3.62	3.62	3.69						
2000	3.94	3.94	3.67						
2001	4.16	4.16	3.76						
2002	4.23	4.23	4.10						
2003	4.47	4.47	4.98	0	0.480	4.47	3.76	5.19	
2004	6.43	6.43	6.11	1	0.548	6.43	5.71	7.15	
2005	7.64	7.64	7.96	1	0.622	7.64	6.92	8.36	
2006	8.68	7.84		0	0.687	7.84	7.12	8.56	
2007	9.26	8.42		0	0.722	8.42	7.70	9.14	

Imports Higher under Congress Leadership

Lab Practice 10 Indicators with Time Series: Impact of Terrorism and Military Strike on Oil Prices

Rolls-Royce is facing several decisions that hinge on future oil prices. They require a forecast of oil prices over the next 4 years, through 2013. They are particularly interested in learning
- The degree to which shocks from terrorism affect oil prices
- The influence of U.S. military strikes in oil rich regions on oil prices

The terrorist incident of September 11, 2001, altered business and economic performance and may have reduced oil prices in 2001 and in the following year, 2002. U.S invasion and occupation of Iraq is thought to have reduced oil prices in 2003 through 2007.

Consultants had developed a multiple linear regression model of oil prices from past economic growth in rapidly developing regions, represented by *GDP in China*. The dataset **Lab Practice 10 oil exploration.xls** contains time series of *World oil prices$_t$* and past *China GDP$_{t-3}$* for years 1998 through 2013.

Build a model of World oil prices, including

1. *An indicator of terrorism influence in 2001 through 2002*
2. *An indicator of U.S. invasion of Iraq influence in 2003 through 2007*
3. *Past Chinese GDP$_{t-3}$*

1. Since this is a time series model, assess the model Durbin Watson statistic to determine whether or not unaccounted for trend or cycles remain.

   DW: _____

   Residuals _____ contain _____ possibly contain   OR   ___ are free of unaccounted for trend or cycles.

2. Is your model valid? Y or N

3. Recalibrate and then write your model equations for *World oil price*:

   - During **baseline years 1998–2000** and **2008–2013**

   - Following terrorism in years **2001–2002**

   - Following U.S. invasion and occupation of Iraq in years **2003–2007**

4. What is your prediction for oil prices in 2012? _____

5. If there were to be a terrorism incident in 2011, similar to 9/11, what would be the estimated impact on oil prices in 2012?

Year	Predicted world oil price ($/barrel)		Expected influence of terrorism incident in 2011 ($/barrel)
	No terrorism incident 2011	Terrorism incident in 2011	
2012			

To assess the impact of terrorism on *World oil prices*, plot *predicted oil prices* by year for years 1998 through 2013 assuming:
- No terrorism incident in 2009 through 2013
- A terrorism incident in 2011, whose impact would last through 2013

6. If there were another U.S. invasion and occupation of an oil rich region, perhaps of Iran, in 2011 through 2013, what would be the estimated impact on oil prices in 2012?

Year	Predicted world oil price ($/barrel)		Expected influence of US occupation in 2011 & 2012 ($/barrel)
	No US occupation 2011	US occupation in 2011 & 2012	
2012			

To assess the impact of a U.S. invasion in 2011 through 2013, add to your plot a third series, *predicted oil prices* by year for years 1998 through 2013 assuming a similar U.S. invasion and occupation of an oil rich region, perhaps of Iran, in 2011 through 2013.

<u>Attach or embed your plot.</u>

7. Which shock would have the greater impact on World oil prices in 2012:

____ Terrorism   OR   ____U.S. occupation of an oil rich region?

Lab 10-1 ANOVA and Regression with Indicators: Global Ad Spending

Advertising Age[a] recently published global ad spending in 2008 by the top 100 advertisers. P&G was Number One, spending $9.7B. P&G management is reviewing the allocation of advertising across global regions and wants to know

1. Whether ad spending by top advertisers differs across global regions, and if it does, which global regions are more intensely targeted by top advertisers

2. Whether top advertisers' ad spending differs, and if it does, where the top spenders are headquartered

They have identified the top advertisers headquartered in eight countries and seven global regions as the sample for comparison. Sample data are in **Lab 10 Global Ad Spending.xls**.

Use ANOVA to determine whether average spending differs across top firms headquartered in the eight countries and whether average spending differs across the seven global regions.

1. What hypotheses are you testing?

2. Does spending differ across the top eight firms' headquarters? Y or N

   Evidence: _____

3. Does spending differ across global regions? Y or N

   Evidence: _____

4. Which matters more in accounting for differences in ad spending: firm headquarter's country or global region targeted?

   ___Country headquartered ___region targeted   Evidence: _____

Use multiple regression with indicators to identify country firm headquarters and global regions where spending is greater.

First make a PivotTable and PivotChart to identify baselines for country of headquarters and targeted global region.

5. In which headquarters country was top firm spending least in 2008? _____

[a]*Advertising Age graciously agreed to provide data for this problem.*

6. In which global region was spending least in 2008? _____

7. Does ad spending differ in at least one country of headquarters or targeted region?

    Y or N   Evidence: _____

8. What percent of the differences in ad spending can be explained by differences in headquarters

    country or global region targeted? _____

9. In which headquarters countries did top advertisers spend more than others? _____

10. In which global regions was spending more than others? _____

11. Which headquarters countries fit the model pattern predictions? _____

Add an indicator of spending in the headquarters home global region.

12. Identify the headquarters countries were firms spend more in global regions other than home

    _____

13. Illustrate the differences that you identified.

Lab 10-2 The H–D Buell Blast

The Buell Motorcycle Company is an American motorcycle manufacturer based in East Troy, Wisconsin, founded by ex Harley–Davidson engineer Erik Buell. H–D bought controlling shares of Buell in 1998 and began selling Buell cycles. Buell became a wholly owned subsidiary of H–D in 2003. Harley–Davidson assumed that new riders who learned on a Buell would later trade up to a H–D.

The Blast. Buell engines were designed to be fuel efficient, and the single cylinder Blast, introduced in 2000, achieved 84 mpg. Body parts of the Blast were made from the same plastic that is used to make the outside of golf balls, to protect Blast parts when dropped.

Buell Blasted. In the economic recession of 2008–2009, lower priced Buell motorcycles began cannibalizing Harley–Davidson bike sales. Keith Wandell, appointed as CEO in 2009, began to question the fit of Buell with Harley–Davidson, mentioning "Erik's racing hobby" and asking "why anyone would even want to ride a sport bike." His team concluded that the adrenaline sport bike segment would encounter high competition, offering low profits, while the cruiser segment could provide high returns.

On October 15, 2009, Harley–Davidson announced the discontinuation of the Buell product line as part of its strategy to focus on the Harley–Davidson brand. The last Buell motorcycle was produced on October 30, 2009.

Harley–Davidson is considering the return of the Blast, and Management would like to know what impact the Blast would have on future revenues, should it be reintroduced.
There is some possibility that H–D may sell Buell. Management would like an estimate of the annual future contribution to revenues that could be attributed to the Buell acquisition.

In 2006, H–D changed its stock ticker from HDI to HOG. Management would like to assess the impact of this move on revenues. Some believe that the ticket change motivated some investors to purchase Harleys.

Harley-Davidson Buell Blast.xls contains annual observations on H–D revenues from 1996 to 2009, as well as demographic data thought to drive revenues.

Add indicators for
   Blast availability, 2000 through 2009
   Buell acquisition in 2003
   Ticker change to *HOG* in 2006
and then build a model to identify revenue drivers and to forecast revenues.
Hide the two most recent datapoints. To more easily identify demographic drivers, run regression
with the three indicators and then plot the residuals.

1. In the most years, the residuals are increasing. Identify the exceptional years in which
    residuals are decreasing.

    _____

2. Which demographic variable shows the same pattern in the last 4 years of the data?

    _____

3. Which demographic variable shows a pattern exactly opposite of the residuals in the last 4
    years of data?

    _____

   Use correlations between residuals and the variables with same or opposite patterns to choose a
   driver to add to the model.

4. Do either the indicators or demographics (or both) influence revenues? Y or N

5. Have you accounted for trend, cyclicality, shocks, and shifts with your model?

                                                 ___N ___Maybe ___Y
6. Is your model valid? Y or N

7. Recalibrate and then plot your fit and forecast for years 1996 through 2012.

8. If H–D were to reintroduce the Blast, how much could revenues in 2012 be expected to increase?

    _____

9. If H–D were to sell Buell, what potential revenues could the buyer expect in 2012?

    _____

10. What additional annual revenues can H–D attribute to the HOG ticker?

    _____

Assignment 10-1 Conjoint Analysis of PDA Preferences

Dell is considering introduction of a new PDA which would be sold at a competitive price through Wal-Marts. New product development managers believe that customers would choose brightly colored Dell PDAs at competitive prices.

Choose four attributes of PDAs that you believe to be influences on college students' preferences. Identify three alterative options for each attribute and fill in the orthogonal array table, below, to make nine hypothetical PDAs. You may use whichever attributes and attribute levels you believe matter to college students. Those shown below are used to illustrate the orthogonal design.

Hypothetical PDA	Brand	Color	Keypad	Price
1	Dell	Silver	Standard	$150
2	Dell	White	QWERTY	$250
3	Dell	Lime green	Touch screen	$350
4	Apple	Silver	QWERTY	$350
5	Apple	White	Touch screen	$150
6	Apple	Lime green	Standard	$250
7	Palm	Silver	Touch screen	$250
8	Palm	White	Standard	$350
9	Palm	Lime green	QWERTY	$150

Rate the nine hypothetical PDAs using a scale from 1 ("undesirable") to 9 ("very desirable"). Ask two friends or classmates to rank the nine hypotheticals also.

Enter your ratings in the **Assignment 10-1 Dell PDA conjoint.xls.** The file contains 27 rows, 9 rows for each person in your sample, and 7 columns, *customer, hypothetical PDA, brand, color keypad, price,* and *rating*. Change the labels to match the attributes and attribute levels that you chose.

Identify the baseline PDA, and then make eight indicator variables to designate options other than baseline.

Run a regression to find the preferred PDA configuration, the *part worth utilities*, and the relative *importances* of attributes.

Deliverables: Write a paragraph to management, summarizing your results, with recommendations for the new product development team.

Attach a copy of your regression sheet with a table and plots of *part worth utilities* and a table of *attribute importances*.

CASE 10-1 Modeling Growth: Procter & Gamble Quarterly Revenues

Procter & Gamble revenues are growing, as the company's managers innovate and forge into new markets and as the company acquires complementary businesses. Procter & Gamble management want to quantify the impact on revenues of the acquisition of Gillette late in 2005. They have asked for a model that quantifies quarterly revenue drivers, including the Gillette acquisition, which can also be used to forecast.

The Gillette Acquisition. Procter & Gamble acquired Gillette in 2005. The first quarter of the combination is December 2005. Revenues in that quarter were nearly $4 billion greater than in the preceding quarter.

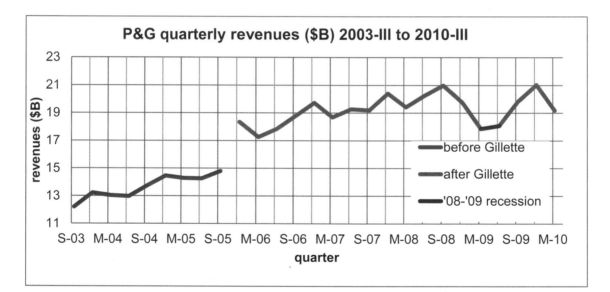

Inertia. Procter & Gamble manufactures and markets packaged goods, many of which could be viewed as necessities. Consumers who choose P&G brands tend to be brand loyal, and choose P&G brands repeatedly. Management would like to know how strong this loyalty response is. For each new dollar in revenues, what proportion can be expected a year later?

Some of the products in P&G's portfolio are seasonal. Laundry detergent, for example, is seasonal, because families wear more layers of clothing in colder seasons. Consequently, quarterly revenues tend to be highest in the fourth quarter each year. One convenient way to account for seasonality is to include an inertia component, such as past year revenues, as a predictor. Past year revenues are, thus, useful for estimating loyalty, and also because they account for seasonality.

2008–2009 recession. P&G management saw revenues falling below forecasts the first two quarters of 2009. Management believed that the disappointing results were the result of the economic recession of 2008–2009 and the related financial crisis of 2009. In the U.S., GDP actually lost value in the first two quarters of 2009. Once decision makers became aware that the economy was officially in recession, major changes occurred. Companies downsized, and consumers were forced to reduce spending. Management wants to confirm that slowed revenues in quarters I and II of 2009 were due to recession and not some firm specific problem. They also want to confirm that revenues had begun to recover and would regain their previous momentum.

Build a time series model of P&G revenues, including the *Gillette acquisition*, the *2008–2009 recession* and the past year *revenues*$_{q-4}$.

Add an indicator of the *Gillette boost*, equal to zero in quarters before December 2005 and equal to one in December 2005 and quarters after.

Allow for the impact of the 2009 recession by adding an indicator *2008–2009 recession,* equal to one in the first two quarters of 2009 and equal to zero in the other quarters.

Make an inertia component, past year *revenue*$_{q-4}$, by copying *revenues*$_q$, and then shifting the lagged inertia column four quarters. Past year revenue from September 2002 will appear in the *revenue*$_{q-4}$ column in the same row as *revenue*$_q$ in September 2003. (Your regression will begin with data from September 2003, as a consequence.)

Be sure to exclude the two most recent quarters, December 2009 and March 2010, to build your model. Then you will be able to test its validity for forecasting.

Assess the Durbin Watson *DW* statistic to decide whether or not your model has accounted for trend, cycles, and seasonality in the quarterly data.

Validate your model and then add the two most recent quarters and recalibrate.

Sensitivity analysis to find expected response under alternate scenarios.

Find forecasts with:
1. The *Gillette* indicator set to zero to determine what *revenues* would have been had the acquisition not been made
2. The *2008–2009 recession* indicator set to zero to determine what *revenues* would have been had the recession not occurred

Deliverables.

1. Write your model equations for
   (a) The baseline quarters before the *Gillette acquisition*
   (b) Quarters after the *Gillette acquisition*
   (c) Quarters during the *2008–2009 recession*

2. What is the margin of error in your forecasts?

3. What percent of each dollar of new revenue can the management expect to come from repeat sales to loyal customers 1 year later?

4.   What is the *95% prediction interval* for revenues in March 2011?

5.   What is the expected percent increase in *revenues* in March 2011, relative to revenues in March 2010?

6. Make a table to show
    a) How much the Gillette acquisition has enhanced *revenues* in each of the quarters since December 2005
    b) The percent of *revenues* contributed by Gillette relative to what *revenues* would have been without Gillette in each of the quarters since December 2005

7. Make a table to show
    a) *Revenue* lost in each of the first two quarters of 2009 due to the *2008–2009 recession*
    b) The percent reduction from expected *revenues* had there been no recession

8. Illustrate your model fit and sensitivity analysis with a scatterplot of
    a) *Revenue predictions*, September 2003 through March 2011
    b) *Actual revenues*
    c) *Revenue predictions* without the *Gillette acquisition* from December 2006 through March 2011
    d) *Revenue predictions* without the *2008–2009 recession* from March 2009 through March 2011

Case 10-2 Store24 (A): Managing Employee Retention* and Store24 (B): Service Quality and Employee Skills**

Read the cases and identify
1.   The performance variables that the management is seeking to improve
2.   The five decision variables under management control that could be used to improve performance (note: hints in the case titles)

Build two models linking the five decision variables to performance, based on the case and using case data in **Case 10-2 Store24.xls.**

In the short term, the management cannot control all of the drivers of performance. (For example, store locations are fixed in the short term.) It will be important to account for both the five controllable and the six uncontrollable drivers in order to accurately assess the influence of decision variables.

Write a memo to the management presenting the results of your models.

*Harvard Business School case 9602096
**Harvard Business School case 9602097

Be sure to include a brief description of the sample that you used.

Include one regression model equation, and a plot showing one of your key results.

If one or more indicator variables are significant in your model, show your equation for both values of the indicator(s). Use standard format to present your equations, including variable names, superscripts for coefficients and *R Square* significance levels, and a hat to denote predicted dependent variable values. <u>Explain your equation and plot—explain what readers should see when they look at your equation and plot.</u>

Report any other results that might influence decision making.

Attach exhibits for additional plots referred to in your memo.

Briefly note other considerations that might matter, but which your model ignores.

Use Times New Roman 12 pt font.

Come to class prepared to present your model and explain what it means to the management.

Chapter 11
Nonlinear Multiple Regression Models

In this chapter, nonlinear transformations are introduced, which expand multiple linear regression options to include situations in which marginal responses are either increasing or decreasing, rather than being constant. We explore Tukey's Ladder of Powers to identify particular ways to rescale variables to produce valid models with superior fit.

11.1 Consider a Nonlinear Model When Response Is Not Constant

To decide whether to use a nonlinear model, first rely on your logic:

- Do you expect the response, or change in the dependent performance variable, to be constant, regardless of whether a change in an independent variable is at minimum values or at maximum values? Linear models assume constant response.

- Is the dependent variable limited or unlimited?

Linear models are unlimited. If your dependent variable couldn't be negative because it is measured in dollars, purchases, people, or uses, a nonlinear model is logically more appropriate.

After consulting your logic, plot your data, then fit a line as well, and examine the residuals. You will see just how well a linear model fits.

11.2 Tukey's Ladder of Powers

Tukey offered a simple heuristic to quickly suggest ways to rescale variables when residuals from linear regression would be either skewed or heteroskedastic. Scales are chosen that reduce skewness of both independent and dependent variables. Models built with variables that have been rescaled to reduce skewness will be nonlinear.

If a variable is positively skewed, as is the variable on the left in Fig. 11.1, shrinking it by rescaling in square roots, natural logarithms, or inverses (reciprocals) will *Normalize*. Square roots are lower absolute power, .5, than inverses, –1, and are less radical. Natural logarithms are moderate, making a bigger difference than square roots and a smaller difference than inverses.

When a variable is negatively skewed, as is the variable on the right in Fig. 11.1, expanding it by rescaling to squares or cubes will *Normalize*. The higher power, cubes, will make a bigger difference.

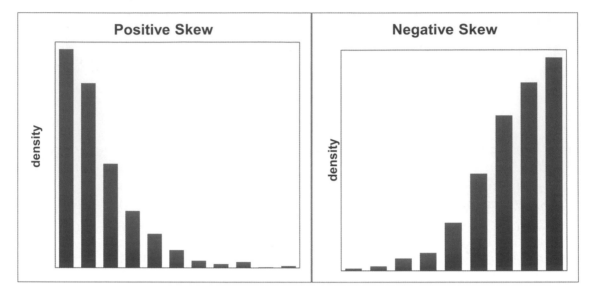

Fig. 11.1 Positively and negatively skewed variables

Moving from the center up or down the *Ladder of Powers*, Fig. 11.2, changing the power more, changes the data and its skewness more. More skewness calls for more adjustment.

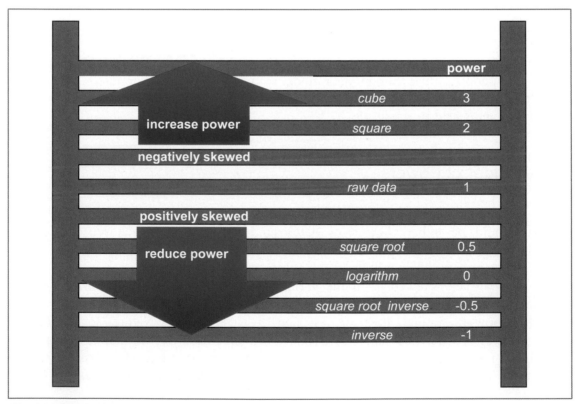

Fig. 11.2 Tukey's ladder of powers

11.3 Rescaling *y* Builds in Synergies

Jointly, two drivers may make a larger difference than the sum of their individual influences. For example, advertising levels may be more effective when sales forces are larger. The impact of population growth in a country may influence imports more if growth in GDP has been relatively high. When the dependent variable is rescaled, the model becomes multiplicative, which produces synergies between predictors. With this potential benefit of improved fit and validity, comes the cost of transforming predictions in rescaled units back to the original units.

Example 11.1 Executive Compensation

The Board of a large corporation in the financial industry is courting a new CEO candidate. To more precisely craft their offer, they would like to be able to relate executive compensation to performance in the industry. They have asked for a model relating executive compensation to firm sales, profits, and returns in similar large corporations. *Forbes* has published a dataset containing executive compensation, firm performance, and demographics from a sample of 500 large corporations. Using this dataset, we build a model to help the board more confidently quantify their offer.

Board members believe that executives from larger, more profitable firms earn more, and that older, more experienced executives are better compensated. They also believe that there may be noticeable differences across industries. We include in the model

- Revenues in billion (B) dollars
- Profits in billion (B) dollars
- Percent return over 5 years
- Executive age in years
- Indicators to distinguish industries

Complete data on these measures are available for 434 firms in six major industries: computers, energy, financial, food, health, and utilities. The best paid executives are compensated well beyond most. Consequently, approximately 10% of the total compensation packages are outliers within each of the six industries and are excluded, leaving a sample of 402 CEOs of large corporations.

Four of the five continuous variables are positively skewed, as Table 11.1 and Fig. 11.3 illustrate. A relatively small proportion of executives are better compensated, and a relatively small proportion of firms have *higher revenues, profits*, and *5 year returns. Age* is approximately *normally* distributed.

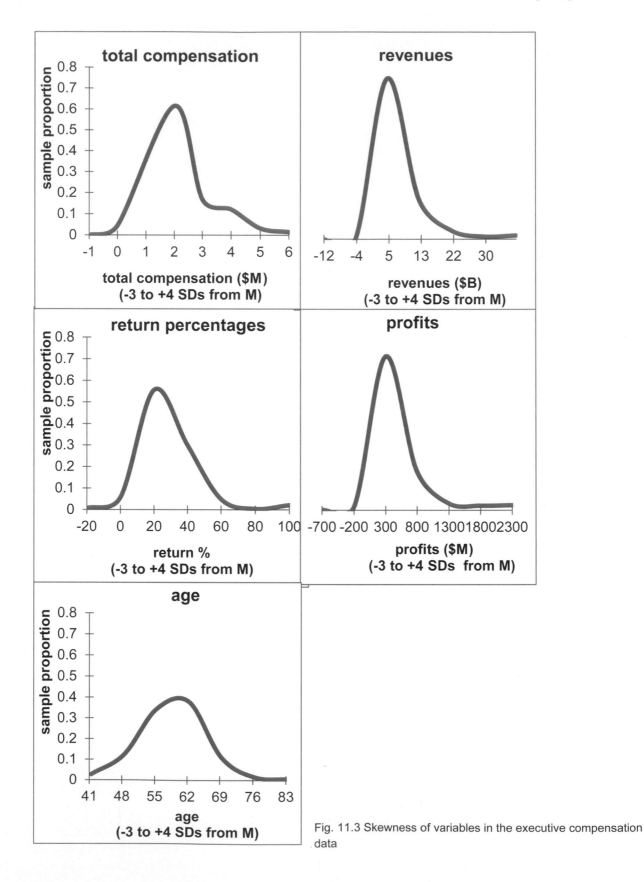

Fig. 11.3 Skewness of variables in the executive compensation data

Table 11.1 Skewness of executive compensation variables

	Total compensation ($M)	Revenues ($B)	5 year return %	Profits ($M)	Age
Skewness	1.5	6.1	3.0	4.4	− .2

To *Normalize* positively skewed total compensation, either the square roots or natural logarithms, shown in Fig. 11.4, reduce skewness to the Normal range, −1 to +1. Revenues are more positively skewed, and the square roots, shown in the left panel of Fig. 11.5, aren't enough correction. The natural logarithms, shown in the right panel of Fig. 11.5, are needed to reduce the positive skew to the Normal range.

With profits and 5 year return, square roots and natural logarithms are not options, because some firms reported negative profits and negative returns. The available option for positively skewed variables with negative values is to invert, scaling in inverses, shown in Fig. 11.6.

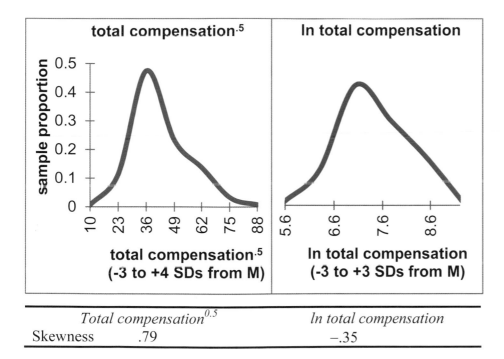

Total compensation$^{0.5}$	*ln total compensation*
Skewness .79	−.35

Fig. 11.4 Rescaled total compensation

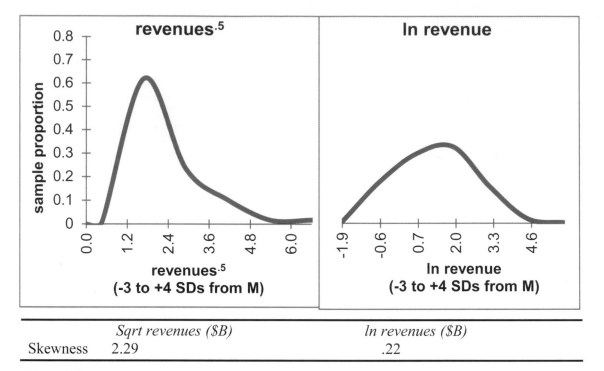

	Sqrt revenues ($B)	ln revenues ($B)
Skewness	2.29	.22

Fig. 11.5 Rescaled revenues

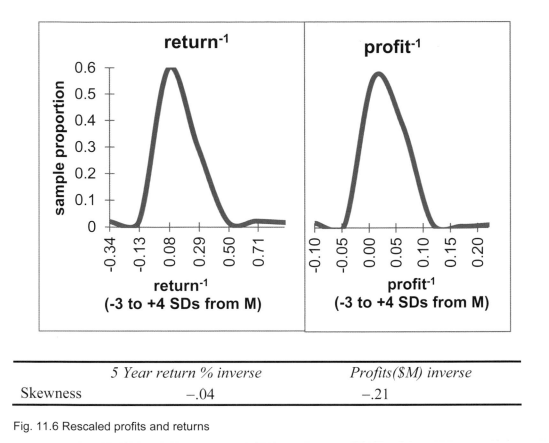

	5 Year return % inverse	Profits($M) inverse
Skewness	−.04	−.21

Fig. 11.6 Rescaled profits and returns

Inverses are fairly drastic and sometimes produce peaked distributions where most datapoints are close to the mean. We will retain the original scales of profits and 5 year return percentage.

The nonlinear multiple regression model results are shown in Table 11.2.

Table 11.2 Executive compensation driven by industry, firm performance, and executive age

SUMMARY OUTPUT

Regression statistics

R Square	.362				
Standard error	.334				
Observations	402				

ANOVA	df	SS	MS	F	Significance F
Regression	9	24.9	2.76	24.7	.0000
Residual	392	43.8	.11		
Total	401	68.7			

	Coefficients	Standard error	t Stat	p Value	Lower 95%	Upper 95%
Intercept	.151	.148	1.0	.31	−.141	.443
Computers	.422	.064	6.6	.0000	.296	.548
Energy	.243	.071	3.4	.0007	.104	.381
Financial	.356	.054	6.6	.0000	.250	.461
Food	.149	.066	2.3	.02	.020	.278
Health	.400	.068	5.9	.0000	.266	.534
Ln revenues (B$)	.128	.021	6.2	.0000	.087	.168
5 Year return %	.0034	.00094	3.6	.0004	.0015	.0052
Profits(M$)	.00013	.000040	3.1	.002	.000046	.0002
Age	.0097	.00253	3.8	.0001	.0047	.015

From regression output, the nonlinear model equation is

$$TotalCom\hat{p}ensation(\$M)^{.5} = .15 + .42^a\ computers + .24^a energy + .36^a financial$$

$$+ .15^b food + .40^a health + .13^a ln(revenues(\$B))$$

$$+ .0034^a return\% + .00013^a profit(\$M) + .0097^a age$$

R Square: 36%[a].

[a]*Significant* at .01 or better
[b]*Significant* at .02

This equation is in square roots. To see the equation in the original scale of million dollars, we square both sides, rescaling the dependent variable back:

$$TotalCom\widehat{pens}ation(\$M) = [\ .15 + .42^a\ computers + .24^a energy + .36^a financial$$

$$+ .15^b food + .40^a health + .13^a ln(revenues(\$B))$$

$$+ .0034^a return\% + .00013^a profit(\$M) + .0097^a age]^2.$$

Variation in industry, firm performance, and executive age accounts for 36% of the variation in CEO compensation. Better performing firms pay their executives more. Older, more experienced executives earn more, and compensation is higher in the computer, financial, and health industries, lower in energy and food industries, and lowest in the baseline industry, utilities.

11.4 Sensitivity Analysis Reveals the Relative Strength of Drivers

Driver influence in a multiplicative model under contrasting scenarios. When the dependent variable is rescaled to build a nonlinear model, the model is multiplicative. The impact of each of the drivers depends on values of all of the other drivers. Predicted compensation values can be compared for contrasting scenarios, such as those that are linked to lower compensation (younger executives from smaller, worse performing firms) and those that are linked to higher compensation (older executives from larger, better performing firms).

As an example, to identify the impact of *return %* on executive compensation in the Financial industry, *Computers, Energy, Food, Health* indicators are set to 0 and the *Financial* indicator is set to 1. Since all driver coefficients are positive, other lower driver values (lower revenue and profit and younger executive age) would reduce compensation, and the impact of *return %* will be lower than in the contrasting case in which revenue and profit are higher and an executive is older. For both of these contrasting scenarios, find predicted compensation at varying levels of *return %* to find its impact. Predicted executive compensation by *return %* in the Financial industry, for contrasting scenarios, is illustrated in Fig. 11.7:

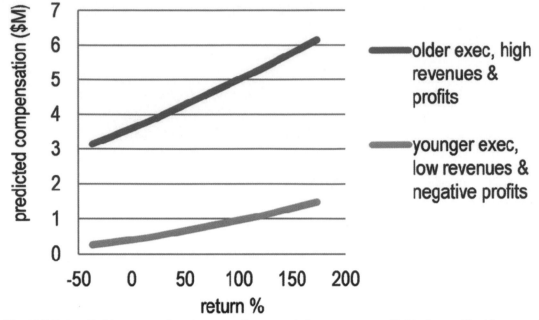

Fig. 11.7 Return% drives executive compensation more in larger, more profitable firms with older executives

Larger compensation packages are associated with greater returns. For older executives managing larger, more profitable firms, *return %* could make a three million dollar difference in compensation in the financial industry. *Return %* matters less for younger executives in smaller, less profitable firms, but could make a difference of one million dollars.

The impact of industry indicators is similarly multiplicative. There are differences in the impact of each driver across industries. To compare the influence in compensation due to a driver, across industries, create multiple industry scenarios, and, within each, allow one driver to vary from the minimum to maximum values, setting the remaining drivers to their median values for each industry. (To acknowledge skewness in the data, the medians are used as the base for comparison, rather than the means.) As an example, Fig. 11.8 compares predicted executive compensation in utility, financial, and computer industries by each of the drivers.

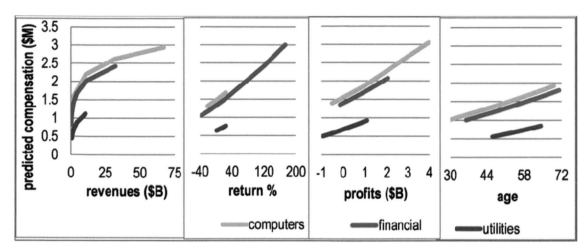

Fig. 11.8 Drivers' Influences are multiplicative

Each driver will be more influential in industries where compensation is higher, which are identified by larger indicator coefficients. A driver will also be more influential in industries where there is a larger range of values. Firm revenue is more influential in the computer industry, where average compensation is higher. Revenue is also more influential in the computer industry because the range in firm revenues is greater than in other industries.

In the financial industry, the large range in returns drives a large difference in executive compensation. Because of this relatively large range in returns, returns are more influential than other drivers in the financial industry.

Total compensation response to revenues increases at a *decreasing* rate. *Executive compensation* differences are greater for firm *revenue* differences among smaller firms than among larger firms: differences in *revenues* influence *executive compensation* more when *revenues* are lower.

Total compensation response to the remaining drivers is increasing. *Executive compensation* differences are greater among more profitable firms with higher profits and among older executives.

Results of the analyses are summarized in the memo to the Board:

MEMO

Re: Executive Compensation Driven by Firm Performance and Age
To: The Board
From: James Melton, Director, Econometric Analysis
Date: January 2011

Analysis of 402 executive compensation packages offered by firms surveyed by Forbes Magazine reveals that industry, firm returns, revenue, and profitability, and executive experience are key drivers.

Compensation Model. Using *Forbes* data from 402 of firms in six broad industries, a model linking industry, executive age and firm performance measures with compensation was built.

Model Results. Industry, executive age, firm revenues, profits, and return percent over 5 years account for 36% of the variation in compensation.

Executives in the financial industry are better rewarded by an average of $400K than those in food, energy, or utilities, but paid an average of $400K less than those in computer or health sectors.

Aside from industry differences, older, more experienced executives and those heading larger, more profitable firms with higher returns are better compensated.

Firm return percent and revenue are the strongest drivers of compensation. A difference of 200% in return percents, the sample range, makes an expected difference of $1M to $3M in compensation. A difference of $32B in revenues, the sample range, also makes an expected difference in compensation of $1–$3M.

Thirty-five additional years in experience, the sample range, can add $.5–$1.6M to expected compensation packages. A difference in profit of $2B, the sample range, can increase compensation by $.4–$1.3M.

Conclusions. More experienced, more successful executives in financial firms are better rewarded, particularly for achieving greater firm returns.

Other considerations. Past compensation and compensation offered by close competitors may also drive compensation, and were not considered in this model.

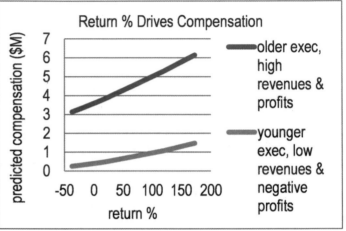

$$\hat{Comp}(\$M) = [.15 + .36^a financial + .42^a computers + .40^a health + .15^b food + .24^a energy + .0034^a return + .00013^a profit(\$M) + .13^a ln(revenue(\$B)) + .0097^a]^2$$

11.5 Gains from Nonlinear Rescaling Are Significant

To see the gain from building a nonlinear model, compare results with those from a simpler linear model. The linear model of total compensation using the same variables and sample is

$$Total\hat{C}ompensation(\$M) = .59^a + 1.2^a computers + 1.0^a health + .74^a energy$$

$$+ .64^a financial + .60^a food + .00077^a profit(\$M)$$

R Square: 26%[a].
[a]*Significant* at .01 or better

In the linear model, firm *revenues* and *5 year return percentage* are not significant and have been removed, accordingly. The remaining predictors, industry indicators, firm *profits*, and executive *age*, account for 26% of the variation in executive *compensation*. Relying on a linear model, the board would ignore the particularly important links among firm *revenues*, firm *return percentage* over 5 years, and *total compensation* reducing potential performance incentives.

Comparing residuals from the nonlinear and linear models, shown in Fig. 11.9, the nonlinear model residuals are less skewed and better satisfy multiple linear regression assumptions.

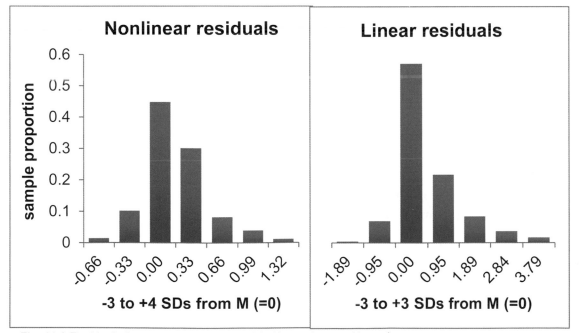

Fig. 11.9 Residuals from the nonlinear model (left) are closer to Normal

11.6 Nonlinear Models Offer the Promise of Better Fit and Better Behavior

It is a challenge to think of an example of a truly linear (constant) response. Responses tend to be nonconstant and nonlinear. The fifth dip of an ice cream is less appetizing than the first. Consumers become satiated at some point, and beyond that point, additional consumption is less valuable. Adding the twentieth stock to a portfolio makes less difference to diversification than adding the third. A second ad insertion in a magazine enhances recall more than a tenth ad insertion. As a consequence of nonconstant, changing marginal response, nonlinear models tend to offer the promise of superior fit and better behaved models, with more nearly random residuals. Nonlinear models do carry the cost of transformation to and back from logarithms, square roots, inverses, or squares. In some cases, a linear model fits data quite well and is a reasonable approximation. Thinking logically about the response that you've set to explain and predict, and then looking at the distribution and skewness of your data and your residuals, will sometimes lead you toward the choice of a nonlinear alternative.

Tukey's Ladder of Powers can help quickly determine the particular nonlinear model that will fit a dataset best. When a variable is positively skewed, rescaling to square roots, natural logarithms, or inverses often reduces the positive skew. Negatively skewed variables are sometimes Normalized by squaring or cubing. The amount of difference corresponds to the power: square roots with power .5 are less radical than inverses with power -1 and squares with power 2 are less extreme than cubes with power 3.

Excel 11.1 Rescale to Build and Fit Nonlinear Regression Models with Linear Regression

Executive Compensation
Executive compensation, including salary, stock options, and bonuses, probably depends on the industry, executive age (reflecting experience), and company performance. Company performance measures include revenues, profits, and 5 year return percentage.

    The fewer exceptional executives are probably compensated more, thus we expect total executive compensation to be positively skewed. Because unsuccessful firms exit markets, we expect company performance measures to be positively skewed, as well.

Data for 402 firms surveyed by *Forbes Magazine* are in **Excel 11.1 Executive Compensation.xls.**

Assess skewness and choose scales: Use Excel's **SKEW(array)** function to assess skewness of total compensation (M$). Then fill in skewness of revenues ($B), profits (M$), return %, and age.

	F404	▾	f_x	=SKEW(F2:F403)		
⊿	A	B	C	D	E	F
1	Wide Industry	total compensation ($MM)	revenues ($B)	Profits (MM$)	Age	Return % Over 5 Yrs
398	Utility	0.416	1.80	166	61	11
399	Utility	0.398	0.87	95.3	55	12
400	Utility	0.391	2.47	-942.6	58	1
401	Utility	0.324	0.54	84.5	55	11
402	Utility	0.314	2.07	297.2	55	15
403	Utility	0.280	1.08	107.2	60	14
404	skew	1.54	6.13	4.43	-0.16	3.04

Total compensation and the three firm performance measures are positively skewed. Executive *age* is approximately Normal.

To see the skewness, make histograms for *total compensation*, *revenues*, *profits*, *return percentage*, and *age*.

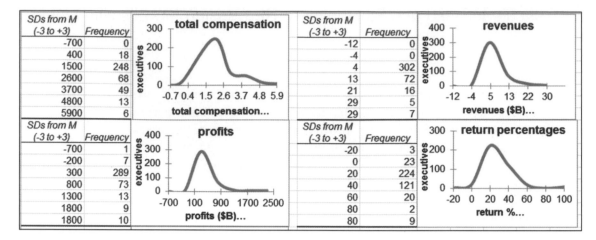

To *normalize* the positively skewed variables, we shrink. For *total compensation* and *revenues*, which are never zero and never negative, consider the square roots and the natural logarithms, which have powers .5 and 0 on Tukey's Ladder.

Add four columns and then make *sqrt total compensation* ($M), *ln total compensation* ($M), *sqrt revenues* ($B), and *ln revenues* ($B) using Excel functions **sqrt(***array***)** and **ln(***array***)**:

	D2		fx	=SQRT(B2)		
	B	C	D	E	F	G
1	total compensation ($M)	revenues ($B)	sqrt total compensation ($M)	ln total compensation ($M)	sqrt revenues ($B)	ln revenues ($B)
382	0.617	4.28	0.79	-0.48	2.07	1.45
383	0.589	0.61	0.77	-0.53	0.78	-0.50
384	0.587	1.64	0.77	-0.53	1.28	0.50
385	0.587	2.33	0.77	-0.53	1.53	0.85
386	0.580	1.91	0.76	-0.54	1.38	0.65

For *profits* and *5 year return*, which are sometimes negative, we cannot use either square roots or logarithms. Consider the inverse of both of these, which has power −1.

Add two columns and then make *profit* (M$) *inverse* and *5 year return % inverse*.

H2			f_x	=J2^(-1)	
	H	I	J	K	L
1	profit (M$) inv	return % inv	Profits (M$)	Age	Return % Over 5 Yrs
382	0.002	0.063	626.4	61	16
383	0.013	0.071	80	63	14
384	0.005	0.071	188.5	50	14
385	0.005	0.077	215.8	63	13
386	0.006	0.077	177.4	56	13

Fill in *skewness* of the six new columns.

C404				f_x	=SKEW(C2:C403)			
	C	D	E	F	G	H	I
1	revenues ($B)	sqrt total compensation ($M)	ln total compensation ($M)	sqrt revenues ($B)	ln revenues ($B)	profit (M$) inv	return % inv
399	0.87	0.63	-0.92	0.93	-0.14	0.010	0.083
400	2.47	0.63	-0.94	1.57	0.91	-0.001	1.000
401	0.54	0.57	-1.13	0.73	-0.62	0.012	0.091
402	2.07	0.56	-1.16	1.44	0.73	0.003	0.067
403	1.08	0.53	-1.27	1.04	0.07	0.009	0.071
404	6.13	0.79	-0.35	2.29	0.22	-0.04	-0.21

The skewness of the square roots and natural logarithms of *total compensation* are within the Normal range, −1 to +1.

Skewness in the square roots of *revenues* remains positive and greater than 1, and skewness in the natural logarithms is close to 0, within Normal range.

Change the square root of *revenues* to a higher root, such as the fifth root, to reduce skewness to Normal.

	F404	▼	f_x	=SKEW(F2:F403)

	D	E	F
1	sqrt total compensation ($M)	ln total compensation ($M)	fifth rt revenues ($B)
400	0.63	-0.94	1.20
401	0.57	-1.13	0.88
402	0.56	-1.16	1.16
403	0.53	-1.27	1.01
404	0.79	-0.35	0.88

Reset the histogram mean and standard deviation to make histograms of *sqrt total compensation*, *ln total compensation*, *fifth root revenues*, *ln revenues*, *profit inverse*, and *return inverse*.

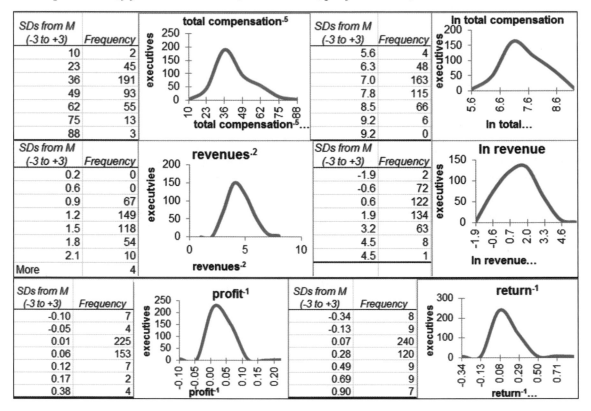

The *profit* and *return* inverses produce peaked distributions with most executives in the two middle bins. This limits their importance as drivers, so the original scales for *profits* and *returns* are used.

Use the square roots of *total compensation* with the natural logarithms of *revenues*.

Add indicators: To account for industry differences in executive compensation, add industry indicators. There are six industries represented in the dataset. It will simplify interpretation if we choose the industry with the lowest average executive compensation for our baseline. Coefficient estimates for the five other industry indicators will reflect the average difference from the least well compensated baseline.

Find average *total compensation* by industry with a PivotTable.

Select cells in the *wide industry* and *total compensation* columns; then use shortcuts to request a Pivot Table **Alt NVT**.

Drag *Wide Industry* to the **ROWS** and *total compensation* ($M) to **DATA**.

	A	B
1	Drop Report Filter Fields Here	
2		
3	Sum of total compensation ($M)	
4	Wide Industry	Total
5	ComputersComm	121.003222
6	Energy	62.15787
7	Financial	213.074231
8	Food	72.785626
9	Health	74.887457
10	Utility	46.23456
11	Grand Total	590.142966

PivotTable Field List

Choose fields to add to report:
- ☑ **Wide Industry**
- ☑ **total compensation ($M)**

Drag fields between areas below:

▽ Report Filter	▦ Column Labels

▦ Row Labels	Σ Values
Wide Industry ▾	Sum of total c... ▾

Use shortcuts to change the PivotTable cells to averages.

Alt JTG to **Summarize value field by average**.

Executives in the *utility* industry are least well compensated, on average.

Designate *utility* as the baseline industry using indicators for each of the remaining five.

In the *Executive compensation* sheet, select all of the data cells, and then use shortcuts to sort the dataset.

Alt dAta **S**ort, My data has **H**eaders with **Sort by** *Wide industry*.

It is good modeling practice to order the independent variables so that the indicators come first. The indicators modify the intercept, and this will make interpretation of results easier.

Add five columns following *sqrt total compensation* for indicators, *computers & communication, energy, financial, health*, and *food*, filling cells with one for the indicator column which matches *Wide industry*, with zeros in the remaining indicator columns.

	A	B	C	D	E	F	G	H	I
1	Wide Industry	total compensation ($M)	revenues ($B)	sqrt total compensation ($M)	computers & communication	energy	financial	food	health
315	Health	2.086	5.79	1.44	0	0	0	0	1
316	Health	2.066	7.48	1.44	0	0	0	0	1
317	Health	1.845	2.38	1.36	0	0	0	0	1
318	Health	1.826	2.26	1.35	0	0	0	0	1
319	Health	1.747	2.47	1.32	0	0	0	0	1

Run regression using the rescaled variables and indicators.

SUMMARY OUTPUT

Regression Statistics

Multiple R	0.602
R Square	0.362
Adjusted R Square	0.347
Standard Error	0.334
Observations	402

ANOVA

	df	SS	MS	F	ignificance F
Regression	9	24.9	2.76	24.7	1.52E-33
Residual	392	43.8	0.11		
Total	401	68.7			

	Coefficients	Standard Error	t Stat	P-value	Lower 95%	Upper 95% v
Intercept	0.151	0.148	1.0	0.3092	-0.141	0.443
computers & communication	0.422	0.064	6.6	0.0000	0.296	0.548
energy	0.243	0.071	3.4	0.0007	0.104	0.381
financial	0.356	0.054	6.6	0.0000	0.250	0.461
food	0.149	0.066	2.3	0.0232	0.020	0.278
health	0.400	0.068	5.9	0.0000	0.266	0.534
ln revenues ($B)	0.128	0.021	6.2	0.0000	0.087	0.168
Profits (M$)	0.000126	0.000040	3.1	0.0020	0.000046	0.000206
Return % Over 5 Yrs	0.00337	0.00094	3.6	0.0004	0.00153	0.00522
Age	0.00970	0.00253	3.8	0.0001	0.00473	0.01466

To assess the *Normality* of the residuals, find the residual skewness and make their histogram.

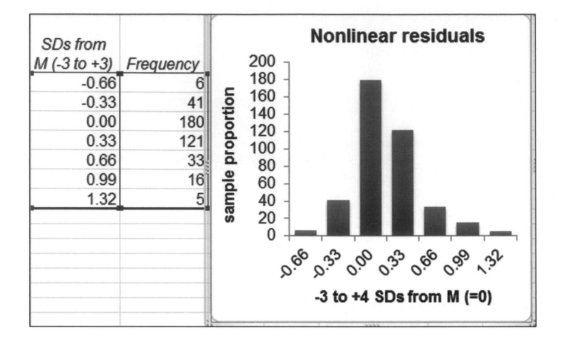

SDs from M (-3 to +3)	Frequency
-0.66	6
-0.33	41
0.00	180
0.33	121
0.66	33
0.99	16
1.32	5

The residuals are approximately *Normal*.

To see *predicted compensation* values, first find *predicted sqrt total compensation* (M$) using the regression equation.

fx   =$O$2+$O$3*E2+$O$4*F2+$O$5*G2+$O$6*H2+$O$7*I2+$O$8*J2+$O$9*K2+$O$10*L2+$O$11*M2

E	F	G	H	I	J	K	L	M	N	O
computers & communication	energy	financial	food	health	In revenues ($B)	Profits (M$)	Return % Over 5 Yrs	Age	predicted sqrt total compensation ($M)	Coefficients
1	0	0	0	0	1.06	-35.8	40	29	1.118978925	0.151
1	0	0	0	0	2.83	1022	39	54	1.717851442	0.422
1	0	0	0	0	0.65	14.8	4	53	1.184571321	0.243
1	0	0	0	0	-1.29	81.2	41	48	1.021610201	0.356
1	0	0	0	0	-0.48	125.6	6	66	1.187695333	0.149
1	0	0	0	0	-0.31	81.7	38	62	1.272472737	0.400
1	0	0	0	0	2.37	1435.2	18	52	1.620853215	0.128
1	0	0	0	0	4.21	3974	14	59	2.229671714	0.000126
1	0	0	0	0	0.23	114.7	44	50	1.250144024	0.00337
1	0	0	0	0	0.72	341.6	13	49	1.226722455	0.00970

	fx	=O2^2

N	O
predicted total compensation ($M)	predicted sqrt total compensation ($M)
1.25	1.12
2.95	1.72
1.40	1.18
1.04	1.02
1.41	1.19
1.62	1.27

To rescale back to the original total compensation scale in million dollars, add *predicted total compensation* ($M), squaring *predicted sqrt total compensation* ($M).

Excel 11.2 Consider Synergies in a Multiplicative Model with Sensitivity Analysis

When a dependent variable has been rescaled, the model becomes multiplicative, and the impact of each driver depends on the values of other drivers. To isolate the importance of a driver, compare expected total compensation of two contrasting sets of hypothetical executives, those likely to be compensated less (younger executives in smaller, less profitable firms with lower returns) and those likely to be compensated more (older executives in larger, more profitable firms with higher returns).

Marginal impact of return percent across contrasting scenarios. To determine the difference in compensation driven by differences in firm *return %* in an industry, add 18 new rows to the dataset which describe two sets of 9 hypothetical executives
- From the *same* industry
- Nine who are youngest, from smallest, least profitable firms
- Nine who are oldest, from largest, most profitable firms

Within each set of nine, hypothetical executives are identical in terms of *age*, firm *revenue*, and *profit* and differ only with respect to their firm's *return%*:
- One will lead the most successful firm with the maximum *return%.*
- Three will manage more successful firms with *return%* at the 75, 90, and 95%.
- The fifth will lead a firm with *return %* at the 50%.
- Three will head less successful firms with *return%* at the 5, 10, and 25% in the sample.
- One will head the least successful firm with the minimum *return%.*

Find representative values of predictors. Focus on the *financial* industry, as an example. Rearrange columns so that the indicators and original executive variables are adjacent, followed by *predicted total compensation ($M)*.

D	E	F	G	H	I	J	K	L	M
computers & communication	energy	financial	food	health	revenues ($B)	Profits (M$)	Return % Over 5 Yrs	Age	predicted total compensation ($M)
1	0	0	0	0	2.87	-35.8	40	29	1.25
1	0	0	0	0	16.96	1022	39	54	2.95
1	0	0	0	0	1.91	14.8	4	53	1.40

Add 18 rows following the last data row and fill in the industry indicators, 1 for *financial*, 0 for the others.

For the first set of nine hypothetical executives, input the maximums of *revenues*, *profits*, and *age*, with 95, 90, 75, 50, 25, 10, 5%, and minimum of *return %* using the Excel functions

$$\textbf{MAX}(array), \ \textbf{PERCENTILE}(array, percentile), \text{ and } \textbf{MIN}(array)$$

with arrays including only the rows from the financial industry.

For the second set of nine hypothetical executives, input the minimums of *revenues*, *profits*, and *age*, with 95, 90, 75, 50, 25, 10, 5% and minimum of *return %*.

| K421 | | f_x | =MIN(K99:K250) | | | | | | |

	B	C	D	E	F	G	H	I	J	K	L
1			computers & communication	energy	financial	food	health	revenues ($B)	Profits (MM$)	Return % Over 5 Yrs	Age
404	oldest, largest,	max	0	0	1	0	0	32	2043	173	71
405	most profitable	95%	0	0	1	0	0	32	2043	34	71
406		90%	0	0	1	0	0	32	2043	30	71
407		75%	0	0	1	0	0	32	2043	22	71
408		50%	0	0	1	0	0	32	2043	16	71
409		25%	0	0	1	0	0	32	2043	10	71
410		10%	0	0	1	0	0	32	2043	4	71
411		5%	0	0	1	0	0	32	2043	-2	71
412		min	0	0	1	0	0	32	2043	-37	71
413	youngest,	max	0	0	1	0	0	0.194	-146	173	36
414	smallest, least	95%	0	0	1	0	0	0.194	-146	34	36
415	profitable	90%	0	0	1	0	0	0.194	-146	30	36
416		75%	0	0	1	0	0	0.194	-146	22	36
417		50%	0	0	1	0	0	0.194	-146	16	36
418		25%	0	0	1	0	0	0.194	-146	10	36
419		10%	0	0	1	0	0	0.194	-146	4	36
420		5%	0	0	1	0	0	0.194	-146	-2	36
421		min	0	0	1	0	0	0.194	-◊6	-37	36

Arrange columns so that *predicted total compensation, ln revenues*, and *predicted sqrt total compensation*, follow *age*, and then fill in the hypothetical rows.

Select *ln revenue*, *predicted sqrt total compensation*, and *predicted total compensation* cells in the last data row, then fill down, using the formulas that you entered earlier, with **Shift+dn** and **Cntl+D**.

	O421			f_x	=N421^2								

	B	C	D	E	F	G	H	I	J	K	L	M	N	O
1			computers & communication	energy	financial	focd	health	revenues ($B)	Profits (MM$)	Return % Over 5 Yrs	Age	ln revenues ($B)	predicted sqrt total compensation (M$)	predicted total compensation (M$)
404	oldest,	max	0	0	1	0	0	32	2043	173	71	3.47	2.48	6.15
405	largest,	95%	0	0	1	0	0	32	2043	34	71	3.47	2.01	4.05
406	most	90%	0	0	1	0	0	32	2043	30	71	3.47	2.00	3.99
407	profitable	75%	0	0	1	0	0	32	2043	22	71	3.47	1.97	3.88
408		50%	0	0	1	0	0	32	2043	16	71	3.47	1.95	3.80
409		25%	0	0	1	0	0	32	2043	10	71	3.47	1.93	3.72
410		10%	0	0	1	0	0	32	2043	4	71	3.47	1.91	3.64
411		5%	0	0	1	0	0	32	2043	-2	71	3.47	1.89	3.56
412		min	0	0	1	0	0	32	2043	-37	71	3.47	1.77	3.14
413	youngest,	max	0	0	1	0	0	0.194	-146	173	36	-1.64	1.21	1.47
414	smallest,	95%	0	0	1	0	0	0.194	-146	34	36	-1.64	0.74	0.55
415	least	90%	0	0	1	0	0	0.194	-146	30	36	-1.64	0.73	0.53
416	profitable	75%	0	0	1	0	0	0.194	-146	22	36	-1.64	0.70	0.49
417		50%	0	0	1	0	0	0.194	-146	16	36	-1.64	0.68	0.47
418		25%	0	0	1	0	0	0.194	-146	10	36	-1.64	0.66	0.44
419		10%	0	0	1	0	0	0.194	-146	4	36	-1.64	0.64	0.41
420		5%	0	0	1	0	0	0.194	-146	-2	36	-1.64	0.62	0.39
421		min	0	0	1	0	0	0.194	-146	-37	36	-1.64	0.50	0.25

Illustrate the marginal response. To see this expected compensation response to differences in *return %*, rearrange columns so that *predicted total compensation* ($K) follows *return %* and then make a scatterplot of the first nine hypothetical executives' *predicted total compensation* ($M) by *return %*).

Add the second series of nine hypothetical executives' *predicted total compensation* ($M) by *return %*).

Choose a style, reset axes minimum, maximum, and major units, and add axes.

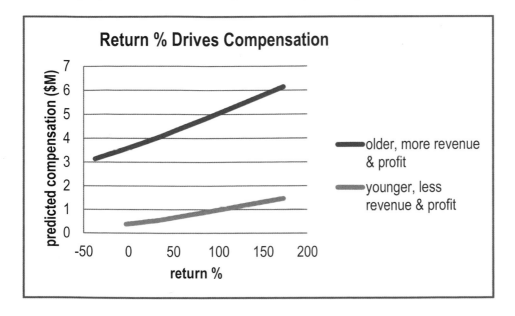

The driver impacts depend on industry, and the industry indicators multiply their influences in the multiplicative model. To see this, compare two industries, financial and food.

Add 18 new hypothetical executives, 9 in the financial industry, and 9 in the food industry.

For the first set of nine, let *return %* take the maximum, 95, 90, 75%, median, 25, 10, 5%, and minimum values among executives in the financial industry. Set the remaining drivers to their medians, also using only the financial rows.

Drag down *ln revenues*, *predicted sqrt total compensation*, and *predicted compensation*.

	M448		fx	=LN(I448)									

	B	C	D	E	F	G	H	I	J	K	L	M	N	O
1			computers & communication	energy	financial	food	health	revenues ($B)	Profits (MM$)	Return % Over 5 Yrs	Age	ln revenues ($B)	predicted sqrt total compensation (M$)	predicted total compensation (M$)
440	Financial	max	0	0	1	0	0	32.20	96	16	56	3.47	1.56	2.43
441	industry	95%	0	0	1	0	0	7.30	96	16	56	1.99	1.37	1.88
442		90%	0	0	1	0	0	4.42	96	16	56	1.49	1.31	1.70
443		75%	0	0	1	0	0	1.84	96	16	56	0.61	1.19	1.42
444		50%	0	0	1	0	0	0.68	96	16	56	-0.39	1.07	1.14
445		25%	0	0	1	0	0	0.37	96	16	56	-0.99	0.99	0.98
446		10%	0	0	1	0	0	0.25	96	16	56	-1.38	0.94	0.88
447		5%	0	0	1	0	0	0.24	96	16	56	-1.43	0.93	0.87
448		min	0	0	1	0	0	0.19	96	16	56	-1.64	0.91	0.82

Repeat this process to fill in comparable rows for the nine hypothetical food executives:

	O457		fx	=N457^2									

	B	C	D	E	F	G	H	I	J	K	L	M	N	O
1			computers & communication	energy	financial	food	health	revenues ($B)	Profits (MM$)	Return % Over 5 Yrs	Age	ln revenues ($B)	predicted sqrt total compensation (M$)	predicted total compensation (M$)
449	food	max	0	0	0	1	0	50.6	114.9	14	57	3.92	1.41	2.00
450	industry	95%	0	0	0	1	0	22.7	114.9	14	57	3.12	1.31	1.72
451		90%	0	0	0	1	0	16.0	114.9	14	57	2.77	1.27	1.61
452		75%	0	0	0	1	0	11.1	114.9	14	57	2.41	1.22	1.49
453		50%	0	0	0	1	0	5.3	114.9	14	57	1.66	1.13	1.27
454		25%	0	0	0	1	0	2.5	114.9	14	57	0.93	1.03	1.06
455		10%	0	0	0	1	0	1.4	114.9	14	57	0.34	0.96	0.92
456		5%	0	0	0	1	0	1.1	114.9	14	57	0.13	0.93	0.86
457		min	0	0	0	1	0	0.5	114.9	14	57	-0.66	0.83	0.69

Plot each of the two series of *predicted total compensation* by *return %*.

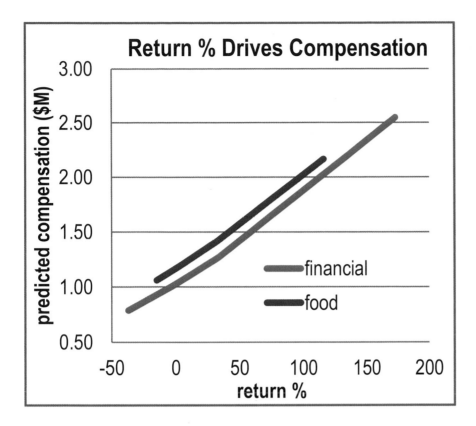

To compare the marginal influences of drivers across industries, repeat this process, adding 18 hypothetical executives whose firms vary with respect to *revenues*, with half of each set in the financial sector and half in food.

Find the marginal differences which *return %* makes in each of the two industries by finding the differences in *predicted compensation* between executives from firms with the maximum and minimum *return %*:

	N439			f_x	=M431-M439							
	C	D	E	F	G	H	I	J	K	L	M	N	O
1	total compens ation	computers & communication	energy	financial	focd	health	revenues ($B)	Profits (MM$)	Return % Over 5 Yrs	Age	predicted total compensation (M$)	difference	due to
422	max	0	0	1	0	0	0.68	96	173	56	2.55		
423	95%	0	0	1	0	0	0.68	96	34	56	1.27		
424	90%	0	0	1	0	0	0.68	96	30	56	1.24		
425	75%	0	0	1	0	0	0.68	96	22	56	1.18		
426	50%	0	0	1	0	0	0.68	96	16	56	1.14		
427	25%	0	0	1	0	0	0.68	96	10	56	1.09		
428	10%	0	0	1	0	0	0.68	96	4	56	1.05		
429	5%	0	0	1	0	0	0.68	96	-2	56	1.01		return %
430	min	0	0	1	0	0	0.68	96	-37	56	0.79	1.76	Financial
431	max	0	0	0	1	0	5.3	114.9	116	57	2.16		
432	95%	0	0	0	1	0	5.3	114.9	33	57	1.42		
433	90%	0	0	0	1	0	5.3	114.9	24	57	1.35		
434	75%	0	0	0	1	0	5.3	114.9	19	57	1.31		
435	50%	0	0	0	1	0	5.3	114.9	14	57	1.27		
436	25%	0	0	0	1	0	5.3	114.9	3	57	1.19		
437	10%	0	0	0	1	0	5.3	114.9	8	57	1.22		
438	5%	0	0	0	1	0	5.3	114.9	-8	57	1.11		return %
439	min	0	0	0	1	0	5.3	114.9	-15	57	1.06	1.10	Food

Find the difference in *predicted compensation* due to *revenues* in both industries, for comparison:

	N457			*fx*	=M449-M457							
C	D	E	F	G	H	I	J	K	L	M	N	O
	computers & communication	energy	financial	food	health	revenues ($B)	Profits (MM$)	Return % Over 5 Yrs	Age	predicted total compensation (M$)	difference	due to
440 max	0	0	1	0	0	32.20	96	16	56	2.43		
441 95%	0	0	1	0	0	7.30	96	16	56	1.88		
442 90%	0	0	1	0	0	4.42	96	16	56	1.70		
443 75%	0	0	1	0	0	1.84	96	16	56	1.42		
444 50%	0	0	1	0	0	0.68	96	16	56	1.14		
445 25%	0	0	1	0	0	0.37	96	16	56	0.98		
446 10%	0	0	1	0	0	0.25	96	16	56	0.88		
447 5%	0	0	1	0	0	0.24	96	16	56	0.87		revenues
448 min	0	0	1	0	0	0.19	96	16	56	0.82	1.61	Financial
449 max	0	0	0	1	0	50.6	114.9	14	57	2.00		
450 95%	0	0	0	1	0	22.7	114.9	14	57	1.72		
451 90%	0	0	0	1	0	16.0	114.9	14	57	1.61		
452 75%	0	0	0	1	0	11.1	114.9	14	57	1.49		
453 50%	0	0	0	1	0	5.3	114.9	14	57	1.27		
454 25%	0	0	0	1	0	2.5	114.9	14	57	1.06		
455 10%	0	0	0	1	0	1.4	114.9	14	57	0.92		
456 5%	0	0	0	1	0	1.1	114.9	14	57	0.86		revenues
457 min	0	0	0	1	0	0.5	114.9	14	57	0.69	1.31	Food

Make plots of *predicted compensation* by *revenues* using the same dependent variable y axis scale to make comparison easy.

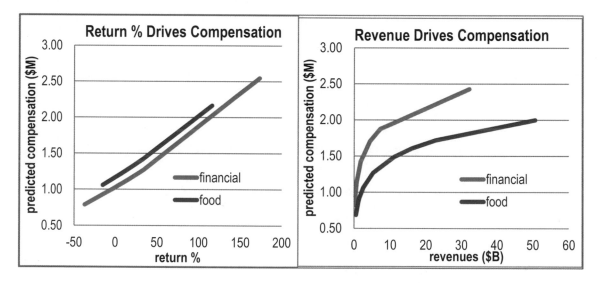

Lab Practice 11

The board of a firm in the communications and computer industry would like to know whether executive compensation packages in their industry are tied to firm performance or executive age. **Lab Practice 11 Forbes Comm.xls** contains data on the largest firms in the industry. Follow the steps in **Excel 11.1** and **Excel 11.2** to build a model of executive compensation for the board. Because all firms are in the same industry, you will not need to add industry indicators.

Which variables are positively skewed? _____

Which variable is negatively skewed? _____

Which scale, roots, natural logarithms, or inverses, *Normalizes* each of the positively skewed variables?

_____

Does rescaling the negatively skewed variable to squares make the variable more *Normal*?

       Y or N

Write your model equation in million dollars ($M) of *compensation*:

Make a table of the marginal impacts of the two significant drivers.

_____

Make scatterplots to illustrate the marginal impacts of the two most important drivers on *compensation* and attach to your lab practice worksheet.

Lab 11 Nonlinear Hybrid Sales

Ford executives have asked for a forecast of hybrid sales in the United States over the next 24 months, through February 2012. They are particularly interested in knowing whether hybrid sales are driven by higher gas prices, and impacts of the Cash for Clunkers program of July and August 2009 and Toyota's recalls of Prius hybrids in January and February 2010.

Gas Prices: Higher gas prices have traditionally spurred hybrid sales months later. Executives believe that the delay may be as long as 24 months, inasmuch as car purchases typically involve research, trade ins, and financing arrangements.

Cash for Clunkers: The Cash for Clunkers program was designed to stimulate sales of fuel-efficient cars, including hybrids. Ford executives would like to know what the program's boost amounted to in the 2 months when it was available, July and August 2009.

Toyota Prius Recalls: Following recalls of other Toyota models, Prius hybrid recalls were announced in January and February 2010. Ford expects that the recalls dampened enthusiasm for hybrids, and that the impact of the recalls is likely to last several years.

Build a valid model of monthly hybrid sales to provide Ford executives with a 24 month forecast and to quantify the impacts of past gas prices, the Cash for Clunkers program, and Prius recalls. Be sure to include a slow season indicator for months when hybrid purchases are unusually low each year.

Lab 11 Prius Recalls.xls contains data on *hybrid sales $(K)_m$*, and past *gas prices (cents per gallon)$_{m-24}$*, for months in the period June 2008 through May 2010.

Gas prices have been lagged. (For example, *gas price$_{m-24}$* in June 2008 is *gas price$_{June\ 2006}$*.)

1. Which of the variables is positively skewed? _____

2. Scales that Normalize the skewed variable: ___sqrt ___ ln ___inv ___ sqrt inv

3. Explain what it means that your model is valid, why it matters, how you know that your model is valid, and provide the evidence that you used to form your conclusion.

4. Write your model equations for the five scenarios of interest using standard format with two or three significant digits:
   (a) Baseline months
   (b) Slow season months
   (c) Cash for Clunkers months
   (d) Months during and after Prius recalls during slow season
   (e) Months after Prius recalls

5. Create a scatterplot to illustrate your model fit and forecast for months June 2008 through February 2012.

6. Explain why you included 95% prediction intervals in your illustration of model fit and forecast.

7. Quantify your forecast:

   (a) What sales do you forecast for

   February 2011: ____ to ____ February 2012: ____ to ____

   (b) Using actual sales in February 2010 as your baseline, what percentage of past year

   sales can be expected in February 2011? ____

   (c) Using expected sales in February 2011 as your baseline, what percentage of past year

   sales is expected in February 2012? ____

8. Quantify the difference in sales over the 2 month period, July 2009 through August 2009, due to the Cash for Clunkers program:

   (a) In this 2 month period, what would total expected sales have been had there been no Cash for Clunkers program? ____

   (b) In this 2 month period, what was the total expected sales given the Cash for Clunkers program? ____

   (c) Using total expected sales had there been no Cash for Clunkers program (a) as your baseline, what percentage was gained because of the Cash for Clunkers program? ____

9. Quantify the difference in sales in the months of the Prius recalls, January and February 2010:

   (a) In this 2 month period, what would total expected sales have been had there been no Prius recalls? ____

   (b) In this 2 month period, what were total expected sales given the recalls? ____

   (c) Using total expected sales had there been no Prius recalls (a) as your baseline, what percentage was lost because of the recalls? ____

10. How much difference in hybrid sales would an increase in past gas prices from $2.00 to $3.00 make 2 years later in future months? _____

11. How much difference in hybrid sales would an increase in past gas prices from $3.00 to $4.00 make 2 years later in future months? ____

12. Create a scatterplot for future months to illustrate the impact of gas prices.
    Note: You do not need to include 95% prediction intervals or actual sales in this graph. Expected sales will be sufficient.

13. Explain why you used expected sales in hypothetical months to answer Questions 8 through 13 instead of actual sales.

CASE 11-1 Global Emissions Segmentation: Markets Where Hybrids Might Have Particular Appeal[8]*

Carbon emissions policies are being watched carefully by Ford Motor Company. Ford executives believe that major markets for new hybrid models will arise in developing countries where increased economic productivity and a growing population stimulate demand for vehicles.

To reduce carbon emissions, the Kyoto Protocol came into effect February 16, 2005, with 141 countries signing on, including every major industrialized country, except the United States, Australia, and Monaco. The Protocol stipulates conditions for systematically reducing carbon emissions. Some of the world's biggest and fastest growing polluters, including China and India, have not signed the Kyoto Protocol. Because they are considered developing countries, they are outside the protocol framework. Yet the publicity about the Kyoto Protocol has heightened interest in carbon emissions reductions (CERs). A number of countries have publicized their expected CERs, shown in the table below.

Expected average annual CERs from registered projects by host party

China	46,500,229	Colombia	414,205	Costa Rica	162,515
Brazil	15,846,288	El Salvador	360,268	Dominican Republic	123,916
India	15,534,244	Ecuador	357,900	Sri Lanka	109,619
Korea	12,362,308	Nicaragua	336,723	Israel	101,617
Mexico	5,566,398	Guatemala	279,694	Panama	96,469
Chile	2,183,123	New Guinea	278,904	Nepal	93,883
Argentina	1,765,007	Philippines	247,885	Bolivia	82,680
Malaysia	1,682,653	South Africa	225,446	Cyprus	72,552
Indonesia	1,557,100	Morocco	223,313	Jamaica	52,540
Nigeria	1,496,934	Honduras	205,251	Cambodia	51,620
Egypt	1,436,784	Peru	199,265	Moldova	47,343
Pakistan	1,050,000	Armenia	197,832	Fiji	24,928
Tunisia	687,573	Bangladesh	169,259	Mongolia	11,904
Viet Nam	681,306	South Africa	225,446	Bhutan	524

Source: Clean Development Mechanism (CDM), cdm.unfcc.int, 10 Feb 07.

Ford executives have asked you to identify drivers of *carbon emissions*. They would like to know, specifically, how important the influences of *GDP*, *population*, and *oil* production are in the global regions, and they are particularly interested in knowing whether emissions are noticeably higher in the BRICKs (Brazil, Russia, India, China, and South Korea). Ford will

---

[8]This example is a hypothetical scenario using actual data.

use that information to promote the manufacture and marketing of their hybrid models in targeted BRICK countries.

Case 11-1 Global Carbon Emissions.xls contains data from 129 countries with measures of *carbon emissions* (metric tons)$_{2007}$, *GDP* ($B)$_{2007}$, *population* (M)$_{2007}$, *barrels of of crude oil produced per day* (K)$_{2007}$.

Build a model of carbon *emissions* to provide Ford with answers.

1. Which variables are positively skewed?

2. Which scale(s) Normalizes each positively skewed variable, considering square roots and logarithms?

   For rescaled variable(s) that continue to show skew, experiment with higher roots to improve your model.

3. Identify outliers (after rescaling).

4. Write your model equations in the original scale of carbon *emissions* for *BRICKs* and other countries.

5. Fill in the table below comparing the marginal driver impacts on emissions:

	BRICKS		Other	
Other drivers:	Max	Min	Max	Min
GDP				
Population				
Oil				

Because the model is multiplicative, find response under both "max" and "min" scenarios by setting the drivers to maximums or minimums for the segment. For each cell, create two scenarios, one with other drivers at maximums for the segment and one with other drivers at minimums for the segment, and then find the difference between predicted emissions across the two.

6. Attach a scatterplot showing the marginal impact of one of the drivers on carbon *emissions*

   - In both BRICK and other segments
   - Under conditions where emissions are likely to be higher (the other drivers at segment maximums) and conditions where emissions are likely to be lower (the other drivers at segment minimums)

   Your scatterplot will contain four series, two for each of the segments, under conditions where emissions are likely to be either higher (solid lines) or lower (dashed lines).

   Show five hypothetical values of the driver in each series, adding a total of 20 hypothetical countries (= four series × five hypothetical driver values) to your data file.

7. Write a paragraph to summarize your key results and their implications for Ford management.

8. Ford managers need a forecast of demand for hybrids in each of the BRICKS in order to choose one or two to target initially. Data on hybrid sales in the BRICKS are not available. Describe how hybrid demand might be estimated from available data.

Chapter 12
Indicator Interactions for Segment Differences or Changes in Response

In this chapter, indicator interactions with predictors are introduced. Adding this type of interaction to models allows us to capture differences in response between segments or changes in response following structural changes or shocks. Indicator interactions alter partial slopes, in the way that indicators alter intercepts.

12.1 Indicator Interaction with a Continuous Influence Alters Its Partial Slope

At times, segment average response levels, the intercepts and responses to an influence, the partial slopes, differ. Two segments may respond differently to an influence. In marketing, segmentation is a basic principle. Customer segments respond differently to prices, advertising, and product characteristics.

In time series models, a structural shift may alter the partial response to a continuous influence. The impact of economic productivity on business performance may differ by party leadership. Consumers may become more sensitive to prices during a recession. In such cases, where segment responses differ, or structural shifts alter responses, we add one or more interactions, each equal to the product of an indicator and a continuous predictor.

To model differences between two segments' responses to a driver X, designate one of the segments as the baseline, add an indicator for the second segment, and make a new interaction variable which is the product of the indicator and the driver X:

$$\hat{Y} = b_0 + b_1 Segment_1 + b_2 X + b_3 Segment_1(X).$$

To model change in response following a structural shift, we add an indicator of the structural shift and make a new interaction variable that is the product of the shift indicator and the driver X_t:

$$\hat{Y}_t = b_0 + b_1 Shift_t + b_2 X_t + b_3 Shift_t(X_t).$$

When the indicator is zero, representing baseline segment response in a cross sectional model, or baseline response before a structural shift in a time series model, the equations are

- In a cross sectional model:

$$\hat{Y} = b_0 + b_1(0) + b_2 X + b_3(0)X$$
$$= b_0 + b_2 X.$$

- In a time series model:

$$\hat{Y}_t = b_0 + b_1(0) + b_2 X_t + b_3(0)X_t$$
$$= b_0 + b_2 X_t.$$

When the indicator is one, representing a second segment's response in a cross sectional model, or response following a structural shift in a time series model, the equations become

- In a cross sectional model:

$$\hat{Y} = (b_0 + b_1(1)) + (b_2 + b_3(1))X$$
$$= (b_0 + b_1) + (b_2 + b_3)X.$$

- In a time series model:

$$\hat{Y}_t = (b_0 + b_1(1)) + (b_2 + b_3(1))X_t$$
$$= (b_0 + b_1) + (b_2 + b_3)X_t.$$

The indicator alters the average level, by adjusting the intercept from b_0 to $b_0 + b_1$, and the indicator interaction alters the response to variation in the predictor, by adjusting the partial slope from b_2 to $b_2 + b_3$.

Example 12.1 Gender Discrimination at Slams Club

A disgruntled Slam's Club employee resigned and decided to sue the firm on grounds of gender discrimination. She alleges that Slams Club pays female employees less than male employees. The Slam's Club Board asked a consultant, Morey Furless, to build a model to assess gender discrimination.

Slams Club executives admitted that women were encouraged to work part time and focus on their roles as homemakers, rather than pursuing long term careers. They maintain that women are paid equally to men in similar positions. Following the meeting with executives, Morey made a note to be sure to include level of responsibility in the model.

From a random sample of 220 employee records, Morey built a model of salaries, including level of responsibility, years of experience, and an indicator for gender. Because it is possible that the value of responsibility and gain from experience each differs across the genders, interactions between the gender indicator and these two continuous variables were included.

Examine skewness of the model variables to choose scales: Examining the distributions of responsibility, experience, and salary, shown in Fig. 12.1, Morey found that salary, responsibility, and experience were positively skewed.

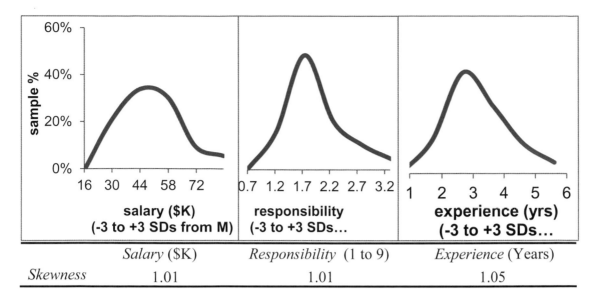

Fig. 12.1 Distribution of responsibility, experience, and salary

To reduce positive skew, we shrink, rescaling to square roots or natural logarithms. The natural logarithms better Normalize *salary*, but are too extreme for *responsibility* and *experience.* The square roots of *responsibility* and *experience* Normalize without overcorrecting. These are shown in Fig. 12.2.

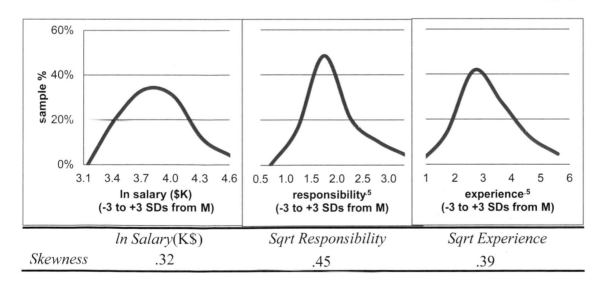

Fig. 12.2 Rescaled variables

When a dependent variable is rescaled, the resulting multiplicative model features built in synergies. With rescaled *salary*($K), this salary model will feature built in synergies among gender, *responsibility*, and years of *experience*. Regression results are shown in Table 12.1.

Table 12.1 Gender differences in the value of responsibility and experience at Slam's Club

SUMMARY OUTPUT

Regression statistics

R Square	.827				
Standard error	.125				
Observations	220				

ANOVA	*df*	*SS*	*MS*	*F*	*Significance F*
Regression	5	15.87	3.17	204.6	.0000
Residual	214	3.32	.02		
Total	219	19.19			

	Coefficients	*Standard error*	*t Stat*	*p Value*	*Lower 95%*	*Upper 95%*
Intercept	2.738	.055	49.4	.0000	2.629	2.847
Male	−.087	.073	−1.2	.2376	−.231	.058
Responsibility$^{0.5}$	.270	.031	8.6	.0000	.208	.332
Experience$^{0.5}$	.208	.037	5.6	.0000	.135	.282
Male ×ₙ responsibility$^{0.5}$	.253	.023	11.2	.0000	.208	.297
Male × experience$^{0.5}$	−.172	.025	−6.8	.0000	−.222	−.122

The *male* indicator is not significant, although the interactions between *male* and *responsibility* and *experience* are significant, so *male* remains in the model. An interaction cannot be included without its parent components, the indicator and the *main effect*, because the interaction is relative to the baseline main effect.

Morey's model is

$$\ln(\hat{sal}ary(\$K)) = 2.74^{a} - .087 \, male + (.27^{a} + .25^{a} \, male) \, responsibility^{0.5}$$
$$+ (.21^{a} - .17^{a} \, male) \, experience^{0.5}$$

R Square: .83[a].
[a]*Significant at .01*

To rescale back to the original thousand dollars, the exponential function is used to undo the natural logarithms:

$$\exp(\hat{ln(sal}ary(\$K)) =$$
$$\hat{sal}ary(\$K) = \exp[\, 2.74^{a} - .087 \, male + (.27^{a} + .25^{a} \, male) \, responsibility^{0.5}$$
$$+ (.21^{a} - .17^{a} \, male) \, experience^{0.5} \,].$$

By setting *male* to zero, the model for women can be written as

$$\hat{sal}ary(\$K) = \exp[\ 2.74^a\ +.27^a\ responsibility^{0.5} + .21^a\ experience^{0.5}\ ]$$

and by setting *male* to one, the model for men can be written as

$$\hat{sal}ary(\$K) = \exp[\ 2.74^a -.087 +(.27^a +.25^a)\ responsibility^{0.5} + (.21^a -.17^a)\ experience^{0.5}]$$
$$= \exp[\ 2.65 +.52\ responsibility^{0.5} + .036\ experience^{0.5}\ ].$$

The interaction between gender and experience: At the median level of *responsibility*, 3, women benefit more than men from increasing *experience*, illustrated in Fig. 12.3. Women with 10 years of experience can expect to be paid about $12,000 more than women with 5 years of experience. Women with 20 years of experience can expect to be paid about $11,000 more than women with 15 years of experience. Gains from *experience* are greater for women with less experience.

Fig. 12.3 Salaries (K$) by experience, responsibility, and gender

Experienced men with median *responsibility* are rewarded less than experienced women. Men with 10 years of experience can expect to be paid about $3,000 more than men with 5 years of experience. Men with 20 years of experience can expect about $2,000 more than men with 15 years of experience. Gains from *experience* are greater for men with less experience. Comparing the rewards expected for increasing experience from 5 to 10 years, women can expect a salary increase of $12,000 whereas men can expect just $3,000.

The interaction between gender and responsibility: In Fig. 12.3, among employees with median years of *experience* women are paid more than men in positions of lower *responsibility*, although men gain more from promotion. Women in middle management (*responsibility* level 5) can expect to be paid about $7,000 more than staff (*responsibility* level 3), and women in upper management (*responsibility* level 7) can expect to be paid only about $7,000 more than middle management. Men in middle management can expect to earn about $11,000 more than staff, and men in upper management can expect to be paid about $11,000 more still.

Men's and women's response curves are not parallel. Men benefit more from increased *responsibility*. Men can expect to gain an average of $6,000 from promotion to level 4 from level 3; a woman can expect to gain an average of about $4,000 from a similar promotion.

Morey was confident that Slam's Club executives would be relieved with his model results, which are summarized in the following memo.

MEMO

Re: Women are Paid More than Men at Slam's Club
To: The Board
From: Morey Furless, Morey Furless Consulting Associates
Date: June 2007

Analysis of a random sample of 220 Slam's Club employee salaries reveals that women are paid more than men, and the difference is greater among more experienced workers.

Salary Model. Using data from 220 randomly selected employee records, a model linking salary, employee responsibility level and tenure was built.

Model Results. Gender, level of *responsibility*, and employee tenure account for 83% of the variation in salaries.

On average, male and female employees are paid equally, though women are paid more for greater tenure.

Women with 10 years tenure earn an average of $6,000 more than men with the same tenure.

Level of responsibility also drives salaries. Middle management workers (*responsibility* level 5) can expect to be paid $4–$13K more than staff (*responsibility* level 3).

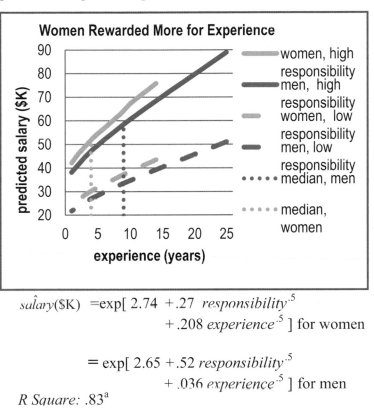

$$\hat{salary}(\$K) = \exp[\ 2.74 + .27\ responsibility^{.5}\\ + .208\ experience^{.5}\ ]\ \text{for women}$$

$$= \exp[\ 2.65 + .52\ responsibility^{.5}\\ + .036\ experience^{.5}\ ]\ \text{for men}$$

R Square: .83[a]

Men do benefit more from promotion to higher levels of *responsibility*. A man can expect to gain an average of $11–$13K from promotion to level 5 from level 3; a woman can expect to gain an average of about $4–$8K from a similar promotion.

Conclusions. Slam's Club does not discriminate against women. Female employees are paid more than men for their years of loyal service.

Limitations. This model does not explore issues related to equal opportunities for promotion to greater responsibility levels. Responsibility is a major driver of salaries. In the case that more men hold positions with higher responsibility levels, this could be considered discriminatory against women.

Example 12.2 Car Sales in China

Every major car manufacturer is watching China closely. As China's GDP grows rapidly, more and more Chinese consumers are buying cars. Some of those cars are imports manufactured outside China and some are the products of joint ventures between Chinese and American partners. Some cars produced in China are exported, particularly to other Asian countries where labor costs are higher. We build a model of car sales in China based on a leading indicator, past year Chinese car production, and political leadership shifts in China. We include two indicators to represent changes in car sales from the baseline years 1994 through 1997, when Deng led China and set import export policy:

- For the period 1998 to 2002, *after Deng*, to represent Third Generation leadership following Deng's death in 1997.
- For the period 2003 through 2011, *Fourth Generation,* to represent current leadership.

Political leadership probably affects car sales response to car production, since imports and exports are either encouraged or discouraged by particular administrations. For this reason, indicator interactions between *after Deng* and *Fourth Generation* with past year *Chinese car production* will be included.

Data contain time series of annual observations from 1994 through 2006 on *car sales in China* (K_t), past year *Chinese car production* ($K)_{t-1}$, indicators for Third Generation leadership *after Deng$_t$*'s death, and *Fourth Generation$_t$* leadership. Both continuous variables, car sales and car production, shown in Fig. 12.4, are positively skewed, suggesting that we shrink each by rescaling to roots, natural logarithms, or inverses.

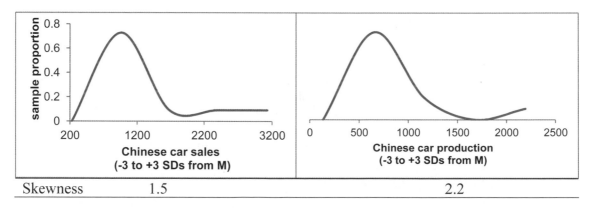

| Skewness | 1.5 | 2.2 |

Fig. 12.4 Skewed dependent and independent variables

The natural logarithms of *Chinese car sales$_t$* and the sixth root of past year *Chinese car production$_{t-1}$*, shown in Fig. 12.5, reduce skewness.

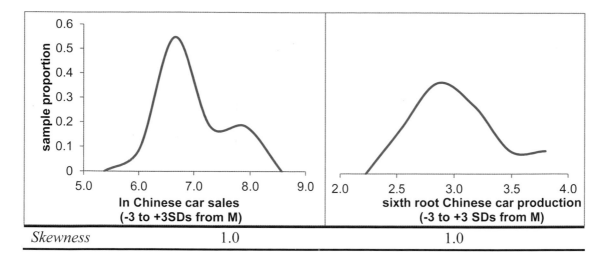

Fig. 12.5 Rescaled variables are less skewed and approximately Normal

Because the dependent variable will be rescaled, the model will feature built in synergies between predictors. The interaction terms will be products of rescaled independent variables and indicators. The model correctly forecasts car sales in China during the two most recent held out years, 2005 and 2006. Those two recent years were then included, and the model was recalibrated. Multiple regression results are shown in Table 12.2.

Table 12.2 Leadership and car production drive chinese car sales

SUMMARY OUTPUT						
Regression statistics						
R Square	.994					
Standard error	.076					
Observations	13					
ANOVA	*Df*	*SS*	*MS*	*F*	*Significance F*	
Regression	5	7.26	1.45	251	.0000	
Residual	7	.040	.0058			
Total	12	7.30				

	Coefficients	*Standard error*	*t Stat*	*p Value*	*Lower 95%*	*Upper 95%*
Intercept	1.19	1.12	1.1	.32	−1.47	3.84
After Deng $_t$	−7.30	1.91	−3.8	.007	−11.82	−2.77
Fourth Generation $_t$	4.39	1.27	3.5	.01	1.40	7.39
Chinese car production $(K)^{1/6}_{t-1}$	1.92	.44	4.4	.003	.89	2.95
After Deng $_t$ x *Chinese car production* $(K)^{1/6}_{t-1}$	2.51	.69	3.6	.008	.87	4.15
Fourth Generation $_t$ x *Chinese car production* $(K)^{1/6}_{t-1}$	−1.27	.47	−2.7	.03	−2.37	−.17

From regression output, we can write the regression equation for the three distinct periods. In each case, we find the exponential function of both sides of the equation to rescale back to car sales in units:

- 1994–1997, the baseline years, with all indicators set to zero:

$$ChineseCar\hat{S}ales(K)_t = exp[\ 1.2 + 1.9\ Chinese\ Car\ Production\ (K)^{1/6}_{t-1}].$$

- 1998–2002, Third Generation leadership after Deng's death, before Fourth Generation leadership, with the *after Deng* indicator set to one:

$$ChineseCar\hat{S}ales(K)_t = exp[\ 1.2 - 7.3 + (1.9 + 2.5)\ Chinese\ Car\ Production(K)^{1/6}_{t-1}]$$
$$= exp[\ -6.1 + 4.4\ Chinese\ Car\ Production(K)^{1/6}_{t-1}].$$

- 2003–present, under Fourth Generation leadership, with the *Fourth Generation* indicator set to one:

$$ChineseCar\hat{S}ales(K)_t = exp[\ 1.2 + 4.4 + (1.9 - 1.3)\ Chinese\ Car\ Production(K)^{1/6}\ _{t-1}]$$
$$= exp[\ 5.6 + .6\ Chinese\ Car\ Production(K)^{1/6}\ _{t-1}].$$

Comparing intercepts, with a given level of car production, car sales would be (and have been) highest in recent years under *Fourth Generation* leadership. The impact of growth in car production is positive in all periods, but particularly strong in the periods after Deng's death, 1998–2002. A scatterplot of the model fit in Fig. 12.6 illustrates the changing patterns of car sales in China.

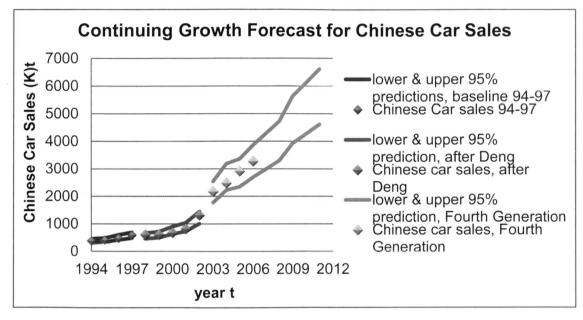

Fig. 12.6 Growth in car sales in China

Residual analysis: The nonlinear model residuals, in Fig. 12.7, are approximately *normally* distributed:

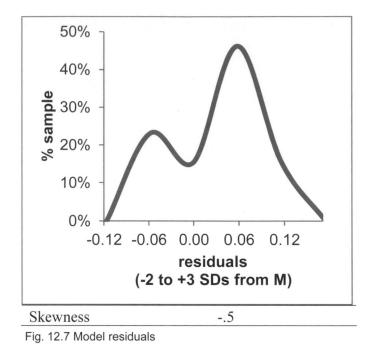

Skewness	-.5

Fig. 12.7 Model residuals

Sensitivity analysis: Fewer cars produced in China are being sold in China under Fourth Generation leadership. To see the impact of Fourth Generation leadership and its interaction with past year *Chinese car production*, we can compare predictions for years 2003 through 2006 under an alternate scenario of continued Third Generation leadership (after Deng's death). Setting the *Fourth Generation* indicator to 0 and the *after Deng* indicator to 1 for those 9 years, the predictions suggest that Chinese car sales might have been higher under continued Third Generation leadership, as Fig. 12.8 and Table 12.3 illustrate. (This comparison assumes that Chinese car production would not differ under Third Generation leadership, a questionable assumption.)

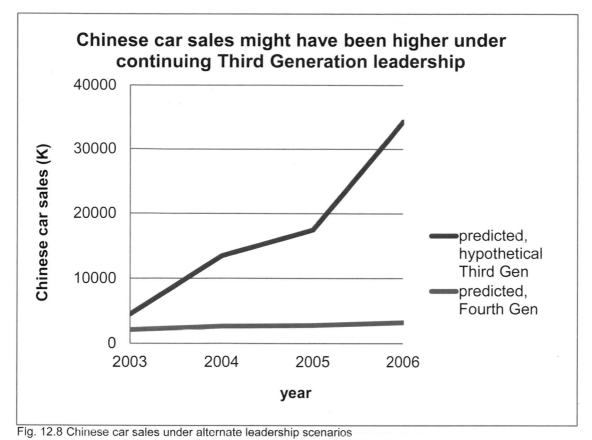

Fig. 12.8 Chinese car sales under alternate leadership scenarios

Table 12.3 Growth in Chinese car sales under alternate leadership

Predicted car sales in China (M) 2003–2011			
Year	Fourth Gen (1)	Third Gen hypothetical (2)	Fourth Gen change (3)=(1)–(2)	% Change (3)/(2)
2003	2.1	3.2	−1.1	−34%
2004	2.7	15.3	−12.6	−82%
2005	2.8	22.0	−19.2	−87%
2006	3.2	56.5	−53.3	−94%

12.2 Indicator Interactions Capture Segment Differences or Structural Differences in Response

Segment responses can be expected to differ. Price discrimination and product differentiation strategies acknowledge this. By incorporating indicator interactions into our models, we add realism. Interactions also allow us to quantify differences in response across segments, improving the value of our results to decision makers.

In time series, structural shifts and shocks sometimes alter both the average level of response and the degree of response to changes in predictors. Adding interaction terms to models improves validity and predictive capability. Interaction terms also allow us to assess differences or changes in response to independent variables in a model. Backcasts can be made to determine the impact of a structural change or shock, and then to estimate what the response would have been had the structural change or shock not occurred. Forecasts can be made to determine the impact of similar shocks or changes in the future. Interaction terms increase the realism and value of models.

Excel 12.1 Add Indicator Interactions to Capture Segment Differences or Changes in Response

Car Sales in China

Build a model of car sales in China, including the leading indicator, past year Chinese car production, and indicators of political leadership shifts, Deng's death in 1997 and Fourth Generation leadership in 2003. Include interactions between the *after Deng* and *Fourth Generation* indicators and past year car production to allow for differences in import and export policies due to leadership.

Data in **Excel 12.1 China Car Sales.xls** contain time series of annual observations from 1994 through 2005 on *Car sales in China (K)$_t$*, *Chinese car production (K) $_{t-1}$* (past year), and indicators for Third Generation leadership *after Deng,* and *Fourth Generation* leadership.

Assess skewness to choose variable scales: To build a more valid model, rescale to reduce skewness of *Chinese car sales (K)* and *Chinese car production (K)$_{t-1}$*, incorporating nonlinear, nonconstant response.

Find *skewness,* the *mean,* and *standard deviation* of *Chinese car sales* and *Chinese car production* using data for years 1994 through 2004, holding out the two most recent years in order to later validate the model.

	A	B	C	D	E
1	year	Chinese car sales (K) t	after Deng t	Fourth Generation t	Chinese car production (K) t-1
2	1994	377	0	0	230
3	1995	411	0	0	248
4	1996	478	0	0	321
5	1997	589	0	0	382
6	1998	600	1	0	488
7	1999	607	1	0	507
8	2000	672	1	0	565
9	2001	814	1	0	605
10	2002	1309	1	0	704
11	2003	2171	0	1	1091
12	2004	2479	0	1	2019
13	2005	2914	0	1	2316
14	2006	3272	0	1	3260
15	2007		0	1	4180
16	2008		0	1	5220
17	2009		0	1	7600
18	2010		0	1	9000
19	2011		0	1	10500
20	skew	1.52			2.19
21	M	955			651
22	SD	726			514

To see the distributions, make histograms of *Chinese car sales* and *Chinese car production*.

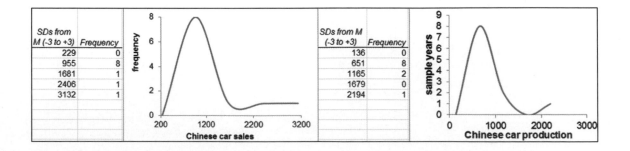

Both variables are positively skewed. To *Normalize* positively skewed variables, we shrink, rescaling to roots, natural logarithms, or inverses.

Add new columns for rescaled variables, and then make *sqrt Chinese car sales* $(K)_t$, a *ln Chinese car sales* $(K)_t$, and *Chinese car sales* (K) *inverse*.

Make rescaled production variables, also: *sqrt Chinese car production* $(K)_{t-1}$, *ln Chinese car production* $(K)_{t-1}$, and *Chinese car production* $(K)_{t-1}$ *inverse*.

Fill in skewness values for the new rescaled variables:

	A	B	C	D	E	F	G	H	I
1	year	Chinese car sales (K) t	Chinese car production (K) t-1	sqrt Chinese car sales (K) t	ln Chinese car sales (K) t	Chinese car sales (K) inv t	sqrt Chinese car production (K) t-1	ln Chinese car production (K) t	Chinese car production (K) inv t
13	2005	2914	2316	54.0	8	0.00034	48.1	7.75	0.00043
14	2006	3272	3260	57.2	8	0.00031	57.1	8.09	0.00031
15	2007		4180				64.7	8.34	0.00024
16	2008		5220				72.2	8.56	0.00019
17	2009		7600				87.2	8.94	0.00013
18	2010		9000				94.9	9.10	0.00011
19	2011		10500				102.5	9.26	0.00010
20	skew	⬦2	2.19	1.3	0.94	-0.10	1.5 ▣	0.77	0.57
21	M	955	651	29.27	6.65	0.00150	24.16	6.27	0.0022
22	SD	726	514	10.41	0.64	0.00074	8.58	0.64	0.0012

The natural logarithms and inverses of both sales and past production are approximately Normal (with skewness between -1 and $+1$).

The square roots improve skewness, although they are more skewed than Normal. Replace the square roots with higher roots: *tenth root Chinese car sales* $(K)_t$ and *sixth root Chinese car production* $(K)_{t-1}$:

A	B	C	D	E	F	G
year	Chinese car sales (K) t	Chinese car production (K) t-1	tenth root Chinese car sales (K) t	ln Chinese car sales (K) t	Chinese car sales (K) inv t	sixth root Chinese car production (K) t-1
2002	1309	704	2.0	7	0.00076	3.0
2003	2171	1091	2.2	8	0.00046	3.2
2004	2479	2019	2.2	8	0.00040	3.6
2005	2914	2316	2.2	8	0.00034	3.6
2006	3272	3260	2.2	8	0.00031	3.9
2007		4180				4.0
2008		5220				4.2
2009		7600				4.4
2010		9000				4.6
2011		10500				4.7
skew	1.52	2.19	1.0	0.94	-0.10	1.0
M	955	651	1.95	6.65	0.00150	2.86
SD	726	514	0.13	0.64	0.00074	0.31

The higher roots are approximately Normal, with skewness of 1.

To compare and inform your choices, make histograms of the scale choices that produce Normal skewness: *Tenth root, Ln,* and *inverse* of *Chinese car sales*, and *sixth root, ln, and inverse* of past *Chinese car production.*

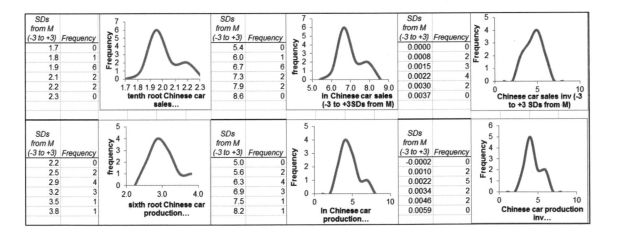

All of the rescaled variables are viable options. The natural logarithms of sales and the sixth roots of production produce a well behaved model and are used in this example.

Add indicator interactions. To model varying car sales response to increasing car production by leadership regime, make interactions between the indicators, *after Deng,* and *Fourth Generation* with *sixth root Chinese car production* $(K)_{t-1}$ by multiplying each indicator column with the *sixth root Chinese car production* $(K)_{t-1}$ column:

f_x	=E2*G2						
C	**D**	**E**	**F**	**G**	**H**		**I**
Chinese car production (K) t-1	In Chinese car sales (K) t	after Deng t	Fourth Generation t	sixth root Chinese car production (K) t-1	after Deng x sixth root Chinese car production (K) t-1		Fourth Gen x sixth root Chinese car production (K) t-1
230	6	0	0	2.5	0		0
248	6	0	0	2.5	0		0
321	6	0	0	2.6	0		0
382	6	0	0	2.7	0		0
488	6	1	0	2.8	2.81		0
507	6	1	0	2.8	2.82		0
565	7	1	0	2.9	2.88		0
605	7	1	0	2.9	2.91		0
704	7	1	0	3.0	2.98		0
1091	8	0	1	3.2	0		3.21
2019	8	0	1	3.6	0		3.56
2316	8	0	1	3.6	0		3.64
3260	8	0	1	3.9	0		3.85
4180		0	1	4.0	0		4.01
5220		0	1	4.2	0		4.16
7600		0	1	4.4	0		4.43
9000		0	1	4.6	0		4.56
10500		0	1	4.7	0		4.68

Run the regression, excluding the two most recent years, 2005 and 2006, to later validate the model.

SUMMARY OUTPUT

Regression Statistics	
Multiple R	0.996
R Square	0.992
Adjusted R Square	0.984
Standard Error	0.082
Observations	11

ANOVA

	df	SS	MS	F	Significance F
Regression	5	4.033	0.807	120.9	3.281E-05
Residual	5	0.033	0.007		
Total	10	4.066			

	Coefficients	Standard Error	t Stat	P-value	Lower 95%	Upper 95%
Intercept	1.19	1.20	1.0	0.3696	-1.91	4.29
after Deng t	-7.30	2.06	-3.5	0.0164	-12.58	-2.01
Fourth Generation t	5.27	1.65	3.2	0.0242	1.03	9.51
sixth root Chinese car production (K) t-1	1.92	0.47	4.1	0.0094	0.72	3.12
after Deng x sixth root Chinese car production (K) t-1	2.51	0.74	3.4	0.0199	0.60	4.42
Fourth Gen x sixth root Chinese car production (K) t-1	-1.54	0.57	-2.7	0.0441	-3.01	-0.06

Assess autocorrelation: Inasmuch as this is a time series model, assess residual autocorrelation to see whether the leading indicator, *past year Chinese car production*, the indicators of shifts due to political leadership, and the interaction have successfully accounted for trend and cycles in *Chinese car sales.*

Next to the residual column in the regression output sheet, add the Durbin Watson statistic to check for unaccounted for trend or cycles, and plot the residuals.

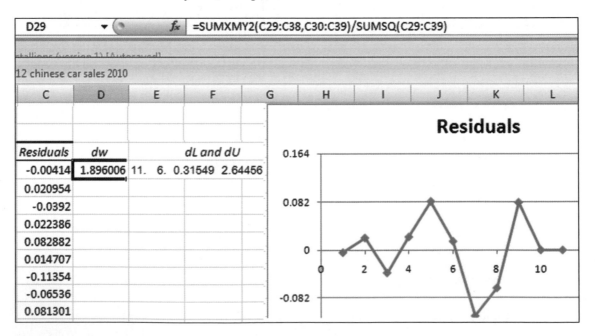

The Durbin Watson statistic 1.90 falls in the gray area, between the lower and upper critical values, dL = .32 to dU = 2.64, suggesting that there may be unaccounted for trend, cycles, shocks, or seasonality. The residuals all fall within two standard errors of the mean of zero, and there is no obvious pattern, so we proceed to assess model validity.

Validate the model: With the model coefficient estimates make *predicted ln Chinese car sales* $(K)_t$.

With the *critical t* value and model *standard error*, make the *margin of error*, *lower* and *upper* 95% *predicted inverse* bounds, which are in natural logarithms.

	K2		f_x	=$N$2+$N$3*E2+$N$4*F2+$N$5*G2+$N$6*H2+$N$7*I2							
E	F	G	H	I	K	L	M	N	O	P	
after Deng	Fourth Generation	sixth root Chinese car production	after Deng x sixth root production	Fourth Generation x sixth root production	predicted ln Chinese car sales				critical		
t	t	(K) t-1			(K)	lower ln	upper ln	coefficients	t	s	me
0	0	2.474719	0	0	5.937213469	5.72727	6.147157	1.187536	2.570582	0.081671	0.20994
0	0	2.5059352	0	0	5.99712619	5.787183	6.207069	-7.29778			
0	0	2.6161077	0	0	6.208577993	5.998635	6.418521	5.269381			
0	0	2.6930907	0	0	6.35632999	6.146387	6.566273	1.91928			
1	0	2.8055933	2.8055933	0	6.313467366	6.103524	6.52341	2.508914			
1	0	2.8239003	2.8239003	0	6.394534383	6.184591	6.604478	-1.53711			

Compare *ln Chinese car sales* (K) in 2005 and 2006 with predictions to confirm that the two values fall within the 95% prediction intervals.

	D13		f_x	=LN(B13)		
2 chinese car sales 2010						
A	B	C	D	L	M	
year	Chinese car sales (K) t	Chinese car sales (K) t	ln Chinese car sales (K) t	lower ln	upper ln	
2004	2479	7.8156	7.8155746	7.605632	8.025518	
2005	2914	7.9774	7.98	7.64	8.06	
2006	3272	8.0932	8.09	7.72	8.14	

The model correctly forecasts held out cars sales in 2005 and 2006.

Recalibrate, including data from 2005 and 2006.

SUMMARY OUTPUT

	Regression Statistics
Multiple R	0.997
R Square	0.994
Adjusted R Square	0.990
Standard Error	0.076
Observations	13

ANOVA

	df	SS	MS	F	ignificance F
Regression	5	7.26	1.45	251.2	9.77E-08
Residual	7	0.04	0.01		
Total	12	7.30			

	Coefficients	Standard Error	t Stat	P-value	Lower 95%	Upper 95% ›
Intercept	1.19	1.12	1.1	0.3249	-1.47	3.84
after Deng t	-7.30	1.91	-3.8	0.0066	-11.82	-2.77
Fourth Generation t	4.39	1.27	3.5	0.0104	1.40	7.39
sixth root Chinese car production (K) t-1	1.92	0.44	4.4	0.0031	0.89	2.95
after Deng x sixth root Chinese car production (K) t-1	2.51	0.69	3.6	0.0085	0.87	4.15
Fourth Gen x sixth root Chinese car production (K) t-1	-1.27	0.47	-2.7	0.0294	-2.37	-0.17

Update forecasts: Copy the recalibrated coefficients and paste over the validation coefficients in the original worksheet to update forecasts.

Update the *critical t* by updating the error degrees of freedom.

Change the standard error *s* to the recalibrated value.

M	N	O	P
oefficient	critical t	s	me
1.19	2.3646	0.076	0.1798
-7.30			
4.39			
1.92			
2.51			
-1.27			

Updating the *critical t* and standard error will update the margin of error and the 95% prediction interval bounds.

Rescale to thousands of cars: The forecasts are in logarithms. To rescale back to thousands of cars, use the Excel function **exp( )**.

Make three new columns, *predicted Chinese car sales* (K)$_t$ *and* 95% *lower* and *upper predicted* (K), by using the exponential functions of *predicted ln Chinese car sales* (K) and lower and upper 95% predicted logarithms.

			M2	f_x =EXP(J2)	
J	**K**	**L**	**M**	**N**	**O**
predicted ln Chinese car sales (K) t	lower 95% ln	upper 95% ln	predicted Chinese car sales sales (K) t	lower 95%	upper 95%
5.94	5.76	6.12	379	317	454
6.00	5.82	6.18	402	336	482
6.21	6.03	6.39	497	415	595
6.36	6.18	6.54	576	481	690

Illustrate the fit and forecast: To see the fit and forecasts, make a scatterplot of actual car sales 95% prediction intervals in original units.

Move the prediction intervals to columns between *year* and *Chinese car sales* (K)$_t$.

Plot each of the distinct periods as a separate set of three series for sales, lower and upper prediction interval bounds, beginning with the two prediction interval columns in baseline years, 1994 to 1997.

year	lower 95%	upper 95%	Chinese car sales (K) t	Chinese car production (K) t-1	In Chinese car sales (K) t	after Deng t	Fourth Generation t
1994	317	454	377				
1995	336	482	411				
1996	415	595	478				
1997	481	690	589				
1998	461	661	600				
1999	500	716	607				
2000	629	901	672				
2001	726	1040	814				
2002	1008	1445	1309				
2003	1777	2546	2171				
2004	2225	3188	2479				
2005	2347	3363	2914				
2006	2695	3862	3272				
2007	2996	4292					

Select the chart and use shortcuts, **Alt JCE** to **Add** the remaining series.

For baseline years 1994 through 1997, add one series, *Chinese car sales* (K).

For years 1998 through 2002 *after Deng's* death, add three series, *lower*, *upper*, and *Chinese car sales* (K).

For years 2003 through 2011, *Fourth Generation,* add the three prediction intervals and actual sales series.

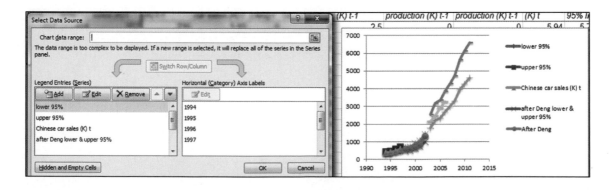

Customize style, design, markers, font, and scales.

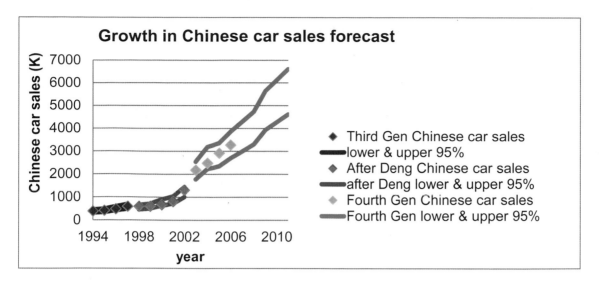

Sensitivity analysis: To estimate the impact of *Fourth Generation leadership*, relative to Third Generation leadership *after Deng's death*, make prediction sales for years 2003 through 2006 under the alternate scenario of continuing *Third Generation leadership*.

First, copy *predicted China car sales* (K) $_t$ for years 2003 through 2006, based on the actual change in leadership in 2003, and paste into a new column, removing formula references, for later comparison: **Alt HVSU.**

Set up the hypothetical scenario of continuing Third Generation leadership after Deng.

In years 2003 through 2006, change after Deng to 1 and Fourth Generation to 0.

Plot the new column containing the original predicted values and the predicted Chinese car sales column, which now reflects continuing Third Generation leadership, by year.

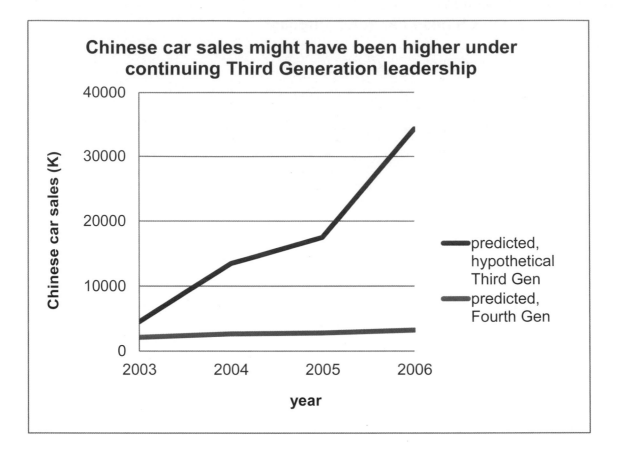

Lab Practice 12 Car Sales in India

An American car manufacturer is considering a joint venture in India where cars would be manufactured for sale to the growing Indian population and Asian markets. Management believes that in India, the leading indicator, population growth, will drive car sales in the next 5 years.

It is also believed that structural shifts from changes in leadership affect both the demand for cars and also the proportion of cars produced that are exported, rather than sold in India. A noticeable structural shift occurred in 1991, following the death of Gandhi. The Congress Party controlled leadership after Gandhi's death until Gandhi's BJP party again gained control in 1997. Congress took back leadership in 2004.

Follow the steps in **Excel 12.1** to build a time series model of *car sales in India*, with **Lab 12 India Car Sales.xls** including

- An indicator of *Congress* leadership to represent the major shifts in economic policy, equal to 1 in years 1991–1996 and 2004–present
- One or more interactions between this indicator and the continuous variables in the model
    - *Past year Indian car production*
    - *Indian population*

Assess skewness: Which variables are skewed? _____

Assess autocorrelation: Is your model free of autocorrelation?_____
(Assess autocorrelation. If *DW* is greater than dL, you do not need to add variables.)

Validate your model, then recalibrate.

Write your model equations

- For the baseline BJP leadership

- For leadership under *Congress*

Forecast: What are *Indian car sales* expected to be in 2010, with 95% confidence?_____

Illustrate your fit and forecast: Make a scatterplot of 95% *lower* and *upper predicted sales* through 2010 with *actual sales* through 2004 to illustrate your model fit and forecast. Plot the distinct leadership periods as separate series:

- *Leadership under BJP Party* 1983 *through* 1990
- *Leadership under Congress* 1991 *through* 1997
- *Leadership under BJP Party* 1998 *through* 2003
- *Leadership under Congress* 2004 *through* 2010

Sensitivity analysis: Make a table to compare *Indian car sales* in 2008 through 2010 under the alternative scenario of BJP leadership from 2008, including the percent increase or decline under BJP leadership, relative to *Congress* leadership.

Add to your scatterplot 95% *lower* and *upper predicted sales* through 2010 given BJP leadership in 2008 through 2010.

Attach a printout of your scatterplot to your lab practice worksheet.

Lab 12 Identifying Global Markets II

Harley–Davidson would like to identify the most promising global markets for motorcycle sales. Consultants built a model of global motorcycle sales identifying population density and GDP as the drivers. The consultants' model accounted for 39% of the variation in motorcycle sales across countries.

Managers believed that GDP per capita was also a driver. They reasoned that as consumers grew wealthier in developing markets, some would, for the first time, purchase a car, instead of a motorcycle. In the case of the BRICK countries, Brazil, Russia, India, China, and South Korea, they felt that GDP per capita would be a particularly important influence.

The managers were also unconvinced that motorcycle sales were driven by population density and GDP at a constant rate. They believed that sales were more likely to exhibit increasing marginal response to increasing GDP per capita, particularly in the BRICKs.

Build a model to identify the drivers of motorcycle market potential, allowing for a differences in driver impacts in BRICK countries. **Lab 12 Global Moto II.xls** contains measures of *Motorcycle sales*, *GDP*, *per capita GDP*, and *population density* in 2009 for 20 countries with the highest motorcycle sales.

Motorcycle sales, GDP, and population density are skewed. Rescale each of these so that each is closer to Normal.

Build indicator interactions between the BRICK indicator and each of the three potential drivers.

Using the BRICK indicator, the three potential drivers, and the three indicator interactions, build a model of motorcycle sales.

1. Identify outliers after rescaling:_____

2. Which economic and population variables drive motorcycle sales?

___ *GDP* ___ *per capita GDP* ___ *population density*

3. Present your regression equations for *BRICK* and other countries.

4. Is the impact of *per capita GDP* greater in the *BRICKs*, as managers suspected?   Y   N

5. Find the difference in *expected motorcycle sales* due to each of the drivers by comparing the maximum and minimum, under otherwise favorable and unfavorable scenarios in each of the two global segments:

| | Difference in *expected motorcycle sales* due to driver | | | |
| | In *BRICKs* | | In Other countries | |
	Unfavorable	Favorable	Unfavorable	Favorable
GDP				
Per capita GDP				
Population density				

6. Illustrate the impact of the two most influential drivers using the same scale for the *y* axis on both plots. Show favorable and unfavorable scenarios in *BRICK* and Other segments as four separate series in both plots.

7. Compare predicted sales with actual sales in all countries in the sample to identify two markets with the greatest unrealized potential that Harley–Davidson should target.

CASE 12-1 Explain and Forecast Defense Spending for Rolls-Royce

Sales to defense contractors are critical to Rolls-Royce growth and profitability. The executives know from experience that the defense business depends critically upon government defense spending, which is influenced by political leadership, global conflict, and the nation's productivity. Ralph Roy, Senior Assistant to the Director of Corporate Planning, has built a model of defense spending, which he must soon present to executives. He has asked you to review his model and suggest improvements.

Indicators and drivers of defense spending: Ralph began by interviewing executives to identify defense spending drivers. From these conversations, the list of likely influences included the following:
Economic productivity, measured by the leading indicator, past year productivity, GDP_{q-4}, shifts in defense policy related to conflicts:
- The War on Terror following 9/11
- The invasion of Iraq
- The shift of focus from Iraq to Afghanistan

and political shifts:
- The beginning nine quarters of a new administration.
- The final four quarters of an administration

Scales to reduce skewness: Defense spending was positively skewed. Ralph used cube roots of defense spending.

Ralph included the five indicators of defense policy and political shifts and the leading indicator, past GDP, in his model.

Ralph was pleased that his model accounted for a high proportion of the variation in defense spending across quarters (95%), that his model was significant, and that the five shift indicators and leading indicator were significant. Coefficients for shifts following conflicts and economic productivity had "correct" positive signs. The model correctly forecasts spending levels in the two most recent quarters that had been hidden to fit and validate the model.

Ralph's regression results are in the workbook **Case 12-1 defense spending forecast for Rolls-Royce.xls.**

Ralph is somewhat concerned that he may have left out one or more important variables or interactions, because the plot of his residuals (on the residuals worksheet) is not pattern free.

1. Is Ralph's model complete? Or should additional variables be added? Document your answer with the appropriate test.

2. Improve Ralph's model, explain how you know whether you have improved Ralph's model, and state your evidence.
   NOTE: You do not need to perfect the model; just improve the model.

3. Write the equations for your improved model in trillions of dollars for spending under four scenarios using proper subscripts, superscripts, and indentations, and rounding to two or three significant digits:

   (a) The baseline, before War on Terror was enacted, in the middle of an administration's term
   (b) Recent quarters, following the shift in focus to Afghanistan, in the beginning quarters of an administration

4. Attach or embed a scatterplot of the 95% *prediction intervals* and *actual defense spending* in hundred billion dollars (T$) through the first quarter of 2011.

5. What quarterly growth in *defense spending* does your model forecast for the first quarter of 2011?

Quarter	Forecast *defense spending* ($T)	
2010 IV		% of Forecast from previous quarter
2011 I		

6. Explain how spending changes from the beginning quarters of an administration to the middle quarters and to the final quarters.

7. Some managers believe that defense spending depends on economic productivity, and that when the economy is growing faster, defense budgets also grow faster the following year. Based on your model results, under what circumstances do you find this contention to be supported? Explain how you used the model to provide evidence for your answer.

CASE 12-3 Pilgrim Bank (A): Customer Profitability and Pilgrim Bank (B): Customer Retention[9]

Framing the Problem. Armed with the information learned from analysis (described in cases (A) and (B)), Pilgrim management has decided to promote online services and online services with billpay to selected customer segments, with the goals of increasing profits and customer retention. (In order to use billpay, a customer must also use online services.)

Identify Potentially Profitable Segments. To assist management, identify customer demographic segments whose use of online services or online services with billpay enhances bank profitability.

Identify Potentially Profitable Segments Who Would Stay with Pilgrim. Online services could potentially either encourage or discourage customer loyalty. Determine which potentially profitable customer segments are
- More likely to continue banking with Pilgrim if offered online services
- Less likely to continue banking with Pilgrim if offered online services

Identify Segments to be Targeted. Describe customers that Pilgrim should target with online services or online services with billpay in order to accomplish their goals of increased profitability and customer retention.

Data and Analysis. Profit data are in **Case 12-3 Pilgrim Profits.xls.** Build a model of profit, including past year profit, indicators for online banking and online billpay, plus customer tenure at the bank, customer age, and customer income.

Customer retention data are in **Case 12-3 Pilgrim Contingency.xls.** Use contingency analysis to test and quantify the association between retention and online services for potentially profitable customer segments.

Memo to Management

Write a memo to Pilgrim Bank management that describes the customer segments which ought to be targeted with online services and with billpay. Use your analysis to explain how these segments were identified.

This memo ought to include one embedded graphic that best illustrates your results and additional graphics in attachments which illustrate each of your key results referred to in your memo, so that they are easily understood by the management. Graphics scales should be in the original units, dollars, years, or retention probability, to help management relate your results to customer data. You should not include any graphics that are not specifically referred to in your text.

---

[9] Harvard Business School Cases 9602095 and 9602103

Chapter 13
Logit Regression for Bounded Responses

In this chapter we introduce *logit* regression which accommodates responses that are *limited* or bounded above and below. For example, the likelihood of trying a new product can neither be negative nor greater than 100%. Market share is similarly limited to the range between 0 and 100%. Indicator 0–1 responses, such as "tried the product or not" and "voted Republican," reflect probabilities, such as the probability of trying a new product, the probability of winning a game, or the probability of voting Republican. In each of these cases, dependent response must be rescaled, acknowledging these boundaries. The odds ratio rescales probabilities or shares to a corresponding unbounded measure. The *logit*, or natural logarithm of an odds ratio, rescales responses, producing an S shaped pattern, which reflects greater response among "fence sitters" with probabilities or shares that are mid range.

13.1 Rescaling Probabilities or Shares to Odds Improves Model Validity

With each response probability π, there is an *odds ratio*, the chance that the response occurs relative to the chance that it does not occur:

$$odds = \pi/(1 - \pi).$$

Response shares, such as market share, also have odds ratios, which reflect percentage of the market owned, relative to the percentage of the market owned by competitors:

$$odds = MarketShare / (100 - MarketShare).$$

Although probabilities and shares are bounded by zero, below, and one or one hundred percent, above, the corresponding odds ratio and its natural logarithm, the logit, are not bounded:

$$Logit = \ln(odds).$$

Rescaling to logits produces an S shaped curve, which, for a probability at .5, or a share at 50%, has a logit of zero. Figure 13.1 illustrates this S shaped scale.

C. Fraser, *Business Statistics for Competitive Advantage with Excel 2010: Basics, Model Building, and Cases*, DOI 10.1007/978-1-4419-9857-6_13, © Springer Science+Business Media, LLC 2012

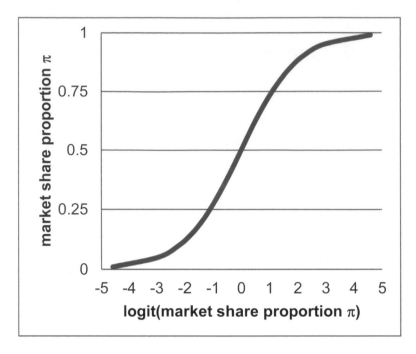

Fig. 13.1 Logits of bounded shares are unbounded

Example 13.1 The Import Challenge[10]

Ford Motors executives were pondering the U.S. car market, where imports had captured the majority of market share. In response to Toyota's successful launch of the hybrid Prius model, Ford had designed and begun selling the hybrid Focus. American cars were known to be less fuel efficient and less reliable than imports, but also less expensive than similar cars designed abroad. What car characteristics drive U.S. car owner satisfaction? Was value enough to sustain share in the U.S. market? Ford executives asked Amanda Arnone, the Director of Quantitative Analysis, to build a model of car owner satisfaction to provide answers.

Consumer Reports (consumerreports.com) routinely collects data on car owners' satisfaction by asking the question, "Would you buy this model again?" Each model's satisfaction rating is the percentage of owners who answered "Yes." Amanda used satisfaction percentages for 33 car models to build the model. She included

- An indicator of whether a car is a *hybrid*
- Fuel economy, *MPG*
- An indicator interaction between *hybrid* and *MPG*
- Lack of power, *seconds* to accelerate from 0 to 60 MPH

Because the proportion of owners of a car who are satisfied, *satisfaction,* is bounded below by zero and above by one, Amanda used the *satisfaction logit* as the dependent variable:

---

[10] This example is a hypothetical scenario using actual data.

$$satis\hat{f}actionLogit_i = \ln\left[\frac{satisfaction_i}{1 - satisfaction_i}\right].$$

MPG, a measure of fuel economy, was skewed. Amanda chose to use the inverses, GPM, gallons per mile, which was approximately Normal.

Regression results from the model **are** shown in Table 13.1.

Table 13.1 Regression of satisfaction logit by car characteristic

SUMMARY OUTPUT						
Regression statistics						
R Square	.68					
Standard error	.31					
Observations	33					
ANOVA	*df*	*SS*	*MS*	*F*	*Significance F*	
Regression	4	5.69	1.42	14.7	.0000	
Residual	28	2.71	.10			
Total	32	8.40				
	Coefficients	*Standard error*	*t Stat*	*p Value*	*Lower 95%*	*Upper 95%*
Intercept	3.71	.42	8.8	.0000	2.85	4.58
Hybrid	2.78	.74	3.8	.0008	1.27	4.29
Seconds	−.19	.04	−4.6	.0001	−.28	−.11
GPM	−18.2	6.42	2.8	.009	−31.3	−5.0
Hybrid x GPM	−77.5	18.6	−4.2	.0003	−115.5	−39.5

The significant and positive coefficient for the *hybrid* indicator suggests that owners of hybrid cars are more likely to be satisfied than owners of conventional cars. However, the negative coefficient for the hybrid × GPM interaction reveals that hybrid owners are less satisfied if fuel economy is lower, requiring more gallons per mile. A greater proportion of owners of all cars are satisfied if a model offers more responsive acceleration (fewer *seconds*). The relative importance of each of the car characteristics is marginal and depends on a car's configuration, as well as whether the car has a conventional or a hybrid engine.

Rescale equations back to satisfaction proportions. The model equation for conventional cars, setting the hybrid indicator to 0, is

$$lo\hat{g}it_i = 3.7 - .19\,seconds_i - 18\,GPM_i.$$

The model for hybrids, with the hybrid indicator set to one, is

$$\hat{logit}_i = 6.5 - .19\,seconds_i - 96\,GPM_i.$$

To see the equations in the original scale of *satisfaction proportion*, first find the *predicted satisfaction odds*, which is the exponential function of the *predicted logits:*

$$\hat{odds}_i = e^{(3.7 - .19\ seconds_i - 18\ GPM_i)} \text{ for conventional models,}$$

and

$$\hat{odds}_i = e^{(6.5 - .19\ seconds_i - 96\ GPM_i)} \text{ for hybrids.}$$

Predicted proportions satisfied are then, for conventional models,

$$\hat{satisfaction}_i = \frac{e^{(3.7 - .19\ seconds_i - 18\ GPM_i)}}{1 + e^{(3.7 - .19\ seconds_i - 18\ GPM_i)}},$$

and, for owners of *hybrids*,

$$\hat{satisfaction}_i = \frac{e^{(6.5 - .19\ seconds_i - 96\ GPM_i)}}{1 + e^{(6.5 - .19\ seconds_i - 96\ GPM_i)}}.$$

Because the dependent variable has been rescaled, the logit model has built in synergies. The value of an improvement in one of the characteristics will be nonconstant and also dependent on the levels of other characteristics. To illustrate the synergies, compare expected satisfaction in response to differences in one of the car characteristics, setting the remaining at best and worst levels.

To see the difference in expected proportion of domestic owners satisfied that *seconds* could make, compare alternate *seconds* for four hypothetical cars:

- Least attractive (minimum MPG, which is maximum GPM) conventional
- Most attractive (maximum MPG, which is minimum GPM) conventional
- Least attractive *hybrid*
- Most attractive *hybrid*

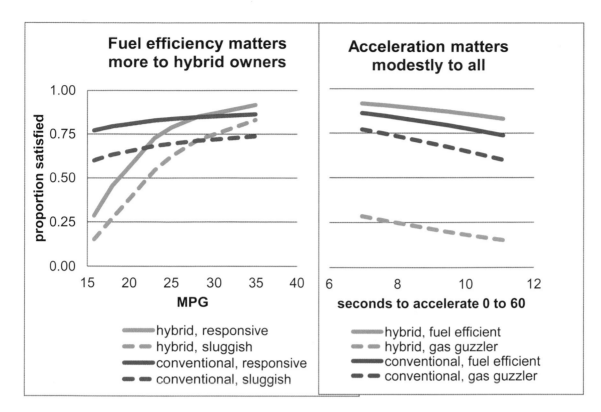

Fig. 13.2 Proportion satisfied by fuel economy and acceleration

Fuel economy: Fuel economy matters more to hybrid owners and, particularly, to hybrid owners of least responsive models. Adding fuel efficiency will be a key to Ford's success in hybrid markets.

Acceleration: Improved acceleration makes a larger difference to owners of the least desirable, least fuel efficient models, shown with dashed lines in Fig. 13.2, and the difference is slightly greater among conventional owners. For Ford, improved response would help to satisfy more, but it would not help enough to satisfy the majority of owners of less fuel efficient hybrids.

The majority of owners of conventional cars remain relatively satisfied, even with poor fuel economy and lack of responsiveness. Hybrid owners, however, require fuel efficiency and are not satisfied with responsiveness alone.

Amanda summarized her model results for Ford executives.

MEMO

Re: Fuel Efficiency Drives Hybrid Owner Satisfaction
To: Ford Strategic Development Executives
From: Amanda Arnone, Quantitative Analysis Director
Date: June 2007

Acceleration increases owner satisfaction modestly. Fuel efficiency increases conventional owner satisfaction modestly and hybrid owner satisfaction dramatically and is essential for hybrid owner satisfaction.

A model of owner satisfaction was built from a representative sample of the proportions of owner satisfied with 33 diverse car models, both hybrid and conventional.

Model results. Differences in fuel source, fuel economy and acceleration account for 68% of the variation in the proportion of car owners satisfied.

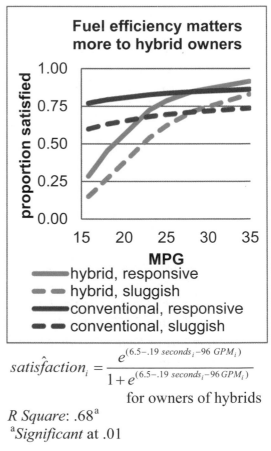

Increasing the fuel efficiency of hybrids has the potential to increase the expected owner proportion satisfied by 60%. Expected satisfaction gain among owners of hybrids with sluggish acceleration is even higher: 70%. Fuel efficiency potentially increases expected conventional owner satisfaction less, about 10%.

Acceleration is a satisfaction driver which partially compensates for lower fuel efficiency. Improving acceleration of fuel inefficient cars would increase the expected percent of conventional owners satisfied by about 20% and the expected percent of hybrid owners satisfied by about 10%.

Conclusions. Owners of hybrids would be more satisfied with more fuel efficient models, though responsiveness also drives satisfaction and partially compensates for less than ideal fuel efficiency.

Other considerations. Recent recalls may affect owner satisfaction. New alternative electric cars may also affect satisfaction. Neither of these potential drivers was considered here.

$$\hat{satisfaction}_i = \frac{e^{(6.5-.19\, seconds_i - 96\, GPM_i)}}{1 + e^{(6.5-.19\, seconds_i - 96\, GPM_i)}}$$

for owners of hybrids

R Square: .68[a]
[a]*Significant* at .01

Example 13.2 Presidential Approval Proportion[11]

The Republican National Committee is planning its 2012 presidential campaign strategy, and the management needs to know what drives public opinion of the president. Some believe that presidential actions which signal defense strength rally public support, whereas others argue that defense references carry costs. The committee is unsure which drives public opinion, the War on Terror and defense strength, or a healthier economy. They suspect that President Bush's declining public opinion may have been linked to war fatalities or to slow growth in wages.

At least three shocks since President Bush's reelection may have induced structural shifts in public opinion.

- In March 2006, President Bush signed the Patriot Act, legalizing government information gathering actions on suspected terrorists.
- In June 2006, the *New York Times* published an article describing illegal government information gathering actions. The White House asked for retraction, and the *New York Times* refused.
- In September 2006, President Bush focused a Labor Day speech on new job creation and designated September 11, 2006, as a day to remember the fifth anniversary of 9/11.

A structural change in political leadership probably also influenced public opinion:

- In November 2006 elections, Democrats gained control of Congress.

Public opinion polls track Americans' approval of the job the president is doing. The Roper Organization (http://www.ropercenter.uconn.edu) publishes results from a number of national polls. **Presidential Approval 13.3.xls** contains the *approval proportions* of 457 polls taken between President Bush's reelection in November 2004 and June 2007.

A consulting firm was retained to build a model of Presidential approval that would identify and quantify drivers and provide short term forecasts. After being briefed by committee representatives, the consultants included:

- An indicator, *Patriot*, following signing into law the Patriot Act
- An indicator, *NYT*, of the *New York Times* article
- An indicator *September 06* of the fifth anniversary of 9/11
- An indicator *Democratic Congress* in 2006 through 2007
- Cumulative military *fatalities* since President Bush's reelection
- A leading indicator of past month average hourly *wage* of American workers

The response variable that the committee was interested in explaining and forecasting is *Proportion Who Approve of the President*. This is a variable bounded below by zero and above by one hundred, so the consultants used the *approval logit* to estimate parameters.

---

[11] This example is a hypothetical scenario using actual data.

Their model was

$$Approval\hat{L}ogit_t = b_0 + b_1 Patriot_t + b_2 NYT_t + b_3 Sept06_t + b_4 DemCongress_t + b_5 fatalities_t$$

$$+ b_6 wage_{t-1}.$$

The model correctly forecasts the two most recent poll results and produced forecasts with a 5% margin of error. Recalibrated results are shown in Table 13.2.

Table 13.2 Logit model of presidential approval

SUMMARY OUTPUT						
Regression statistics						
R Square	.802					
Standard error	.112					
Observations	455					
ANOVA	*df*	*SS*	*MS*	*F*	*Significance F*	
Regression	7	22.8	3.3	259.0	.0000	
Residual	447	5.6	.013			
Total	454	28.4				
	Coefficients	*Standard error*	*t Stat*	*p Value*	*Lower 95%*	*Upper 95%*
Intercept	−6.60	.85	−7.7	.0000	−8.28	−4.92
Patriot	−.19	.02	−10.6	.0000	−.22	−.15
NYT	−.30	.08	−3.7	.0002	−.45	−.14
Sept 06	.20	.03	7.1	.0000	.14	.25
Dem Congress	−.16	.03	−4.7	.0000	−.22	−.09
surge	−.087	.024	−3.6	.0004	−.13	−.04
fatalities (K) *to date*	−.39	.01	−34.2	.0000	−.41	−.36
wage ($)$_{t-1}$	.86	.10	8.2	.0000	.66	1.07
DW: 1.83						

The model accounts for much of the variation, 80%, in *approval logits*.

The Patriot Act, the *New York Times* article alleging government abuses of privacy, Democratic control of Congress, and military fatalities reduce approval. President Bush's September 2006 focus on new jobs, followed by the memorial service commemorating the fifth year anniversary of 9/11, as well as growing wages, enhance public opinion.

The baseline equation, before renewal of the Patriot Act, is

$$Approval\hat{L}ogit_t = -6.60 - .39\ fatalities(K)_t + .86\ wage(\$)_{t-1}.$$

During the 3 months that followed passage of the Patriot Act, the model equation is

$$Approval\hat{L}ogit_t = -6.79 - .39 \, fatalities(K)_t + .86 \, wage(\$)_{t\text{-}1}.$$

After the *New York Times* publication, the equation is

$$Approval\hat{L}ogit_t = -6.90 - .39 \, fatalities(K)_t + .86 \, wage(\$)_{t\text{-}1}.$$

After the fifth 9/11 anniversary, the model equation is

$$Approval\hat{L}ogit_t = -6.40 - .39 \, fatalities(K)_t + .86 \, wage(\$)_{t\text{-}1}.$$

Following the 2006 election, the model equation is

$$Approval\hat{L}ogit_t = -6.76 - .39 \, fatalities(K)_t + .86 \, wage(\$)_{t\text{-}1},$$

and following Bush's presentation of the Surge plan, the equation is

$$Approval\hat{L}ogit_t = -6.69 - .39 \, fatalities(K)_t + .86 \, wage(\$)_{t\text{-}1}.$$

Rewriting the equations as expected odds,

$$Approval\hat{O}dds_t = e^{(-6.60 - .39 \, fatalities(K)_t + .86 \, wage(\$)_{t\text{-}1})}$$

in baseline days before renewal of the Patriot Act,

$$= e^{(-6.79 - .39 \, fatalities(K)_t + .86 \, wage(\$)_{t\text{-}1})}$$

following renewal of the Patriot Act,

$$= e^{(-6.90 - .39 \, fatalities(K)_t + .86 \, wage(\$)_{t\text{-}1})}$$

following the *New York Times* article,

$$= e^{(-6.4 - .39 \, fatalities(K)_t + .86 \, wage(\$)_{t\text{-}1})}$$

following September 2006,

$$= e^{(-6.76 - .39 \, fatalities(K)_t + .86 \, wage(\$)_{t\text{-}1})}$$

following the 2006 election, through 2007, and

$$=e^{(-6.69\ -.39\,fatalities(K)_t\ +.86\,wage(\$)_{t-1})}$$

following the Surge plan speech.

Predicted approval proportions

$$Approval\_\hat{p}roportions_t = [Approval\hat{O}dds_t\ /(1+Approval\hat{O}dds_t)]$$

are shown in Fig. 13.5 by day from President Bush's reelection through June 2007.

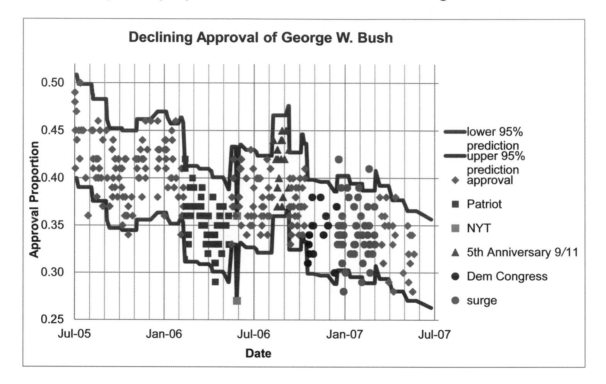

Fig. 13.3 Presidential approval proportion

Predicted Presidential approval was 51% in November 2004, following reelection. The predicted *approval proportion* declined gradually over the next 16 months to 40% in March 2006. Following renewal of the Patriot Act in March 2006, a structural shift in public opinion occurred, reducing approval ratings by an estimated 4% for a 3 month period.

In June 2006, predicted approval was 38%, but dropped briefly to 31% following the *New York Times* article alleging government abuses of privacy. By September 2006, predicted approval was 37%. President Bush's commemoration of the fifth anniversary of 9/11 stimulated a brief structural shift, raising predicted *approval proportions* an estimated 5%.

Before the 2006 election, predicted approval was 38%. With Democratic wins insuring a Democratic Congress, a structural shift reduced approval proportions by an estimated 3%. In January 2007, after President Bush's presentation of the Surge plan for increased troop involvement in Iraq, predicted approval dropped 2–35%. Increasing military fatalities and falling hourly wages brought predicted approval to a low of 31% by July 2007. The margin of error in forecasts is 5%.

The National Committee now has evidence that both the war effort and the domestic economy, in the form of hourly wages, drive public opinion. Democratic control of Congress reduced approval, as well.

13.2 Logit Models Provide the Means to Build Valid Models of Shares and Proportions

When responses are bounded below and above, these limits must be built into models to get accurate pictures of drivers and valid forecasts. Rescaling shares or proportions to odds, and then to their natural logarithms, the logits, provides more valid models. Although both odds and logits are unbounded, the corresponding predicted proportions or shares are bounded below and above, guaranteeing believable forecasts.

Excel 13.1 Regression of a Limited Dependent Variable Using Logits

Proportion Who Would Try Pampers Preemies
Procter & Gamble management believes that their new diaper may attract mothers who were choosing cloth diapers. Natural composition is a known advantage of cloth diapers. Build a model of trial intentions to see whether the importance of natural composition and selected demographics are drivers.

Rescale bounded dependent variables to unbounded logits: In concept test data, **Excel 13.1 Pampers Concept Test.xls**, we have the *trial intentions* of 97 preemie mothers, measured on a 5 point scale ("Definitely Not" = .05, "Probably Not" = .25, "Maybe" = .5, "Probably" = .75, "Definitely" = .95).

From *trial intent*, which is bounded between zero and one, make *trial odds*, the chance of trying to the chance of not trying, and *trial logit*, the natural logarithms of the *trial odds*.

	E2			f_x	=A2/(1-A2)	
	A	B	C	D	E	F
1	trial intention	income (K$)	natural importance	only child	trial odds	trial logit
2	0.05	6	3	1	0.05	-2.94
3	0.05	6	3	1	0.05	-2.94

The concept test measures include household demographics, an indicator of absence of other children in the households, *only child* and *income* ($K), and the importance rating of natural composition, *natural importance*.

Run regression of *trial logit* with *only child*, *income*, and *natural importance*.

	A	B	C	D	E	F	G
1	SUMMARY OUTPUT						
2							
3		*Regression Statistics*					
4	Multiple R	0.590					
5	R Square	0.348					
6	Adjusted R Square	0.327					
7	Standard Error	1.844					
8	Observations	97					
9							
10	ANOVA						
11		*df*	*SS*	*MS*	*F*	*Significance F*	
12	Regression	3	169	56.2	16.5	1.07289E-08	
13	Residual	93	316	3.4			
14	Total	96	485				
15							
16		*Coefficients*	*Standard Error*	*t Stat*	*P-value*	*Lower 95%*	*Upper 95%*
17	Intercept	-0.573	0.363	-1.6	0.11771	-1.294	0.148
18	income (K$)	0.0172	0.0040	4.3	4.9E-05	0.0092	0.0252
19	natural importance	-0.289	0.091	-3.2	0.00203	-0.470	-0.108
20	only child	1.730	0.415	4.2	6.8E-05	0.907	2.553

Sensitivity analysis: To quantify the influence of each driver, find *predicted trial intentions* for hypothetical combinations of the three predictors.

To find the sample ranges, find the 5% and 95%, using the Excel functions **PERCENTILE(***array,.05***)** and **PERCENTILE(***array,.95***)**, and the *median*, using the Excel function **MEDIAN(***array***)** of *Income*(K) and *natural importance*.

B99			f_x	=PERCENTILE(B2:B98,0.05)
	A	B	C	D	
1	trial intention	income (K$)	natural importance	only child	trial odd
96	0.95	160	5	1	
97	0.95	160	6	1	
98	0.95	170	1	0	
99	5%	6	1	0	
100	median	50	3	1	
101	95%	160	8.2	1	

Find the marginal impact of each driver when the other drivers are at most favorable and unfavorable levels.

Natural composition: Add 20 hypothetical mothers to the bottom of the dataset:
  Ten mothers with low income ($K), *6*
    Five with only one child (only child=*1*)
    Five with other children (only child=*0*)
  Ten with high income ($K), *160*
    Five with only one child (only child=*1*)
    Five with other children (only child=*0*)

Within each set of five demographically identical moms, let one of each rate *natural composition*
  Unimportant (natural importance=*1*)
  Not very important (natural importance=*3*)
  Moderately important (natural importance=*5*)
  Important (natural importance=*7*)
  Very important (natural importance=*9*)

	A	B	C	D
1	*trial intention*	*income (K$)*	*natural importance*	*only child*
100	median	50	3	1
101	95%	160	8.2	1
102	low income	6	1	0
103	not only child	6	3	0
104		6	5	0
105		6	7	0
106		6	9	0
107	only child	6	1	1
108		6	3	1
109		6	5	1
110		6	7	1
111		6	9	1
112	high income	160	1	0
113	not only child	160	3	0
114		160	5	0
115		160	7	0
116		160	9	0
117	only child	160	1	1
118		160	3	1
119		160	5	1
120		160	7	1
121		160	9	1

Predicted trial logits: Use the coefficient estimates from your regression output sheet to make *predicted trial logits*, using the regression equation, filling in the column through the 20 hypotheticals.

| G2 | | | f_x | =$H$2+$H$3*B2+$H$4*C2+$H$5*D2 | |

	A	B	C	D	G
1	trial intention	income (K$)	natural importance	only child	predicted trial logit
99	5%	6	1	0	-0.76
100	median	50	3	1	1.15
101	95%	160	8.2	1	1.54
102	low income	6	1	0	-0.76
103	not only child	6	3	0	-1.34
104		6	5	0	-1.91
105		6	7	0	-2.49
106		6	9	0	-3.07
107	only child	6	1	1	0.97
108		6	3	1	0.39
109		6	5	1	-0.18
110		6	7	1	-0.76
111		6	9	1	-1.34
112	high income	160	1	0	1.89
113	not only child	160	3	0	1.31
114		160	5	0	0.74
115		160	7	0	0.16
116		160	9	0	-0.42
117	only child	160	1	1	3.62
118		160	3	1	3.04
119		160	5	1	2.47
120		160	7	1	1.89
121		160	9	1	1.31

Rescale to find predicted trial intentions: Rescale *predicted trial logit* to *predicted odds* using the Excel exponential function, **EXP( ).**

Rescale the *predicted odds* to *predicted trial intentions:*

=predicted odds/(1+predicted odds).

	F102		▼	f_x	=EXP(E102)		
◢	A	B	C	D	E	F	G
1	trial intention	income (K$)	natural importance	only child	predicted trial logit	predicted trial odds	predicted trial intention
102	low income	6	1	0	-0.76	0.47	0.32
103	not only child	6	3	0	-1.34	0.26	0.21
104		6	5	0	-1.91	0.15	0.13
105		6	7	0	-2.49	0.08	0.08
106		6	9	0	-3.07	0.05	0.04
107	only child	6	1	1	0.97	2.64	0.73
108		6	3	1	0.39	1.48	0.60
109		6	5	1	-0.18	0.83	0.45
110		6	7	1	-0.76	0.47	0.32
111		6	9	1	-1.34	0.26	0.21
112	high income	160	1	0	1.89	6.63	0.87
113	not only child	160	3	0	1.31	3.72	0.79
114		160	5	0	0.74	2.09	0.68
115		160	7	0	0.16	1.17	0.54
116		160	9	0	-0.42	0.66	0.40
117	only child	160	1	1	3.62	37.42	0.97
118		160	3	1	3.04	21.00	0.95
119		160	5	1	2.47	11.78	0.92
120		160	7	1	1.89	6.61	0.87
121		160	9	1	1.31	3.71	0.79

Illustrate synergies between predictors: To see the synergies between the importance of natural composition, income, and absence of other children, plot *predicted trial intentions* by *natural importance*, four each of the four sets of demographically identical moms as separate series.

Rearrange columns, moving *predicted trial intentions* right of *natural importance*.

Plot *predicted trial intentions* by *natural importance* for the first set of five hypotheticals, *lower income, only child.*

trial intention	income (K$)	natural importance	predicted trial intention	only child	predicted trial logit	predicted trial odds	Coefficient	trial odds	trial logit
low income	6	1	0.32						
not only child	6	3	0.21						
	6	5	0.13						
	6	7	0.08						
	6	9	0.04						
only child	6	1	0.73						
	6	3	0.60						
	6	5	0.45						
	6	7	0.32						
	6	9	0.21						
high income	160	1	0.87						
not only child	160	3	0.79						
	160	5	0.68						
	160	7	0.54						

Chart: "low income, not only child"

Add the three remaining hypothetical series, choosing a style and design, and adjusting axes.

trial intention	income (K$)	natural importance	predicted trial intention	only child	predicted trial logit	predicted trial odds	Coefficient	trial odds	trial logit
low income	6	1	0.32						
not only child	6	3	0.21						
	6	5	0.13						
	6	7	0.08						
	6	9	0.04						
only child	6	1	0.73						
	6	3	0.60						
	6	5	0.45						
	6	7	0.32						
	6	9	0.21						
high income	160	1	0.87						
not only child	160	3	0.79						
	160	5	0.68						
	160	7	0.54	0	0.16	1.17			
	160	9	0.40	0	-0.42	0.66			
only child	160	1	0.97	1	3.62	37.42			
	160	3	0.95	1	3.04	21.00			
	160	5	0.92	1	2.47	11.78			
	160	7	0.87	1	1.89	6.61			
	160	9	0.79	1	1.31	3.71			

Chart: "Those who value natural less likely to try" — predicted trial intention vs. natural importance (1 to 9); series: high income, only child; high income, not only child; low income, only child; low income, not only child

Find the marginal difference that natural composition makes, given alternate demographics: To quantify the marginal difference that the importance of natural composition makes in expected trial intention, find the *marginal difference in expected trial intention* for each of the four hypothetical demographics.

	E117	▾ ◯	f_x	=D117-D121	
◢	A	B	C	D	E
1	trial intention	income (K$)	natural importance	predicted trial intention	marginal difference due to natural importance
102	low income	6	1	0.32	0.42
103	not only child	6	3	0.21	
104		6	5	0.13	
105		6	7	0.08	
106		6	9	0.04	
107	only child	6	1	0.73	0.52
108		6	3	0.60	
109		6	5	0.45	
110		6	7	0.32	
111		6	9	0.21	
112	high income	160	1	0.87	0.47
113	not only child	160	3	0.79	
114		160	5	0.68	
115		160	7	0.54	
116		160	9	0.40	
117	only child	160	1	0.97	0.19
118		160	3	0.95	
119		160	5	0.92	
120		160	7	0.87	
121		160	9	0.79	

Lab 13 T-Mobile's Plans to Capture Share in the Cell Phone Service Market[12]

Verizon offers competitive cell phone network service in a wide geographic area, buoyed by their acquisition of Alltel.

T-Mobile management believes that with advertising to boost brand equity and improvements in their service, T-Mobile could overtake Verizon. Competitors ATT and Sprint are also making plans.

In 26 cities, samples of cell phone customers were drawn and surveyed. Survey measures included *service provider*, *satisfaction*, service *coverage rating*, *dropped calls rating*, and *static rating*. Ratings were on a 5 point scale, where a higher number indicated better service. In the data file, **Lab 13 tmobile.xls**, are

- *City*
- *Service provider*
- *Proportion of customers satisfied*
- *Coverage rating*
- *Dropped calls rating*
- *Static rating*

Build a model of customer satisfaction for the T-Mobile executives that quantifies the importance of *service provider*, *coverage*, *dropped calls*, and *static*.

Proportion satisfied is a limited dependent variable with values between 0 and 1. Rescale to acknowledge these limits.

Make a PivotChart to compare average customer satisfaction with the service providers.

1. Which service providers achieved greater than average satisfaction? _____

Choose the service with the least satisfied as your baseline.

To incorporate executive judgment, include in your model, interactions among

- *Verizon* and each of the three service ratings
- *T-Mobile* and each of the three service ratings

Fit your model, first removing insignificant indicator interactions, and then removing insignificant variables and indicators. (If an indicator interaction is significant, but either one of the main effects involved in the interaction is not, keep the main effects in the model to support the interaction.)

---

[12] The case is a hypothetical scenario using actual data.

Because each of the service ratings ought to influence satisfaction positively, use one tail *t tests* of the coefficient estimates of the three ratings (by dividing the two tail *p* Values by 2). (For the coefficient estimates of indicators and indicator interactions, we don't know which will be positive and which will be negative. Use a two tail test for coefficients of indicators and indicator interactions.)

Use your coefficient estimates to make *predicted logits*, and then rescale to make *predicted odds*, and *predicted proportion satisfied.*

2. Write your equations for the *predicted satisfaction odds*.

- For *ATT* customers

- For *Sprint* customers

- For *Verizon* customers

- For *T-Mobile* customers

Use a PivotTable to find the average service ratings for each of the service providers.

Add hypothetical services to the data file, comparing *predicted customer satisfaction proportions* across the competing service providers, ATT, *Sprint, T-Mobile,* and *Verizon,* given *current average* service ratings.

3. With current average ratings, which service provider can expect the highest proportion satisfied?

_____

4. If T-Mobile were to improve brand equity to equal Verizon's brand equity, could T-Mobile

expect to have more satisfied customers than Verizon? _____

Add 12 hypotheticals to your dataset:

- ATT service with current *coverage* and *static* ratings, *dropped calls* rating of 2
- ATT service with current *coverage* and *static* ratings, *dropped calls* rating of 3
- ATT service with current *coverage* and *static* ratings, *dropped calls* rating of 4
- ATT service with current *coverage* and *static* ratings, *dropped calls* rating of 5

- Sprint service with current *coverage* and *static* ratings, *dropped calls* rating of 3
- Sprint service with current *coverage* and *static* ratings, *dropped calls* rating of 4
- Sprint service with current *coverage* and *static* ratings, *dropped calls* rating of 5

- T-Mobile service with current *coverage* and *static* ratings, *dropped calls* rating of 3
- T-Mobile service with current *coverage* and *static* ratings, *dropped calls* rating of 4
- T-Mobile service with current *coverage* and *static* ratings, *dropped calls* rating of 5

- Verizon service with current *coverage* and *static* ratings, *dropped calls* rating of 4
- Verizon service with current *coverage* and *static* ratings, *dropped calls* rating of 5

Make a scatterplot showing expected satisfaction proportion by *dropped calls* rating, showing each service provider as a separate series.

5. Which service gains the most from an improvement of one *dropped calls* rating scale point?

_____

6. Toward which should T-Mobile management dedicate more resources:

    ___brand equity, or ___dropped calls?

Assignment 13-1 Big Drug Co Scripts

The leading manufacturer of a popular anti allergy drug would like to know how reformulations affect their share of prescriptions dispensed. Big Drug's major competition comes from generic copycat brands. When the generic competition begins to gain share, Big Drug introduces a reformulation, which sends the generics back to the lab to reformulate their copies. Reformulation is expensive, because it includes research and development, as well as repackaging and reformulating promotional materials.

Semiannual data in **Assignment 13-1 Big Drug Co.xls** include time series of a semiannual counter of time periods, the share of prescriptions dispensed of Big Drug Co's antiallergy drug, and indicators for a major and a minor reformulation.

Build a logit trend model to estimate the impact of reformulations on Big Drug Co's share and to forecast Big Drug Co's share in the next 5 years.

Write a one page memo to Big Drug Co management concerning the impact of reformulations on share and share forecasts for the next 5 years. Embed one figure to illustrate your results. Include in your memo:

- Share estimates had the drug not been reformulated
- Suggested date for Big Drug Cos introduction of Reformulation 3, and recommendations for either a major or a minor reformulation

Assignment 13-2 Competition in the Netbook Market

Dell leads the netbook market with the lightest weight option, although rival HP is a strong second. The managers want to know whether resources ought to be directed to the design of an ultra light weight netbook or to building Dell's existing netbook brand equity.

Build a model using data in **weighty netbooks.xls** to inform Dell management. These data include three recently published estimates of netbook market share by brand and weight.

Make the Acer and Toshiba, the less advertised brands, the baseline.

Note: You will not be able to include indicator interactions because of the small sample size.

1. Write your model equations for Dell and HP using standard format with two or three significant digits.

2. Which is the most significant driver? ___brand or ___weight

3. Which makes a bigger difference on expected market share? ___brand or ___weight

4. Attach one scatterplot that illustrates your conclusion. Show four brand series, with at least four points per series. Use actual weight in lbs for the x axis.

5. If HP were to introduce a lighter, 2.5 lb netbook, what market share could HP expect? _____

6. If Dell were to match HP's brand equity, what market share could Dell expect? ____

7. If Dell were to introduce an ultra light, 2.0 lb netbook, what market share could Dell expect?

   _____

8. Explain your dependent variable scale choice.

Case 13-1 Pilgrim Bank (B): Customer Retention[13]

Framing the Problem: Armed with the information learned previously in cases (A) and (B), Pilgrim management has decided to promote online services and online services with billpay to selected customer segments. (In order to use billpay, a customer must also use online services.) There are two customer segments of interest that must be identified:

1. Customers who are more likely to continue banking at Pilgrim after using online services or online services with billpay. This is the prime segment of most interest to Pilgrim.

2. Customers who are less likely to continue banking at Pilgrim after using online services or online services with billpay. Pilgrim will try to limit online banking in this segment.

Pilgrim could offer online services without billpay or online services with billpay to identifiable customer segments. Identify these customer segments so that online services and promotion can be customized to increase retention.

Data and analysis: Customer retention data are in **Case 13-1 Pilgrim Retention.xls.** Customers who have stayed with the bank have been coded with *retention probability* .95. Customers who left the bank have been coded with *retention probability* .05.

Build a model, focusing on customer *retention probability*, including indicators for online banking and online billpay, indicators for district, customer tenure, age, and income.

Memo to Management: Write a memo to Pilgrim Bank management that presents the results of your model of retention to guide management in targeting customer segments for both online services and online services with billpay.

---

[13] Harvard Business School Case 9602103.

This memo ought to include one embedded graphic that best illustrates your results. You should illustrate each of your key results, so that they are easily understood by management. Attach additional pages with supporting graphics that are referred to in your memo.

You should not include any graphics that are not specifically referred to in your text.

Index